Restoring Floodplains in Europe

Restoring Floodplains in Europe

Policy contexts and project experiences

Edited by Timothy Moss
and Jochen Monstadt

IWA Publishing

IRS
Leibniz Institute for
Regional Development
and Structural Planning

Published by IWA Publishing, Alliance House, 12 Caxton Street, London SW1H 0QS, UK
Telephone: +44 (0) 20 7654 5500; Fax: +44 (0) 20 7654 5555; Email: publications@iwap.co.uk
Web: www.iwapublishing.com

First published 2008

Cover design by www.designforpublishing.co.uk
Index provided by C.M. Graham

British Library Cataloguing in Publication Data
A CIP catalogue record for this book is available from the British Library

Library of Congress Cataloging- in-Publication Data
A catalog record for this book is available from the Library of Congress

ISBN: 1843390906
ISBN13: 9781843390909

Contents

About the contributors

Ross Beveridge

Ross Beveridge (ross.beveridge@ncl.ac.uk) is a doctoral student in the School of Geography, Politics and Sociology at the University of Newcastle. He is currently conducting research on water policy controversies in Germany and Great Britain. Previously he carried out research on innovation and the shifting socio-technical organisation of the English water sector.

Thomas Dworak

Thomas Dworak (thomas.dworak@ecologic.eu) is a policy advisor working primarily on water resource protection, especially the EU Water Framework Directive, and the links between water protection and other environmental issues, including the EU Common Agriculture Policy, soil protection and spatial planning. He is head of the Ecologic office in Vienna, Austria. Before joining Ecologic, Thomas Dworak worked for the Wuppertal Institute for Climate, Environment and Energy in the research group "Eco-Efficiency & Sustainable Enterprises" and the department for climate policy. There, he focused on

 Restoring Floodplains in Europe: policy contexts and project experiences edited by Timothy Moss and Jochen Monstadt. ISBN: 1843390906. Published by IWA Publishing, London, UK.

sustainable water management, organic farming and the sustainable development of information technologies.

Maria Hagemeier

Maria Hagemeier (hagemeier@forst.wzw.tum.de) is a graduate in geography and works as a research fellow at the Technical University of Munich. Under the Chair of Forest and Environmental Policy she is currently working on her PhD within the EU Project FloodScan, which aims to improve and to simplify two-dimensional flood simulation for the identification of floodplains and for the elaboration of flood hazard maps. Her research focuses on risk communication, risk perception and risk management in the field of flood risks.

Francine Hughes

Dr Francine Hughes (f.hughes@anglia.ac.uk) is Senior Lecturer in Ecology and Conservation in the Department of Life Sciences at Anglia Ruskin University, Cambridge. She is a hydro-ecologist whose research interests include the regeneration ecology of European floodplain forests, the design of 'environmental flows' for riparian ecosystems and the development of monitoring and evaluation systems for landscape-scale wetland restoration projects. She was previously a joint co-ordinator, together with Professor Keith Richards, of a series of EU-funded research projects based in the Geography Department at the University of Cambridge, including the project on which this book is based: "Floodplain Biodiversity and Restoration (FLOBAR) 2: Integrated natural science and socio-economic approaches to catchment flow management".

Axel Klaphake

Dr Axel Klaphake (klaphake@imup.tu-berlin.de) works at the Berlin University of Technology (Institute of Landscape Architecture and Environmental Planning). His research focuses on economic aspects of water management, nature protection and regional policies. In the context of several research projects he investigates institutional aspects of river basin management in Europe. In addition, Dr Klaphake addresses integrated water management in developing and threshold countries and cross-border cooperation for the protection of water bodies.

Sylvia Kruse

Sylvia Kruse (sylvia.kruse@uni-lueneburg.de) studied environmental sciences at the University of Lüneburg and worked as a research associate in the Institute for Resource Management, Berlin, on projects on precautionary flood protection and infrastructure development in the context of sustainable regional development. Her special research interests lie in the areas of sustainable urban and regional development, spatial transformation and precautionary disaster management. Her PhD research work (2005-2008) focuses on flood risk management from a socio-ecological perspective and includes extensive fieldwork and multi-method empirical research in the river basin of the Elbe River.

Jochen Monstadt (editor)

Jochen Monstadt (monstadt@yorku.ca) holds a PhD in environmental planning. He works as a senior research fellow at the City Institute at York University, Toronto (CDN) and at the Keston Institute for Infrastructure at the University of Southern California, Los Angeles (USA). Previously he conducted and coordinated research on the transition of energy, water and wastewater systems, innovation management in the utility sectors, river basin management, infrastructure planning, and regional/urban governance and planning in different research institutes: the Swiss Federal Institute for Aquatic Science and Technology (EAWAG), the Leibniz Institute for Regional Development and Structural Planning (IRS), Erkner, and the Institute for Environmental Management and Planning at the Berlin University of Technology.

Timothy Moss (editor)

Timothy Moss (moss@irs-net.de) holds a DPhil in urban history. He is head of the department "Regional institutional change" at the Leibniz Institute for Regional Development and Structural Planning (IRS) in Erkner, near Berlin, where he has coordinated and conducted a number of national and international research projects on institutional dimensions to environmental change in urban and regional contexts. His special research interests are in the areas of infrastructure management, river basin management and the sustainable development of cities and regions in Europe.

Keith Richards

Keith Richards (keith.richards@geog.cam.ac.uk) is Professor of Geography and Fellow of Emmanuel College at the University of Cambridge. A fluvial geomorphologist, his research interests are in river management, river and floodplain restoration, and inter-relationships between hydrological and ecological processes in floodplain environments. Together with Dr Francine Hughes he coordinated the EU-funded research project "Floodplain Biodiversity and Restoration (FLOBAR) 2: Integrated natural science and socio-economic approaches to catchment flow management".

Foreword: *European Commission, Directorate General for Research*

Hartmut Barth

Following a communication to the other European institutions in July 2004 on "Flood risk management – Flood prevention, protection and mitigation", in January 2006 the European Commission proposed a Directive of the European Parliament and of the Council on the assessment and management of floods. This Directive aims to establish a common framework for assessing and reducing the risk which floods within the European Union may pose to human health, the environment, property and economic activity. Since the envisaged prevention and management measures are organised by river basin districts (which may cover several river basins), the Directive is directly linked to the EU Water Framework Directive (WFD), which came into effect in December 2000. The WFD aims to establish by 2015 good chemical and ecological status of all surface, ground, transitional and coastal waters in the European Member States, respecting at the same time cost-efficient approaches.

In support of informed policy development and implementation at European and national levels, research within the EU RTD Framework Programmes has, since the early 1990s, been complementing national research efforts to investigate the structure as well as the biogeochemical and ecological functioning of aquatic and wetland ecosystems. In a second step this understanding of "intrinsic" ecosystem *functioning* has been related to the many **functions** of aquatic – and in particular wetland – ecosystems. These goods and services provided to the general environment and to society comprise the filtering of polluted water, reduction of diffuse loads of nutrients to surface and ground waters (e.g. from agricultural land), buffering against floods and the maintenance of high specific biodiversity (e.g. by acting as a breeding ground for important fish species), to mention just a few. With this research the European Community has contributed substantially to characterising and – to some degree – quantifying the "boxes" of the classical Drivers-Pressures-State-Impacts-Response (DPSIR) model, which is used *inter alia* by the European Environment Agency (EEA) to regularly describe the status of the European environment, and which is often used as the basis for EU environmental policy.

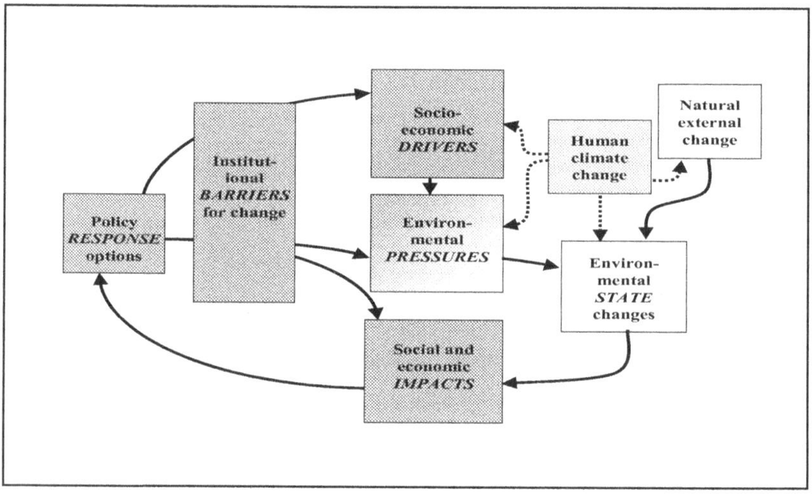

Figure X.1: Extended DPSIR model, by Prof. Laurence Mee (ELME project, 2006, University of Plymouth, unpublished).

Until the end of the 1990s the DPSIR assessment had rarely been translated into an operational scheme, because it ignored the complex institutional setting behind environmental policy implementation. Conflicting interests of the various actors involved in environmental decision-making and in practical environmental planning and management often comprise a significant barrier to environmental change. Figure X.1 demonstrates an improved DPSIR model, which takes into account institutional barriers to change and includes strong feedback mechanisms along the chain of socio-economic drivers, environmental pressures, environmental state changes, social, economic and environmental impacts and policy response options.

In the past only a few degraded or destroyed floodplain wetlands of European importance have been restored successfully. This may well be due – in part, at least – to inadequate consideration of the role of the many laws, rules and regulations governing restoration schemes and the often conflicting interests of the multiple stakeholders involved in an integrated ecosystem management process. This book on the policy contexts and project experiences of restoring floodplain ecosystems is, therefore, extremely timely. It is based on the research project "FLOBAR2" (contract no. EVK1-CT-1999-00031) funded within the 5th RTD Framework Programme and which I supervised as Scientific Officer at the European Commission. The book addresses an important gap in our knowledge about the institutional preconditions to change and, on the basis of experiences drawn from three European Member States, it provides practical guidance to policy makers and river and floodplain wetland ecosystem managers on ways of resolving or coping with institutional problems. This will certainly contribute to improved implementation of the Floods Directive and the EU Water Framework Directive mentioned above, but also to a better understanding of the manifold other conventions and legal frameworks relevant to wetlands and their restoration.

European Commission
Directorate General for Research

Foreword: *European Centre for River Restoration*

Bart Fokkens

Floodplain restoration is an important measure, reconnecting rivers to the adjacent floodplain. It is well known that many floodplains along rivers no longer fulfil their key functions in river ecology and flood prevention. Many major floods along European rivers like the Elbe, Danube, Odra and Rhine have demonstrated that it is beneficial (from an economic point of view) to prevent flood disasters by giving more space to the rivers and by precautionary spatial and landscape planning rather than just continuing to heighten the dykes. Today, programmes exist in various European countries in which issues of sustainable water and river management are used as a guiding principle and more appropriate approaches to flood prevention have been developed.

The European Centre for River Restoration (ECRR) is a network organisation whose vision is to enhance ecological river restoration as an integral part of sustainable water management throughout Europe, connecting people and organisations working on river restoration. The ECRR places particular emphasis on ecological restoration. River restoration is understood as an important

© 2008 IWA Publishing. *Restoring Floodplains in Europe: policy contexts and project experiences* edited by Timothy Moss and Jochen Monstadt. ISBN: 1843390906. Published by IWA Publishing, London, UK.

component of sustainable water management. The proceedings of our three international river restoration conferences can be found on our website. These comprehensive conference proceedings on floodplain biodiversity and restoration clearly show the policy change in floodplain restoration in Europe over the last three decades. Although there is still a lot to do, initial projects are presented which demonstrate how the complex challenges of large-scale, integrated floodplain restoration have been addressed successfully. In the 1990s early attempts were made and projects were carried out on a local and regional scale to restore or, at least, rehabilitate floodplains in Europe. In order to regain the flood protection function of a catchment area, however, an integrated approach is increasingly necessary.

The ECRR joins other organisations in emphasising the value of the catchment and/or integrated approach. The latter is the method of the future despite the fact that projects may take longer to get started; the multiple objectives are more important than the time schedule. In some projects it has already been shown that although the planning process was time-consuming, the implementation phase was accelerated as there were only few objections, or none at all. Working with strong partnerships and looking for various funding sources might not be very easy in the beginning but often produces good results in the end. For example, we can identify many cases of recent flood prevention schemes which combine flood protection with land-use change, nature development, mineral extraction and spatial planning (see examples in the UK and The Netherlands, with their 'space for water' and 'room for the river' concepts, and along the rivers Elbe, Danube, Loire and Rhine).

Enjoy reading this book on policy contexts and project experiences in floodplain restoration and be inspired for your next restoration project. And don't forget to inform the ECRR so that it can disseminate information on your river restoration projects and/or activities.

ECRR

EUROPEAN CENTRE FOR RIVER RESTORATION

www.ecrr.org – ecrr@cirf.org

CIRF

CENTRO ITALIANO PER LA RIQUALIFICAZIONE FLUVIALE

Preface

This book represents – so far as the authors are aware – the first systematic analysis of the policy contexts of floodplain restoration in a number of EU Member States coupled with in-depth comparative analysis of experiences of floodplain restoration schemes in these European countries. It presents the findings of the social science module of the interdisciplinary research project “Floodplain Biodiversity and Restoration 2: Integrated natural science and socio-economic approaches to catchment flow management”, funded by the European Commission under the Framework 5 Research Programme between 2000 and 2003 (no. EVK1-CT-1999-00031). The “FLOBAR2” project was coordinated by the University of Cambridge (UK), which, together with project partners at the University of Grenoble (France), The Centre National de Recherche Scientifique, Toulouse (France), the University of Umea (Sweden) and the University of Lethbridge (Canada), conducted the natural science modules in the project. The social science module was the responsibility of the Leibniz Institute for Regional Development and Structural Planning (IRS), Erkner (Germany).

The editors would like to take this opportunity of expressing their gratitude to the European Commission for providing generous funding for the FLOBAR2 project, in particular to Hartmut Barth, the project officer in charge, for his continuous support. We are grateful also to the European Centre for River Restoration, particularly to Ute Menke, for kindly providing financial support for the production of this book. We thank all the researchers on the FLOBAR2 team for providing such a supportive and creative setting for interdisciplinary work between natural and social scientists – a rare experience indeed. We are particularly indebted to the coordinators of the project, Keith Richards and Francine Hughes, for giving us the opportunity to learn about the challenges of restoring floodplains. Their interest in, and encouragement of, our work as well as their tireless and professional leadership of the project was instrumental behind our highly productive and enjoyable collaboration. We would also like to thank all those involved in producing this book. Beyond the authors this applies in particular to Manuela Siener, Thomas Weith, Ulrike von Schlippenbach, Andreas Uhrlau, Tina Swierczynski and Ross Beveridge, who contributed significantly to our research and provided valuable inputs. We are also grateful to Irene Wilson for her excellent translation work, Jackie Chilton, Britta Kastens and Barbara Jantzen for transcriptions and Petra Geral, Martina Leppler and Bernd Schulze for the layout and graphics. Alan Click, Michelle Jones and their team from IWA Publishing have always been very supportive and shown great patience in waiting for the final product. Last but not least, we would like to take the opportunity to thank our numerous interview partners for sharing their knowledge and insights on the policy and practice of floodplain restoration with us. We hope that their voices echo throughout the text.

Timothy Moss, Erkner
Jochen Monstadt, Toronto

July 2007

List of Abbreviations

[France (F), Germany (D), United Kingdom (UK)]

APPB	Prefectoral Decree for Biotope Conservation (F)
BauGB	Federal Building Code (D)
BAW	Federal Waterways Engineering and Research Institute (D)
BMBF	Federal Ministry for Education and Research (D)
BMU	Federal Ministry for Environment, Nature Conservation and Nuclear Safety (D)
BMVBW	Federal Ministry for Transport, Building and Housing (D)
BNatSchG	Federal Nature Conservation Act (D)
BUND	German Federation of Environment and Nature Conservation (D)
BWaldG	Federal Forestry Act (D)
CAP	Common Agricultural Policy
CC	Countryside Commission (UK)
CFMP	Catchment Flood Management Plan (UK)
CPRE	Council for the Protection of Rural England (UK)
COFAGAR	Private Forest Cooperation (F)
CREN	Regional Nature Conservationists (F)
CRFP	Regional Commission for Forests and Forest Products (F)
CSP	Administration for Fishing and Fishing Surveillance (F)

CTE	Territorial Contracts of Exploitation (F)
DATAR	Delegation of Spatial Planning and Regional Action (F)
DCLG	Department for Communities and Local Government (UK)
DDAF	Departmental Administration for Agriculture and Forests (F)
DDE	Departmental Administration for Public Works (F)
Defra	Department of Food, Rural Affairs and Agriculture (UK)
DIREN	Regional Environmental Administration (F)
DPF	Public Fluvial Domain (F)
DRAC	Regional Administration for Culture (F)
DREF	Regional Administration of Public Works (F)
EA	Environment Agency (UK)
EC	Environmental Code (F)
ECRR	European Centre for River Restoration
EDF	French Power Company (F)
EN	English Nature (UK)
EPALA	Public Agency for the Management of the Loire River and its Tributaries (F)
ERDF	European Regional Development Fund
ESA	Environmentally Sensitive Areas (UK)
EU	European Union
FDP	Departmental Fish Federation (F)
FPR	Floodplain restoration
FWAG	Farming and Wildlife Advisory Group (UK)
GDR	German Democratic Republic
GIS	Interdisciplinary Research Group (F)
GREBAR	Management Group of Dams in the Seine Basin (F)
HWSchG	Federal Flood Control Act (D)
ICE	Institution of Civil Engineers (UK)
ICPE/IKSE	International Commission for the Protection of the Elbe
ICPO/IKSO	International Commission for the Protection of the Oder
ICPR/IKSR	International Commission for the Protection of the Rhine
ICZM	Integrated Coastal Zone Management
IDB	Internal Drainage Board (UK)
IFEN	French Institute for the Environment (F)
IIBRBS	Interdepartmental Organisation for Flood and Drought Protection (F)
IRMA	Interreg Rhine-Meuse Activities
IWRM	Integrated Water Resources Management
LAGS	State Agency for Large Protected Areas in Brandenburg (D)
LAMP	Levels and Moors Partnership (UK)
LAWA	Inter-state Working Group on Water (D)
LIFE	European Financial Instrument for the Environment

LOADDT	Framework Act on Spatial Planning and Sustainable Regional Development (F)
LOADT	Framework Act on Spatial Planning and Regional Development (F)
LwG	Federal Agriculture Act (D)
MATE	Ministry for Spatial Planning and for the Environment (F)
MAFF	Ministry of Agriculture, Fisheries and Food (UK)
MKRO	Inter-ministerial Conference on Regional Planning (D)
NE	Natural England (UK)
NFU	National Farmers Union (UK)
NGO	Non-governmental organisation
NMP	Nature Midi Pyrénées (non-governmental organisation) (F)
NRA	National Rivers Authority (UK)
OECD	Organisation for Economic Cooperation and Development
OIEau	International Office for Water (F)
ONZH	National Observatory of Wetlands (F)
ORF	Orientations for Regional Forests (F)
PAPI	Flood Prevention Action Programme (F)
PCP	Parrett Catchment Project (UK)
PER	Natural Risk Exposure Plans (F)
PCWMSAP	Parrett Catchment Water Management Strategy Action Plan (UK)
PNRZH	National Research Programme in Wetlands (zones humides) (F)
PPG	Policy Planning Guidance (UK)
PPR	Foreseeable Natural Risk Prevention Plans (F)
RDR	Rural Development Regulation
REFS	Rural Environmental Facilitation Service (UK)
ROG	Federal Regional Planning Act (D)
RRP	River Restoration Project (UK)
RSPB	Royal Society for the Protection of Birds (UK)
SAC	Special Areas of Conservation (UK)
SAGE	Water Management Plan (F)
SDAGE	Master Plan for Water Management (F)
SDRIF	Regional Urban Development Plan (F)
SMEAG	Interdepartmental and Interregional Organisation for Water Management on the Garonne (F)
SPA	Special Protection Areas (UK)
SRADT	Regional Planning Documents (F)
SSSI	Sites of Special Scientific Interest (UK)
WFD	Water Framework Directive
WHG	Federal Water Act (D)

WLMAP	Somerset Levels and Moors Water Level Management Action Plan (UK)
WUF	Wise Use of Floodplains Project
WWF	World Wide Fund for Nature
ZNIEFF	Natural Zone of Ecological, Floral and Faunal Interest (F)

Part A

INTRODUCING FLOODPLAIN RESTORATION

1

Institutional dimensions of floodplain restoration in Europe: an introduction

Timothy Moss and Jochen Monstadt

1.1 FLOODPLAIN RESTORATION AS A SOCIETAL CHALLENGE

Throughout history there has been a tendency in many industrialised countries to uncouple rivers from their genetic functional landscape, to regulate them and to overexploit land in river basins (Maltby and Blackwell 2005: 133). As well as being hydrologically detached from their rivers of origin, lowland river landscapes with potentially fertile soils but high water tables were brought into arable production by field drainage throughout the nineteenth and early

 Restoring Floodplains in Europe: policy contexts and project experiences edited by Timothy Moss and Jochen Monstadt. ISBN: 1843390906. Published by IWA Publishing, London, UK.

twentieth centuries (Werritty 2006: 16). Furthermore, floodplains[1] have been routinely drained and developed to make easier the construction of routeways across valleys and to reduce the risk of waterborne diseases, commonly associated with wet floodplain backswamps (Maltby and Blackwell 2005: 133). While floodplains and wetlands were traditionally avoided for settlement and commerce in most cases, this aversion disappeared during the twentieth century as unplanned urban growth spilled onto floodplains (Werritty 2006: 17).

However, the negative impact on fluvial hydrosystems of the use of floodplains for agricultural production and human settlements has today become all too apparent, aggravated in many cases by weather extremes associated with global warming. As a consequence of draining wetlands and floodplains for agriculture, of permitting steady encroachment of settlements onto floodplains and extending impermeable surfaces within urban areas, flooding in recent times has become more frequent, extreme, damaging and costly than in the past (Werritty 2006: 17). Furthermore, the alarming degradation of river ecosystems as one of the most species-rich habitats has caused growing concern in the public domain across Europe (Hughes 2003). In Europe and North America, up to 90% of floodplains have been 'cultivated' and ecologically impaired (Tockner and Stanford 2002: 310). They are thus among the most threatened ecosystems.

During the last twenty years there has, however, been growing recognition of the fact that action is required to change the way that floodplains are used and managed. Particularly during the last decade, shifts in problem-solving approaches in a number of relevant policy fields have created new openings for the sustainable management of river environments and for restoring floodplains.[2] Major flooding events across Europe, growing evidence of climate change, the continued loss and deterioration of valuable habitats, inefficiencies in agricultural production and the cost of maintaining flood defence barriers – to name but a few key drivers – are cause for growing concern across Europe. The severity of these problems is prompting policy-makers to critically reappraise many existing institutional regulations and incentives on water and land use

[1] *River floodplains* are defined as areas of low-lying land that are subject to inundation by lateral overflow water from rivers with which they are associated (cf. Tockner and Stanford 2002: 309).

[2] *Floodplain restoration* refers, for the purposes of this book, to the creation of ecosystems typical of floodplains on low-lying land which exhibit a hydrological link between river and land. We thank Bill Adams for formulating this widely accepted definition. To avoid confusion, it should be noted that the term 'restoration' used here, as commonly elsewhere, does not mean restoring river ecosystems to some previous or natural state – which is impossible for all human-modified ecosystems – but rather a form of 'rehabilitation' or 'enhancement' of their ecological functions (cf. Adams and Perrow 1999: 89-92; Eden *et al.* 2000: 259-261; Morris *et al.* 2005: 106).

with a view to improving their effectiveness in combating these problems (Adams *et al.* 2004; Ledoux *et al.* 2005). Giving more space to rivers is part of this canon of new national and European policy responses. At the same time we can observe growing interest amongst local and regional actors in restoring floodplains in order to address a variety of localised issues, ranging from flood risks and loss of biodiversity to agricultural restructuring and rural development. This interest at both strategic and operational levels is combining to create favourable conditions for floodplain restoration schemes which address multiple policy objectives, engage a wide range of stakeholders, seek to fulfil a combination of functions and take more of a catchment perspective to river and land management.

This growing public interest in the potential of floodplain restoration to improve flood protection, enhance the environment and raise life quality can tap into rich veins of scientific knowledge on the functioning of floodplain ecosystems and the regulation of rivers. Since the mid-1980s international scientific communities of ecologists, biologists, hydrologists and geomorphologists have made huge advances in our understanding of how rivers work and how they interact with their natural floodplains (e.g. Petts 1996; Brookes 1996; Ward *et al.* 1999; Hughes and Rood 2001; Hughes 2003; Rohde *et al.* 2006; Tockner and Stanford 2002). These studies highlight that natural floodplains are considered the most biologically productive and diverse ecosystems on earth, as far more species of plants and animals occur on floodplains than in any other landscape unit in most regions of the world (Tockner and Stanford 2002: 310).

In addition to providing important habitats for plants and animals, floodplains can perform a multitude of other functions and provide numerous benefits for humans. They offer potential alleviation to downstream flooding by retaining water, reducing the flow velocity of a river and minimising erosion. They improve human water supplies by filtering out sediments, nutrients and pollutants and by stabilising river flows and groundwater levels. Furthermore, floodplains provide valuable artefacts for socio-economic reproduction, such as crops and timber, and less tangible but equally valuable functions, such as attractive landscapes or opportunities for recreational pursuits. The fish and wildlife habitats help support fishing, nature tourism, hunting and other recreational uses (cf. Maltby and Blackwell 2005: 134-5; WWF 2004: 2-4).

Considering these multiple beneficial hydrological, ecological and socio-economic functions and values of river and floodplain ecosystems, the restoration of floodplains and new approaches in flood prevention oriented towards giving more space to rivers have become important policy issues. It has thus been "widely recognised that a multifunctional, holistic, catchment based approach to the management of rivers and floodplains is required" (Maltby and Blackwell 2005: 140). Natural flood reduction measures and projects on floodplain restoration are thus regarded as "environment-friendly, commonly acceptable, [and] lend themselves well to application in a transnational context, and are more sustainable than the structural measures because they are more flexible" (Kundzewicz and Menzel 2005: 130).

Given this recent surge of support for restoring floodplains among policy and research circles as well as in the public domain, it is astonishing how very few schemes for restoring functional floodplains have been put into practice in Europe to date (cf. Ledoux *et al.* 2005: 130). Whereas river restoration has in recent years become a widely practised form of environmental enhancement in Western Europe, floodplain restoration is still under-developed, being limited largely to a few, small-scale demonstration sites. The next section will examine why this is the case when floodplains represent some of the most species-rich and endangered ecosystems in Europe, offering additional benefits in flood prevention, regional economic development and recreation etc.

1.2 INSTITUTIONAL COMPLEXITIES OF FLOODPLAIN RESTORATION

The difficulties in restoring floodplains lie partly in the human activities which have contributed to the loss of functional floodplains in the past, such as settlement development, agricultural land use, flood defence and navigation. The continued importance of these activities poses strong arguments against allowing floodplains to flood naturally and ecosystems typical of floodplains to develop unimpeded. There are, though, many additional factors which contribute to frustrate schemes to restore floodplains.

Floodplain restoration places considerable demands on project managers and policy makers alike. Involving essentially a reconfiguration of the interaction between a river and adjacent low-lying land, floodplain restoration has far-reaching implications for existing forms of water and land use (on wetlands as the interface between several ecosystems, see Adger *et al.* 2000: 78). It affects a wide range of institutions – or rule systems – designed to secure a variety of private and public goods associated with water and land use. These institutions relate to water protection, nature conservation, flood defence, navigation, recreation, urban and rural development and the protection of historical landscapes, to name but a few of the most significant. A scheme to restore a floodplain requires the successful enrolment of these institutions and organisations in such a way as to create a result acceptable to the principal stakeholders. As institutional arrangements have traditionally stimulated the use of floodplains for settlements, technical flood protection, agricultural land-use etc., the task of restoring floodplains calls for complex institutional reforms in various policy fields and societal domains.

The restoration of functional floodplains requires changes to existing activities on the site of the floodplain itself but also – particularly in the case of larger schemes – along whole reaches of a river and even a whole catchment. Maintaining desirable river flow dynamics demands a catchment perspective to river regulation and cooperation between upstream and downstream users (Adger *et al.* 2000: 87). On this wider scale, floodplain restoration can significantly influence, for instance, levels of flood protection, the navigability of a river reach or the viability of current farming practices. In this way floodplain restoration affects a wide range of stakeholders and interests (Adams and Perrow 1999: 94-95; Adams *et al.* 2004; Turner *et al.* 2000: 13-14; Adger and Luttrell 2000: 78), making it potentially highly controversial. Striking an optimal balance between the multiple services a floodplain can provide, to humans and the wider natural environment, requires agreement between many actor groups spanning a variety of policy fields.

Behind these stakeholders and interests lie institutions designed to protect and provide a variety of private and public goods, ranging from commercial products to rights of access. For each of the policy fields affected by floodplain restoration – primarily water protection, flood defence, nature conservation, recreation, navigation, agriculture, forestry and urban and rural development – complex institutional arrangements have been designed and adapted over the years. Each institutional arrangement comprises a set of codified norms (such as laws, regulations and contractual obligations), planning instruments and funding mechanisms, organised and individual actors responsible for, or to, these rule systems as well as standardised procedures of operation, values and accepted practices of the relevant actors. These institutions span a wide range of often diverse interests, and frequently pursue competing objectives. They can be – and often are – inconsistent, sending contradictory signals (Turner *et al.* 2000: 10).

Managing the interdependence of multiple functions, actors and institutions is, however, not the only major socio-political challenge to floodplain restoration. Considerable uncertainty is generated by the diverse spatial scales and time frames involved (Hughes *et al.* 2007). Spatially, restoring a floodplain requires consideration of the longitudinal connectivity of a floodplain to river uses both up and down stream, as well as of the lateral connectivity to ways in which adjacent land is used (Adams and Perrow 1999: 92-93). Even on site, interventions generally cut across several functional and administrative boundaries. These can relate to the spatial remit of local landowners and farmers, planning authorities, government agencies, protected areas or infrastructure networks (e.g. railways, roads, canals).

Temporally, functional floodplains are characterised by their dependence on flooding events, which are by their nature periodic and unpredictable, and by significant time lags between changes in biotic and abiotic systems (Adams and Perrow 1999: 93). Socio-economically too, the process of restoring a floodplain is framed by diverse time scales, ranging from the payback periods for investments in altered practices of agriculture and forestry to the electoral periods of key public authorities. Floodplain restoration is, therefore, not only highly complex but also highly unpredictable. The institutional challenge is further complicated by the fact that the many institutional arrangements, in particular for nature conservation, exhibit a strong tendency to protect existing conditions rather than encourage change. Those which do pursue change are generally geared towards achieving specific targets rather than creating suitable frameworks for open-ended processes, as is required for functional floodplains. This tendency towards institutional inertia is reinforced by the longevity of the material-physical components of the flood defence infrastructure or settlements on floodplains. Once they are built, it is complicated, time-consuming and in many cases expensive (at least in the short term) to substitute them with alternative technological solutions or to dismantle them.

1.3 STATE OF THE ART AND PURPOSE OF THE BOOK

Considering this institutional complexity, surprisingly little is known about the institutional dimensions of floodplain restoration. Our knowledge of institutions, which can support – or obstruct – the protection of public goods such as water, flood defence and biodiversity has been developing rapidly over the past decade (cf. Breit *et al.* 2003; Young 2002). This knowledge has been widely applied to river basin management as an exemplar of an integrated, strategic approach to managing water and land use at several scales and in a variety of different contexts (Jaspers 2003; Kliot *et al.* 2001; Bernauer 2002; Moss 2003). Floodplain restoration has, however, only rarely been researched from an institutionalist perspective. Exceptions include studies of institutional constraints and complexities (Adams and Perrow 1999; Hodge and McNally 2000) and of competing discourses of floodplain restoration (Adams *et al.* 2004). Overviews of the impact of EU policies on floodplains and their restoration have been published by the World Wide Fund for Nature, on the basis of its "Wise Use of Floodplains" programme (WWF 2000, 2004), and by Maltby and Blackwell (2005) and Dworak and Görlach (2005). Case studies in Europe have been analysed comparatively, but only in brief, by Zöckler (2000), Tunstall *et al.* (1999) and Morris *et al.* (2005). There exist a larger number of studies on economic valuations of the functions, services and values provided by floodplains or wetlands (Gren *et al.* 1995; Turner *et al.* 2000; Adger and Luttrell 2000). Eden *et al.* (2000) apply an Actor Network Theory perspective to analyse the interplay between human and natural processes on the project to restore the River Cole in England. However, there exists no systematic analysis of the policy contexts of floodplain restoration in individual European countries or a detailed comparative analysis of experiences of floodplain restoration schemes in different European countries.

This book is designed to help fill this knowledge gap on the institutional drivers and constraints of floodplain restoration in Europe. Here, we understand drivers to refer to institutional forces conducive to floodplain restoration. These forces – it should be emphasised – are neither given nor static, but are constructed by stakeholders and are, accordingly, subject to change. Drivers can, therefore, range from the problem perceptions of key stakeholders to political reforms or economic incentives. Similarly, constraints need to be conceived not as fixed barriers 'out there' which need to be circumvented, but as institutional forces working against floodplain restoration. Some act as direct counter-forces to specific drivers, some are sector-specific but do not address a specific driver whilst others are of a general nature, affecting a range of policy fields.

With this book we aim to explain why so few projects have been successfully implemented, how recent policy shifts are creating new opportunities for floodplain restoration and what lessons for policy development and project management can be drawn from in-depth analysis of past and present schemes. At a time of growing interest in restoring floodplains as an important component of efforts to improve flood protection, enhance riparian habitats, strengthen catchment management, raise water quality and pursue integrated rural development, we explore the relationship between macro-level policy development and enforcement, and micro-level project design and implementation. We provide a comparative analysis of past and ongoing efforts to restore floodplains in three EU member states (France, Germany and Britain) at both strategic (policy) and operational (project) levels. The book explores the policy contexts of floodplain restoration, focusing on the principal drivers and constraints, analyses six schemes to restore floodplains (two from each country studied) and summarises comparatively the principal findings, providing recommendations for ways of developing policy and designing projects in the future.

The book is targeted at researchers, policy makers, project managers and other practitioners engaged or interested in policy development and project implementation at EU, national and local levels in fields relevant to river and floodplain restoration. It is designed to provide professionals and social scientists with as yet largely undocumented information on, and assessment of, general constraints and drivers of floodplain restoration, illustrated with examples from six detailed case studies. On the basis of this analysis it provides practical recommendations for the design and implementation of floodplain restoration schemes and for ways of improving policy delivery in the future. In particular for the research community, it addresses important issues of governance relating to the integrated management of water and land use in complex, multi-stakeholder contexts and the different ways in which the actors studied strove to exploit institutional drivers and cope with the constraints.

1.4 THE RESEARCH CONTEXT OF THE BOOK

This book is based on original research conducted between 2000 and 2003 under the project "FLOBAR2", funded by the European Union within its 5th Research Framework Programme.[3] FLOBAR2 was an interdisciplinary project of natural and social scientists investigating hydrological, biological and institutional aspects of the restoration of floodplains and their woodlands (on the project as a whole, see Chapter 2 in this volume). The central project objective was to develop scientific guidelines for the allocation of river flows, which benefit floodplain ecosystems while maintaining acceptable levels of flood control. This task culminated in the joint publication "The flooded forest: guidance for policy makers and river managers in Europe on the restoration of floodplain forests" (Hughes 2003). The socio-economic component of FLOBAR2 was conducted at the Leibniz Institute for Regional Development and Structural Planning (IRS), Erkner (Germany). This book – on institutional aspects of floodplain restoration in Europe – is the principal product of the socio-economic investigations on the FLOBAR2 project, updated with supplementary investigations in 2006/07. The research objective was to identify institutional drivers and constraints of floodplain restoration, to analyse how they affected the design and development of restoration schemes and to draw conclusions on ways of promoting floodplain restoration – at both the operational (project) and strategic (policy) levels – on the basis of these observations.

Pursuing a 'funnelled' research strategy, the project team first conducted a review of the institutional framework of river basin management in France, Germany, England and Wales, using a literature and document analysis to describe the status and responsibilities of the relevant organisations, and map their interdependencies and spatial remits. In a second step, it then analysed the institutional mechanisms affecting floodplain restoration in the same countries. The principal policy, legislative and planning documents relevant to floodplain restoration were evaluated, on the basis of substantial documentary material and

---

3 The full project title was "Floodplain Biodiversity and Restoration 2: Integrated natural science and socio-economic approaches to catchment flow management" (EVK1-CT-1999-00031). FLOBAR2 was coordinated by Keith Richards and Francine Hughes of the Department of Geography at the University of Cambridge (UK). The other project partners were: the University of Grenoble (France), The Centre National de Recherche Scientifique, Toulouse (France), the University of Umea (Sweden), the University of Lethbridge (Canada) and the Leibniz Institute for Regional Development and Structural Planning, Erkner (Germany). The project website is http://www-flobar.geog.cam.ac.uk/.

circa 10 face-to-face interviews with experts in each country, in terms of their importance and effectiveness. To reflect the importance of institutional interplay on floodplain restoration, the analyses drew on a wide range of relevant policy fields, from water quantity and quality management to flood protection, nature conservation, forestry management, land-use planning and agricultural production.

Based on a review of relevant floodplain restoration projects in each country, the third step was to conduct in-depth case studies of six schemes of floodplain restoration (two each in France, Germany and England), comprising circa 12 face-to-face interviews and the analysis of extensive documentary material in each case. The selection of suitable case studies was informed by the desire to include on the one hand schemes which were already completed, in order to observe their evolution and assess their impact, and on the other hand schemes which reflect the recent shift in thinking towards a more integrated, catchment-oriented approach to floodplain management.

1.5 STRUCTURE OF THE BOOK

The structure of the book largely follows this funnelled research design, setting microanalyses of individual floodplain restoration schemes in broader institutional contexts, with some notable additions. Prior to the analysis of policy contexts the first section (Block A) sets out the key issues of floodplain restoration in Europe. Following this introductory chapter, Francine Hughes and Keith Richards make the case for restoring functional floodplains on the basis of the results from the FLOBAR2 project as a whole, providing a natural science perspective and linking this to the social science research of this book.

In Block B the focus of attention shifts to the policy contexts of floodplain restoration, first at EU level and subsequently in Germany, France and England and Wales. Thomas Dworak analyses EU policy in the fields of flood prevention, water resource management, agriculture, rural development, nature conservation, urban and regional planning and coastal zone management to identify factors working in favour of, and against, floodplain restoration. The subsequent country analyses identify the principal policy, legislative and planning documents relevant to floodplain restoration, specify references to floodplains and their restoration in these documents, describe the principal organisations and instruments relevant to floodplain restoration and summarise how they work in practice. Comparability between the three chapters – authored by Jochen Monstadt and Timothy Moss (Germany), Jochen Monstadt (France) and Timothy Moss (England and Wales) – is enhanced by having an identical structure.

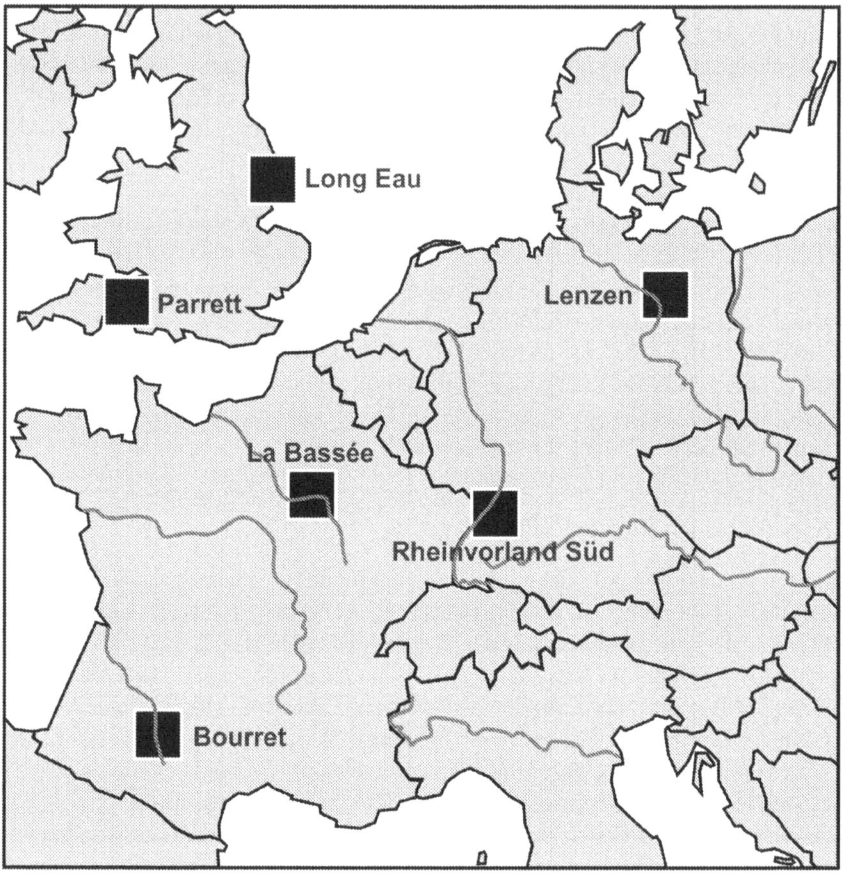

Figure 1.1: Map illustrating the location of the six case studies.

The following two sections of the book are dedicated to case studies of six schemes of floodplain restoration in France, Germany and England (see Figure 1.1). Block C covers three schemes selected as examples of early generation projects which tended to be relatively limited in terms of their objectives, the instruments used, the number of actors involved, the spatial scale of reference and the area of floodplain restored. Maria Hagemeier and Axel Klaphake investigate a small-scale scheme to reconnect an old arm to the river Garonne and restore an alluvial forest – primarily for environmental benefits – at Bourret, in the Midi-Pyrenées Region of France. Sylvia Kruse analyses the Rheinvorland-Süd Project, in Baden-Württemberg, Germany. This was a scheme to improve hydrological and ecological conditions in a section of the

Rhine floodplain near Rastatt, widening ducts and removing structures to allow water to enter the floodplain and dead arms. Timothy Moss looks at the Long Eau Project, near Louth in Lincolnshire, England, a small-scale, completed scheme to set back flood banks for both environmental and flood defence benefits.

The chapters in Block D then address three schemes selected as examples of recent generation projects which are often more ambitious and complex in terms of the issues they address, the actors they enrol, the instruments they apply and their spatial scale of reference. Maria Hagemeier and Axel Klaphake explore a planned, large-scale flood retention scheme at La Bassée on the Seine upstream of Paris, which is designed to provide multiple benefits of improved flood protection, biodiversity and regional development. The German case, authored by Jochen Monstadt, is of a large-scale scheme to set back flood banks along a length of the River Elbe near Lenzen, in Brandenburg, allowing flooding of the floodplain primarily for nature conservation, but also flood defence and regional development benefits. Ross Beveridge and Timothy Moss investigate the Parrett Catchment Project in Somerset, England. This was a project to promote more sustainable techniques of flood management in the whole catchment of the River Parrett, seeking consensus between often divergent flood protection, nature conservation and agricultural production interests.

The principal findings of the book are summarised in the concluding chapter in Block E, drawing on the findings from each of the policy analyses and case studies. Timothy Moss and Jochen Monstadt first appraise the main institutional drivers and constraints to floodplain restoration, substantiated with detailed empirical evidence from the case studies. They then describe and explain the emergent 'windows of opportunity' for floodplain restoration being created as a result of recent shifts in policy objectives relating to a number of relevant policy fields. Subsequently, the authors assess the comparative strengths and weaknesses of the new generation of multi-purpose, catchment-oriented floodplain restoration projects seeking to exploit the emergent 'windows of opportunity'. On this basis the chapter concludes with recommendations for policy development, for project design and implementation and for future research needs.

1.6 REFERENCES

Adams, W. M. and Perrow, M. (1999) Scientific and institutional constraints on the restoration of European floodplains. In *Floodplains: Interdisciplinary Approaches* (eds. S. Marriott, J. Alexander and R. Hey), Geological Society, London, Special Publications, 163, pp.89-97.

Adams, W. M., Perrow, M. R. and Carpenter, A. (2004) Conservatives and champions: river managers and river restoration discourses in the United Kingdom. *Environment and Planning A*, **36**, 1929-1942.

Armbruster, J., Muley-Fritze, A., Pfarr, U., Rhodius, R., Siepmann-Schinker, D., Sittler, B., Späth, V. Trémolières, M., Rennenberg, H. and Kreuzwieser, J. (eds.) (2006) *FOWARA* -

Forested Water Retention Areas: Guideline for decision makers, forest managers and land owners. University of Freiburg, Freiburg.

Adger, W.N. and Luttrell, C. (2000) Property rights and the utilisation of wetlands. *Ecological Economics*, **35**, 75-89.

Bernauer, T. (2002) Explaining success and failure in international river management. *Aquatic Sciences* **64**(1), 1-19.

Böhm, H. R., Haupter, B., Heiland, P. and Dapp, K. (2004) Implementation of flood risk management measures into spatial plans and policies. *River Research and Applications* (Special Issue "Towards Sustainable Flood Risk Management in the Rhine and Meuse River Basins") **20**(2), 255-267.

Breit, H., Engels, A., Moss, T. and Troja, M. (eds.) (2003) *How Institutions Change. Perspectives on Social Learning in Global and Local Environmental Contexts.* Leske + Budrich, Opladen.

Brookes, A. (1996) Floodplain restoration and rehabilitation. In *Floodplain Processes* (eds. M.G. Anderson, D.E. Walling and P.D. Bates), John Wiley, Chichester, Sussex, pp.553-576.

Dworak, T. and Görlach, B. (2005) Flood risk management in Europe - the development of a common EU policy. *International Journal of River Basin Management* **3**(3), 97-103.

Eden, S.; Tunstall, S.M. and Tapsell, S.M. (2000) Translating nature: river restoration as nature – culture. *Environment and Planning D* **18**, 257-273.

Gren, I.-M., Groth, K.-H. and Sylvén, M. (1995) Economic Values of Danube Floodplains. *Journal of Environmental Management* **45**, 333-345.

Hale, B. W. and Adams, M. S. (2007) Ecosystem management and the conservation of river-floodplain systems. *Landscape and Urban Planning* **80**(1-2), 23-33.

Hodge, I. and McNally, S. (2000) Wetland restoration, collective action and the role of water management institutions. *Ecological Economics* **35**, 107-118.

Hooper, B. (ed.) (2005) *Integrated River Basin Governance: Learning from International Experience.* Urban and Regional Planning and Development Series. IWA Publishing, London, UK.

Hughes, F.M.R. (ed.) (2003) *The flooded forest: guidance for policy makers and river managers in Europe on the restoration of floodplain forests.* Department of Geography, University of Cambridge, Cambridge, UK.

Hughes, F.M.R., Moss, T. and Richards, K.S. (2007) Uncertainty in riparian and floodplain restoration. In *Uncertainty in River Restoration* (eds. S. Darby and D. Sears). Wiley, Chichester (in press).

Hughes, F.M.R. and Rood, S.B. (2001): Floodplains. In *Habitat Conservation: Managing the Physical Environment* (eds. A. Warren and J.R. French), John Wiley & Sons Ltd., Chichester, UK, pp.105-121.

Jaspers, F.G.W. (2003) Institutional arrangements for integrated river basin management. *Water Policy* **5** (2003), 77–90.

Kliot N., Shmueli D. and Shamir U. (2001) Institutions for management of transboundary water resources: their nature, characteristics and shortcomings. *Water Policy* **3(3)**, 229-255.

Kundzewicz, Z.W. and Menzel, L. (2005) Natural flood reduction strategies – a challenge. *International Journal of River Basin Management* **3**(2), 125-131.

Ledoux, L., Cornell, S., O'Riordan, T., Harvey, R and Banyard, L. (2005) Towards sustainable flood and coastal management: identifying drivers of, and obstacles to, managed realignment. *Land Use Policy* **22**, 129-144.

Maltby, E. and M.S.A. Blackwell (2005) Managing riverine environments in the context of new water policy in Europe. *International Journal of River Basin Management* **3**(2), 133-141.

Morris, J., Hess, T.M., Gowing, D.J.G., Leeds-Harrison, P.B., Bannister, N., Vivash, R.M.N. and Wade, M. (2005) A framework for integrating flood defence and biodiversity in washlands in England. *International Journal of River Basin Management* **3**(2), 105-115.

Moss, T. (ed.) (2003) *Das Flussgebiet als Handlungsraum: Institutionenwandel durch die EU-Wasserrahmenrichtlinie aus raumwissenschaftlichen Perspektiven*. LIT, Münster.

Moss, T., Klaphake, A., Monstadt, J. and Siener, M. (2003) What prospects for restoration? In *The flooded forest. Guidance for policy makers and river managers in Europe on the restoration of floodplain forests* (ed. F. Hughes), University of Cambridge, Cambridge, pp.71-84.

Petrow, T., Thieken, A. H., Kreibich, H., Merz, B. and Bahlburg, C. H. (2006) Improvements on Flood Alleviation in Germany: Lessons Learned from the Elbe Flood in August 2002. *Environmental Management* **38**(5), 717-732.

Petts, G. E. (1996) Sustaining the ecological integrity of large floodplain rivers. In *Floodplain Processes* (eds. M.G. Anderson, D.E. Walling and P.D. Bates), John Wiley, Chichester, Sussex, pp.535-551.

Puhlmann, G. and Jährling, K.-H. (2003) Erfahrungen mit "nachhaltigem Auenmanagement" im Biospärenreservat "Flusslandschaft Mittlere Elbe". *Natur und Landschaft* **4**, 143-150.

Rohde, S., Hostmann, M., Peter, A. and Ewald, K. C. (2006) Room for rivers: An integrative search strategy for floodplain restoration. *Landscape and Urban Planning* **78**(1-2), 50-70.

Tockner, K. and Stanford, J. A. (2002) Riverine flood plains: present state and future trends. *Environmental Conservation* **29**(3), 308-330.

Tunstall, S.M., Tapsell, S.M. and Eden, S. (1999) How Stable are Public Responses to Changing Local Environments: A 'Before' and 'After' Case Study of River Restoration. *Journal of Environmental Planning and Management* **42**(4), 527-547.

Turner, R.K., van den Bergh, J.C.J.M., Söderqvist, T., Barendregt, A., van der Straaten, J., Maltby, E. and van Ierland, E.C. (2000) Ecological-economic analysis of wetlands: scientific integration for management and policy. *Ecological Economics* **35**, 7-23.

Ward, J.V., Tockner, K. & Schmeier, F. (1999) Biodiversity of floodplain ecosystems: ecotones and connectivity. *Regulated Rivers: Research & Management* **15**, 125-139.

Werritty, A. (2006) Sustainable flood management: oxymoron or new paradigm? *Area* **38**(1), 16-23.

WWF (World Wide Fund For Nature) (2000) Wise Use of Floodplains. Policy and Economic Analysis of Floodplain Restoration in Europe. Opportunities and Obstacles. Report, November 2000.

WWF (2004) Living with floods: Achieving ecologically sustainable flood management in Europe. Policy Briefing.

Young, O. (2002) *The Institutional Dimensions of Environmental Change. Fit, Interplay, and Scale*. MIT Press, Cambridge (MA).

Zöckler, C. (2000) Wise Use of Floodplains – LIFE Environment Project: A Review of 12 WWF River Restoration Projects Across Europe. WWF European Freshwater Programme, Copenhagen.

2

Floodplains in Europe: the case for restoration

Keith Richards and Francine Hughes

2.1 INTRODUCTION

In the 1990s there was an increasing interest in river restoration, both in the UK and in Europe more widely. In the UK between 1994 and 1998, the River Restoration Project undertook recovery of the natural (meandering) morphology and ecology of the channelised River Cole on the Oxfordshire-Wiltshire border, and the River Skerne in County Durham. From 1998, the "Project" was succeeded by a "Centre", which continues to promote the principles of river restoration. The European Centre for River Restoration was established in 1995, and performs a similar function across Europe, "support[ing] the development of river restoration as an integral part of sustainable water management throughout Europe, ensuring that projects will be more cost effective, more likely to succeed, and will encompass multifunctional objectives" (ECRR 2006).

The work of these organisations is likely to become increasingly important as the European Union Water Framework Directive (WFD) becomes operational, and national and federal water management agencies seek to improve aquatic and riparian ecology in order to comply with its objectives.

The focus of the restoration efforts of the 1990s was mainly on the river channel and the aquatic ecology, probably for the pragmatic reason that land ownership constraints inhibited extending the practice laterally. By comparison, therefore, there was a relative lack of focus on floodplain restoration, in spite of (i) the evidence of declining ecological status of floodplains because of direct and indirect effects of both land use and river management; and (ii) the scientific evidence of the integrated functioning of the river and its floodplain, of the role of river corridors in providing a larger-scale structure to the functions of habitat, migration and dispersal, and the spatial framework for meta-populations. To respond to the evidence of decline, and to reverse it where possible and desirable, requires further scientific research into the interdependence of rivers and their floodplains, and improved understanding of how their varying inter-relationships influence river and floodplain properties and functions, such as those involving flood attenuation and biodiversity. This will rectify the imbalance by allowing greater restoration focus on riparian and floodplain environments.

The research reported in this book is a contribution to this effort, and is based on the EC-funded FLOBAR2 project (FLOodplain Biodiversity And Restoration 2: integrated natural science and socio-economic approaches to catchment flow management; 2000-2003). As its full title indicates, this project reflected the inter-disciplinary nature of the research required to support floodplain restoration initiatives. Such initiatives may vary widely in their primary focus on different floodplain functions, depending on local circumstances, implying that valuation and choice on the part of both the affected publics and the responsible institutions constitute part of the research question posed in each case. This chapter discusses the theoretical and practical implications of this inter-disciplinarity and reviews the status of European floodplains, with particular reference to those that could be re-naturalised by allowing riparian, or "wet", woodland to regenerate. It then considers the case for restoration in terms of the potential value of various natural floodplain functions, and the current opportunities for restoration, set against the various institutional constraints.

2.2 THE FLOBAR2 PROJECT AND THE INTER-DISCIPLINARITY OF FLOODPLAIN RESEARCH

The FLOBAR2 project grew out of previous EC-funded research on the controls of declining regeneration of floodplain woodland, and had four primary objectives. The first was to develop scientific guidelines for the application of river flow prescriptions that benefit floodplain ecosystems while maintaining acceptable levels of flood control. The second was to quantify aspects of the relationships between hydrological inputs to a floodplain and plant response, in terms of water consumption, growth and diversity of regeneration strategies. Thirdly, it aimed to investigate the flow resistance associated with woody riparian vegetation using field studies and mathematical and numerical modelling. Finally, it investigated and compared the effectiveness of selected institutional arrangements for restoring floodplain environments at different spatial scales and in different national/local settings. These objectives were addressed in seven Work Packages (WP) that are summarised in Table 2.1 and illustrated in Figure 2.1.

From the beginning, FLOBAR2 was an avowedly inter-disciplinary project. The minutes of the initial meeting of partners includes the co-ordinator's welcome, and states:

> "He stressed the inter-disciplinary nature of the project, and the importance of properly integrated social and environmental science (both because of FP5 priorities, and the aims of FLOBAR2 itself). The Work Package on institutional structures was concerned with the reflexive relationship between science and policy, and the guidelines to be produced for Work Package 1 would need to consider adaptations of scientific priorities to the institutional goals, as well as institutional changes as scientific understanding and needs develop" (Keith Richards, FLOBAR2 Meeting, 11-13 May 2000).

European research funding, since Framework Programme 3 (FP3)[1], emphasised the need to incorporate socio-economic dimensions into research and development projects in many fields. However, this had also been based on an assumption that socio-economic research questions were concerned with downstream analysis of policy and its implementation. Debates about this inter-disciplinarity have increasingly emphasised the importance of upstreaming the socio-economic research so that it constitutes an integral part of a project from the outset, and may even lead to adjustment of research agendas, or even to agendas that are largely determined by explicitly researched and understood social needs, rather than implicit goals driven by what are seen to be natural science or policy imperatives. In the river restoration field, Eden *et al.* (2000) have shown that social preferences and goals are not always consistent with scientific and science policy goals.

---

[1] The European Commission funds Research and Development in the European Research Area through a series of 5-year Framework Programmes; FP3 was from 1990-1994.

Table 2.1: The Work Packages into which the FLOBAR2 project was divided.

WP1	A manual of river flow prescriptions for riparian ecosystems. This developed scientific guidelines on best practices for water allocation to improve the quality of floodplain ecosystems, collating information and expertise from sources including the scientific community and end-users.
WP2	The influence of hydrology on water uptake and growth of woody riparian species. This investigated links between hydrological inputs to floodplains, and the consumption of water by, and growth of, woody riparian species.
WP3	The influence of hydrology and litter accumulation on riparian species diversity. This investigated plant community response to manipulated changes in hydrology and litter dynamics with particular reference to their impacts on biodiversity.
WP4	The influence of hydrology on sex ratios in dioecious* floodplain species. This investigated the influence of hydrology and other environmental variables on sex ratios of members of the dioecious Salicaceae family, and thus the regeneration strategies exhibited by these riparian species.
WP5	The influence of hydrology on sexual and vegetative regeneration strategies in floodplain species. This investigated the influence of hydrological and sedimentological inputs to floodplains on the balance of vegetative and non-vegetative regeneration exhibited by the early successional members of the genus Populus.
WP6	Flow resistance properties of woody riparian vegetation. This investigated the flow resistance of riparian vegetation as it changes during restoration initiatives and in response to flow management in floodplain zones.
WP7	Institutional mechanisms for floodplain management. This assessed the effectiveness of institutions for restoring floodplain environments and enhancing floodplain ecosystems at different scales and in different settings and consistent with scientific guidelines.

*Dioecious plants are species that have male and female reproductive structures on separate plants: examples are the Salicaceae family, which includes two genera, willows (*Salix*) and poplars (*Populus*).

Of course, there is no guarantee that what is referred to as inter-disciplinary actually conforms in practice to what this collaboration demands. Instead, the scientific practice may be closer to multi-disciplinary or cross-disciplinary research, in which researchers from different disciplines contribute in studying a problem, and cooperate and coordinate. Inter-disciplinarity requires teams to engage in joint activity, to share responsibility, and to adapt their research practices mutually. Even in inter-disciplinary research, disciplinary assumptions may persist about legitimate research themes, acceptable methods, and agreed truth-claims and warrants, and this may prevent the emergence of innovative intellectual connections (Brewer 1999). In FLOBAR2, there were deliberate attempts to involve the environmental science members of the team in the social science research, and to interact in developing the outputs for Work Packages 1 and 7. However, this did not extend to revision of the goals of the environmental science Work Packages; it would require a remarkably flexible funding arrangement to permit such reflexivity of research objectives.

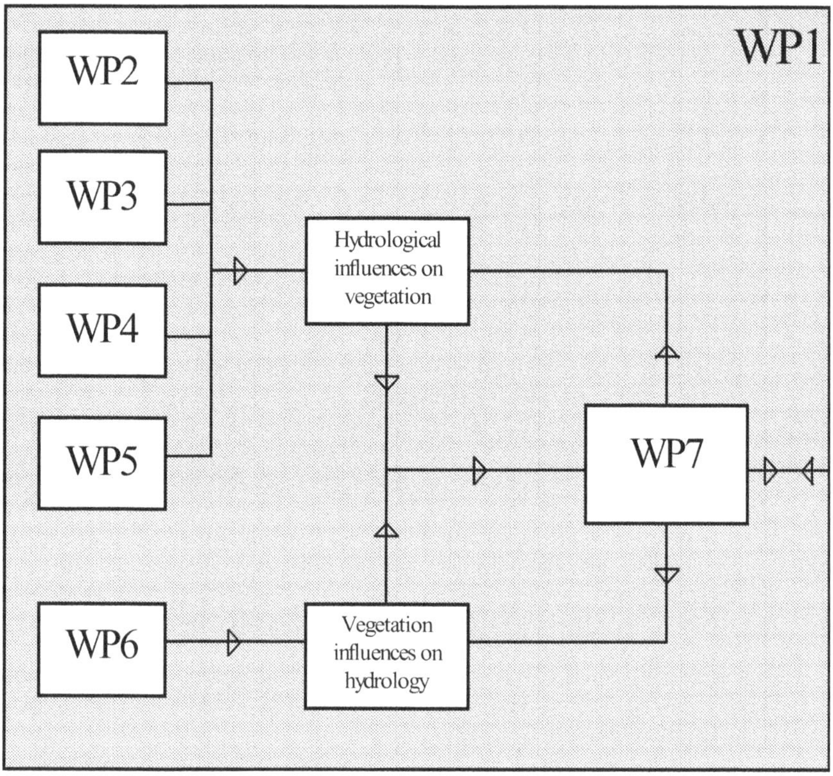

Figure 2.1: The structure of Work Packages (WP) in the FLOBAR2 project. This shows that Work Package 7 integrates the output from the environmental science projects and resolves the potential conflict between the goals of biodiversity gain and flood control.

Instead, therefore, the project was set up to expose a conflict potentially arising because floodplain restoration is viewed positively by ecologists and conservationists (in terms of the potential for biodiversity gain), but negatively by engineers (in terms of weakening flood control). This implies that a resolution of this conflict is required before restoration initiatives become institutionally acceptable goals, and also that it is the responsible management agencies that must achieve this resolution. As implied by the role of Work Package 7 in Figure 2.1, this process is itself part of the research project, as there is no guarantee that existing policies and agencies are structured to facilitate either the resolution, or the restoration that may then follow. The minutes of the initial project meeting show Tim Moss educating the partners in the social science meaning of "institution":

> "Tim outlined the social science definition of "Institutions"; a complex idea, not meaning just an organisation (or agency), but sets of rules governing human actions (politically/socially accepted), value systems, norms and customs (e.g. political, legal, social structures). A key point about formalised institutions is that they should reflect informal social institutions; and institutional analysis is about the web of inter-relations between the formal and the informal; and about dynamics of change in both" (Tim Moss, FLOBAR2 Meeting, 11-13 May 2000).

Thus, both responsible agencies (institutions, *sensu stricto*) and prevailing ideas may limit the opportunities for river restoration in practice; while the kinds of restoration objective and the required underpinning science may need to be tailored to the extant institutional contexts (*sensu lato*) in different national and local circumstances. This means that the value of the scientific Work Packages may vary across Europe, as the formal institutions for river management, and the prevailing emphasis in river management objectives, themselves vary. Noticeably, in the lifetime of the project, that emphasis was undergoing rapid change, with increasing acceptance of the potential for floodplain restoration in appropriate locations within catchments, in order to effect win-win scenarios in which both biodiversity gain and flood management formed joint objectives. A FLOBAR2 workshop of scientists and practitioners took place on 16-17 May 2002 to explore the requirements for the Guidelines document of WP1, and revealed that already by this stage, the UK Environment Agency was increasingly interested in this possibility. This itself shows that the institutional framework is not fixed and given, but itself needs to form part of the research project, about which recommendations may be included in the Guidelines.

The question of inter-disciplinarity is significant both for the identification of research questions, and for the question of the scientific closure required for the framing of reliable knowledge (although these are closely related). Where the inter-disciplinarity is derived from the instrumental demands of policy, it can strongly affect the scientific agenda. This is evident from this example derived from an EC Framework 6 Work Programme Research Topic:

> **Task 4 - Ecological status assessment - filling the gaps:** Development of tools in the implementation of the Water Framework Directive, such as hydromorphological pressures (e.g. issues such as river continuity), testing of the European Fish Index in Mediterranean and central and Eastern eco-regions, and the application of classification schemes in very large rivers.

This specifically defines a research objective as providing a policy implementation tool, and links it to a gap-filling exercise related to the output of a prior research project which developed an index, prosecuted within an existing policy context. This context is deemed settled and unchallengeable, although the European Fish Index is by no means uncontested scientifically or as a policy tool. The tool is a practical descriptor of the ecological status of a river, based on pre-existing scientific knowledge (which defines the choice of variables in the index), and the pre-determined policy agenda (the Water Framework Directive). Neither the scientific methods to be used nor the form of scientific output required are likely to lead to fundamentally new science, and the research agenda is rendered sacrosanct by the policy agenda. The scientific closure (that is, the experimental closure) required to generate reliable and testable knowledge is here defined by these external constraints, rather than being defined by the scientific questions posed.

In an inter-disciplinary context in which the institutions and policies themselves form part of the research project, it is possible for scientific investigation to continue the search for more basic knowledge, as in the projects undertaken for Work Packages WP2 to WP6 (see Figure 2.1). These are the most reductionist Work Package projects in FLOBAR2, and they represent a conventional form of "reliable science", or "Mode 1 science" in the terms outlined by Gibbons *et al.* (1994). In this research, experimental closure is of the kind perceived necessary to generate "reliable" scientific knowledge. However, different combinations of the results of these and other investigations are drawn upon for a different scale of closure, in which the scientific knowledge base required in particular locations and for particular restoration projects is quite specifically selected. Each example of river or floodplain restoration is itself a form of experiment, which relies on a variety of elements of reliable scientific knowledge, depending on the goals for the project set by the responsible agencies, and the particular knowledge required by the geographical contingencies of the location. A form of "robust science" (Irwin 1995) or "Mode 2 science" (Gibbons *et al.* 1994) is therefore required, in which elements of "reliable science" are selected by the agencies involved, and the process whereby they achieve this in a successful restoration is part of the experiment itself. This suggests a hierarchical form of closure, in which conventional "reliable science" effects its closure at a different scale from a new kind of "robust science", which effects closure by selecting certain relevant areas of scientific knowledge for particular purposes and contexts (and ignores others, or rejects them as irrelevant). The case studies of floodplain restoration undertaken in Work Package WP7 in FLOBAR2, and described in this book, perhaps exemplify this latter "robust" form of closure.

2.3 EUROPEAN FLOODPLAINS

Evolution of this kind of robust science and policy is necessitated by the varied circumstances in which management and restoration of rivers and their floodplains may take place. Essentially, adaptive and flexible combinations of science are required, both spatially as the priorities for management differ from one river corridor to another, and temporally as the essentially indeterminate responses of rivers to various interventions demand continual changes of strategy. Formal catchment and river management institutions thus have to select the science they require, depending on circumstances and possibilities. In turn, those formal institutions adapt to the context of informal institutions, such as the ideas, perceptions, expectations and wishes of the publics they serve, while also acting to adapt the latter, both directly and indirectly, through educative processes that may change perceptions and expectations. In Europe, this inter-disciplinary science and policy is nevertheless constrained in both its development and its potential for action by the severely impacted status and character of European floodplains. These issues are discussed in the following sections.

2.3.1 The natural status of European floodplains

Europe includes some very large trans-boundary river systems. The floodplains of both main rivers and tributaries have, with few exceptions, been occupied and used by humans for several thousand years. Most natural floodplains could be characterised as mosaics of forests, grassy clearings and small wetland areas, prior to the Iron Age. There is much archaeological evidence that, as early as the Iron Age and early Roman periods, many floodplains were settled and being cultivated; for example, in the Thames River valley in Britain there are many prehistoric occupation sites. There are also biological remains in the floodplain and the first gravel terrace of the Thames in Oxfordshire that are evidence of grasslands maintained for hay and grazing, many of which persisted well into the 20th Century. Perhaps for this reason, the aim of the restoration of the River Cole (see Section 2.1 above) was to recover its natural, but stable, meandering character, and to restore wet (hay) meadow with a rare snake's head fritillary. Interestingly, consultation with local people revealed opposition to these proposals: a straight incised channel with dry banks was preferred (Eden *et al.* 2000). The River Restoration Project valued a particular form of naturalised floodplain environment, but not necessarily that preferred by its local public, and not even necessarily that favoured by all scientists. Restoration may, therefore, involve somewhat arbitrary choices based on an ecologically-favoured or simply a practicable "reference condition" (see Section 2.5.3 below).

Some legitimisation for the chosen form of restoration is provided because only 1.9% of pollen from Iron Age floodplain sites in the general area of the River Cole shows evidence of remnant floodplain tree and scrub species (Robinson 1978; Peterken and Hughes 1995). However, as many European floodplain trees species are members of the *Salicaceae* family, whose pollen does not preserve, there may have been more trees at this time than is suggested by the pollen evidence. Although floodplain land use as hay meadow was common in Europe as well, tracts of floodplain forest remained until the 18th

Century in many locations in mainland Europe and particularly in the large river systems of Eastern Europe (Petts 1989). Thus, floodplain woodland or forest, of a dynamically changing and spatially complex form, is a potential early reference state for many river corridors, and the following sub-sections discuss some of the characteristics of wooded floodplains in Europe, in principle and in reality. In the UK, wet woodlands on floodplains constitute a threatened natural habitat for which restoration is a target under the UK Biodiversity Action Plan.[2]

2.3.2 Patterns in floodplain ecology

Erosion and deposition are processes that drive a natural propensity for unmanaged rivers to migrate across their floodplains, and thereby to create sites whose physical habitats are suitable for the regeneration of typical floodplain vegetation species and plant assemblages. This results in the vegetation mosaic typified by Figure 2.2 below (see also section 2.5.2). Healthy stands of alluvial or riparian forest incorporate a mixture of forested and non-forested areas in given river corridors, pioneer species on recently-deposited sandy or gravelly sediment, wetland areas in abandoned channels infilled with fine sediment, and more mature, drier woodland on higher areas such as levées or low terraces that are less frequently flooded. In the Drôme River valley in France, the difference between floodplain vegetation mosaics on channelised and non-channelised reaches of a river can be seen in vivid contrast. The former are characterised by a very narrow band of floodplain vegetation and restricted vegetation regeneration patterns, and the latter by a highly mobile channel and a more diverse vegetation mosaic (Barsoum 2002; Dufour 2005).

Across Europe, at a larger spatial scale, there are river reaches (most of them in Eastern Europe, but there are some stretches in Western Europe) that still contain dynamic and self-regenerating floodplain forests. Figure 2.3 illustrates their distribution; this is a map of European floodplain forests, based on data compiled for the FLOBAR2 project by UNEP-WCMC (2000). According to the EU Interpretation Manual of European Habitats (1999) there are five categories of forest to be found along European rivers (Table 2.2). These types reflect Europe-wide variations in the typical forms of alluvial forest, dependent on the climatic and hydrological regimes, and the regional climatic and ecological histories.

---

[2] See http://www.ukbap.org.uk/UKPlans.aspx?ID=4

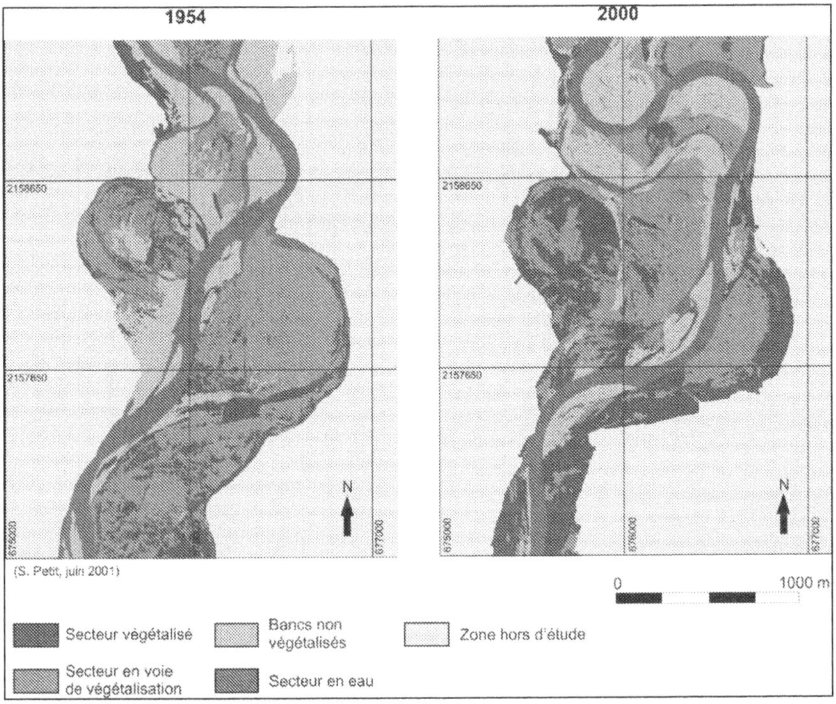

Figure 2.2. The dynamics of the River Allier, near Chatel-de-Neuvre, Allier, France, and the floodplain vegetation patterns that reflect its channel migration. Based on remote sensing and field mapping (from Petit 2001). It is possible to map more detailed vegetation communities to show fine-scale patterns of colonisation, succession and regeneration of the floodplain ecosystem (e.g. Amoros *et al.* 1986).

2.3.3 Threats to European river corridors

During the 18th and 19th Centuries, progress in civil engineering expertise resulted in widespread channelisation in order to improve navigation, to reduce flooding in some urban areas, to reclaim floodplain land for agriculture, and in some northern regions to improve channel form for timber floating (Nilsson *et al.* 2005). These engineering works also allowed greater control over domestic and industrial water supply. Later, dam construction and transfer of water between river basins caused massive changes to the water delivery patterns in rivers downstream of these structures. As a result of this engineering, floodplains became physically isolated from their adjacent rivers, and characteristic hydrological and sedimentary processes and patterns became fundamentally changed in most river systems. The previously typical vegetation mosaics discussed above and in Section 2.5.2, and illustrated in Figure 2.2, became very restricted in extent and floodplain woody species were frequently unable to regenerate (eg.

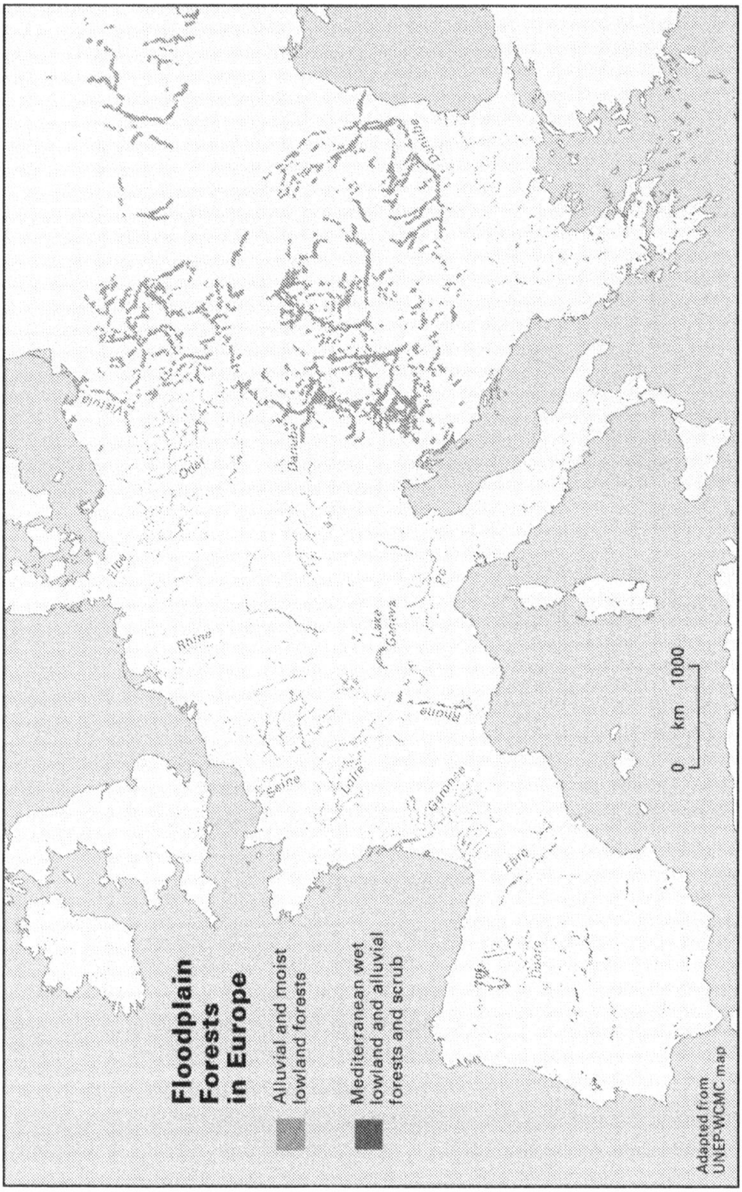

Figure 2.3. Map illustrating European floodplain forests, produced by the United Nations Environment Programme-World Conservation Monitoring Centre for the FLOBAR2 project.

Table 2.2: European riparian habitats (based on Girel et al. 2003).

EU Habitats (1999)	**Description**
91E0 Residual alluvial forests (Boreal, Alpine & temperate Europe)	Alluvial forests (*Alnus glutinosa, Alnus incana, Prunus padus, Fraxinus excelsior, Ulmus glabra*) of temperate and Boreal Europe (lowland, piedmont, montane and sub-montane rivers of the Alps, Pyrenees, Carpathians, Balkans and North Apennine); arborescent galleries of tall willows (*Salix alba, S. fragilis, Alnus, Fraxinus, Populus nigra, Populus alba*) on heavy soils periodically inundated, well-drained and aerated during low-flows.
91F0 Mixed hardwood riparian forests (temperate Europe)	Diverse riparian forests of the middle and lower courses of large rivers (eg, Rhone, Loire, Rhine, Danube, Elbe, Weser, Oder, Vistula), inundated by large floods; mature forests of hardwood trees (*Quercus robur, Fraxinus excelsior, F. angustifolia, Ulmus laevis, U. glabra, U. minor, Prunus avium, P. padus...*) growing on recent alluvial deposits; soils well drained or remaining wet between high-flow periods; the level of the water table determines the dominant species (from shallow to deep water tables: *Fraxinus, Ulmus* or *Quercus*).
92A0 White willow & white poplar galleries (Mediterranean Europe)	Riparian forests of the Mediterranean zone dominated by tall willows (*Salix alba, S. fragilis*) and poplars (*Populus alba, P. caspica, P. euphratica*) (Distribution: France, Greece, Italy, Portugal, Spain).
92B0 Riparian communities on intermittent rivers (Mediterranean Europe)	Relict alder galleries (thermo- and meso-Mediterranean zones) with *Alnus glutinosa, A. cordate, Betula sp., Fraxinus angustifolia, Osmunda regalis* (Distribution: France, Italy, Spain, Portugal).
92C0 Plane & sweet-gum woods (Mediterranean Europe)	Riparian forests and woods dominated by *Platanus orientalis* and *Liquidambar orientalis*; presence of *Salix alba, Alnus glutinosa, Celtis australis, Populus alba, Fraxinus ornus, Cercis siliquastrum*. (Distribution: Greece, Sicily).

Van Urk and Smit 1989). For example, Roux *et al.* (1989) describe the loss of viable floodplain forest along the upper Rhône in France, and Hager and Schume (2001) describe the same phenomenon along the Austrian stretches of the River Danube.

One consequence has been that pioneer species such as the black poplar (*Populus nigra*) and species of willow (*Salix* spp.) now have few opportunities to regenerate, and *Populus nigra* has become an endangered species, with near extinctions in the United Kingdom, Belgium and the Netherlands (Girel *et al* 2003). Genetic conservation of the black poplar is now being encouraged by the European Forest Genetic Resources Programme (Rotach 2001).[3] The genetic status of black poplar is also threatened by hybridisation with cultivated imported species, highlighting a second general problem for floodplain ecology, namely the invasion of river corridors by exotic non-native species that can severely impact upon regeneration of native species. Aggressive exotic herbaceous species such as Japanese knotweed (*Reynoutria japonica*) or Canadian golden-rod (*Solidago canadensis*) compete with native species for light and nutrients. Exotic woody species such as box elder (*Acer negundo*) and black locust (*Robinia pseudacacia*) also often replace local species. These trends are particularly marked where channel dynamism is reduced (Planty-Tabacchi *et al.* 1996), but riparian corridors may also be vulnerable to invasion because of their frequent disturbance (Naiman *et al.* 2005).

Ninety per cent of the original, native, natural alluvial forests have now disappeared from the European landscape. However, through the European Habitats Directive (92/43/EEC 1992) there is official recognition of the need to conserve the residual alluvial forests (those mapped in Figure 2.3). This need has been recognised since at least 1981, when a report produced by the Council of Europe highlighted the very reduced extent of these ecosystems (Yon and Tendron 1981). As well as reduced forest extent, there are concerns over the quality of remaining forests. Measures to 'restore' floodplain forests have been interpreted in widely different ways. In many locations, natural self-regenerating forests have been considered unproductive and replaced with productive forestry plantations (often using hybrid poplars) within the floodplain forest zone. In many European countries state-owned riparian forests have been managed by Forestry Departments and some have been leased to private foresters and farmers who have removed floodplain forest to plant quick-growing exotic trees or crops (Girel *et al.* 2003). For example, in Hungary most of the significant remaining floodplain forests are located along the Danube and Tisza Rivers, but 40% of their area has been converted to forestry plantations and floodplain land continues to be converted to agricultural use (Harazthy 2000). These trends all highlight the importance of a sustained inter-disciplinary research effort to provide the knowledge base to support conservation and restoration initiatives.

---

[3] See www.euforgen.org

2.4 THE CASE FOR RESTORATION

In spite of the evidence of severe decline in European floodplain woodlands, it remains of value to consider the benefits that may derive from restoration and conservation. Floodplains and their habitats are potentially important for many different reasons. Some qualities are intrinsic and hard to value in monetary terms while others offer environmental gains or have more obvious economic values. A high diversity of species, high productivity, high habitat dynamism, and the provision of species refugia and dispersal routes are all good examples of the former, while an ability to store floods or produce timber trees are good examples of the latter. Floodplains are also able to provide valuable functions, services and products although these are not always easy to distinguish. For example recreation on floodplains may be thought of as either a function or a service, and can also be marketed as a product. It is often the case that natural functions and services cannot readily be assigned monetary value, but there is increasing appreciation of the ecosystem services that floodplains and their forests can perform. Some of these are identified below.

2.4.1 Floodwater storage

The essential natural function of a floodplain is to store water and sediment derived from the river basin, especially during floods. Peak flows that cannot be accommodated within the channel spill onto the floodplain. As a result, the flood hydrograph is attenuated in downstream locations. This is because water is stored on the floodplain, and is then released slowly after the peak has passed, reducing the peak discharge and spreading the hydrograph over a longer period. Some water will infiltrate, and this reduces the total hydrograph volume downstream. When floodplains are wooded, the forest contributes to this natural flood retention function by increasing the hydraulic roughness of floodplain areas, slowing the release of water stored on the floodplain surface even more. There is still a need to explore this effect, in order to understand how much additional storage delay is introduced by the frictional effect of woody vegetation, and how extensive floodplain woodland needs to be in order for the effect to be quantitatively significant (Horn and Richards, in press).

It has to be recognised that the effect may also sometimes be negative. Excessive delay in the release of water from floodplain storage may increase the risk of flooding, and reduce the effectiveness of attenuation, in a second hydrograph following soon after the first. Along the Isère River in France, there are floodplain patches and bars within a channelised river, and on these surfaces woody vegetation growth has helped to increase flood water levels by up to a metre, causing increased risk of over-topping of the flood embankments. Furthermore, a traditional engineering objection to woody riparian vegetation is that it provides a source of large woody debris that may cause dangerous blockage of bridges which may lead to sudden surges and increased risk of damage. Thus, the effects of woody vegetation on floodplains can be viewed both negatively and positively, and sharply contrasted stances on this function are documented in the literature (e.g. Adams *et al.* 2004).

Nevertheless, in restoration projects, tree growth on floodplains could provide a service by retaining water on the floodplain longer, and thereby assist in attenuating hydrographs to reduce flood magnitudes downstream. The effect of forest in this respect will vary with the location, size, and shape of a patch. Assessment of the costs and benefits need to be made for each case, and it is necessary to appraise upstream and downstream consequences of such re-naturalisation. This requires assessment of the spatial distribution of floodplain woodland restoration within a catchment management procedure. However, this is very different from a general negative presumption about floodplain woodland, and implies that continuous negotiation is needed between the institutions (*sensu lato*) that support, respectively, conservation and flood management, which requires institutions (agencies) capable of managing this. Restoration may become a more practicable goal as a component of the River Basin Management Plans required by the Water Framework Directive. Although it may seem idealistic to expect such planning to be systematic, given the constraints of land ownership, the opportunity for environmental gain associated with Single Farm Payments and the Farm Stewardship Scheme of the reformed Common Agricultural Policy offer some economic benefit for floodplain restoration.

2.4.2 Groundwater recharge

An additional hydrological service provided by floodplains is as a site of natural groundwater recharge. Over-bank flow onto the floodplain results in infiltration across a broad surface area, and a loss of hydrograph volume. Wooded floodplains encourage this further as a result of their higher roughness, which slows the flow, and also because their roots provide macro-porosity. Ironically, in order to improve agriculture on floodplains it has often been necessary to drain floodplain soils and counter this natural tendency to store water. Control of river flow, for example for hydropower generation, has reduced seasonal flood maxima, resulting in lower levels of recharge that have contributed to the declining potential for regeneration of floodplain woodland. Thus, encouraging renewal of the natural flooding has a dual benefit.

2.4.3 Timber production

Floodplain ecosystems are highly productive, and in some places floodplain forests have been harvested for timber over many centuries, usually by systems of pollarding and coppicing. This means that natural forest on floodplains may represent a productive soft-wood or hard-wood timber resource. However, in many places they have been converted to plantation forestry, such as the extensive poplar plantations along the Garonne River in southern France. This is one of the most obvious ways in which a natural floodplain function can be converted into a marketable, economically-valued product. Conversion of natural floodplain forest to plantation forestry is detrimental to the biodiversity of the floodplain but where floodplain forest plantations replace marginal agricultural lands they can bring biodiversity gains as well as economic benefits to an area (e.g Stanturf *et al.* 2001). The rapid biomass production of a floodplain may also form a natural service in respect of carbon sequestration to offset CO_2 emissions and their global warming impacts. Through carbon trading instruments, it is possible for this service to be valued explicitly.

However, it is likely that realisation of timber production benefits will require that flows are naturalised or managed (see Section 2.5.1), since reduced flow levels and loss of recharge of floodplain sediments may otherwise inhibit growth rates. This is because floodplain soils are inherently "droughty" because of their high permeability, and require regular recharge especially at critical stages of tree life cycles.

2.4.4 Pollution control

An additional role of floodplain forest is that it can contribute to diffuse pollution control, by recycling the nutrients carried in runoff and subsurface flow from adjacent agricultural land. There is uncertainty about the optimal width of riparian buffer strips, about the hydrological pathways that permit buffer strips to strip nutrients effectively, and about the time period over which they maintain their function. However, as an additional beneficial function of floodplain forest, this service adds value to a re-naturalisation initiative, and is another potential environmental gain that can be associated with Farm Stewardship.

From this brief review it appears that the benefits of flood management, groundwater recharge, timber production and management, and pollution control can be mutually interactive, enhancing the opportunity to negotiate for restoration initiatives. However, although conservation and restoration of the remaining forested floodplains in Europe has been embraced as a priority in the EU Habitats Directive, few restoration projects that involve re-establishing some form of natural vegetation on a floodplain include forested habitats among their goals (Moss *et al.* 2003). There are notable exceptions, such as the well-documented Regelsbrunner Au Project (http://www.donauauen.at/) on the River Danube (e.g. Schiemer *et al.* 1999) and efforts to naturalise the species composition of forests through forestry practices, for example in the Gemenc floodplain forests of the Duna-Drava National Park in Hungary. In general, most restoration projects on floodplains have concentrated on developing wet grasslands, reedbeds and aquatic habitats. This is in part because forests are viewed as posing a threat to both flood defence measures, and to structures such as bridges when trees are felled as a result of bank erosion during flood events and then float downstream. Also, conservation and restoration of floodplain ecosystems needs to restore the habitat mosaic described in Section 2.3, by re-instating channel dynamism. This is commonly seen as an undesirable source of instability, causing legal problems over land ownership. A negative view of floodplain forest is also a cultural response to the fact that, in many countries, forests on floodplains disappeared from the landscape too many years ago for people to consider them as a natural component of the floodplain vegetation mosaic, except in rare cases such as the Biesbosch at the head of the Rhine delta in the Netherlands (http://www.biesbosch.org/). Public perception can be a major influence on the direction taken by a restoration project, but must be assessed carefully since restoration goals that are perceived as desirable and the restoration of river functions necessary to achieve some form of ecological integrity may not always be the same (Wohl 2005). Thus, there remain constraints to be overcome, and these are reviewed in more detail in the next section.

2.5 THE PRACTICE OF RESTORATION: REQUIREMENTS AND CONSTRAINTS

2.5.1 The need for, and limitations on, environmental flows

The local-scale approach to river restoration that prevails in Europe makes it difficult to restore systems to an ecologically dynamic and physically mobile state. Catchment-scale and process-driven approaches may become feasible using water allocation methodologies but have not been fully explored in the European context. This is because high population densities and intensive development on floodplain land make it difficult to consider restoration outside the context of site-specific initiatives (Hughes *et al.* 2001; Hughes and Rood 2003). In addition, it is difficult to be strategic and easier to be opportunistic in site selection because long histories of land tenure and highly complex stakeholder relationships make even site-based restoration initiatives difficult to implement, as the case studies in this book illustrate. Nevertheless, there is great merit in considering the possibilities of integrating these site-based schemes with water allocation policies in river basins in order to improve flow arrival patterns at restoration sites. The possibilities for this integration vary greatly across Europe and will ultimately depend on reduced overall demands for water resources, an ethic that will not be embraced readily by either the general public or planning sectors but does require debate. The role that natural scientists can play in informing and directing this debate in the US is discussed by Reuss (2005), who highlights the fact that ecologists can help their planning colleagues choose from among rational choices that balance ecological and human demands, rather than base their decisions on solely economic criteria.

In the specific case of remaining floodplain forests, conservation of variable annual flow patterns is crucial in maintaining their ecological integrity. There are numerous examples of reduced ecological health of these forests along European rivers. For example, on the Danube River in North West Hungary, floodplain forests in the Szigetkoz area have had reduced growth rates and deteriorating health since 1992, when annual floods ceased following construction of the Gabcikova Barrage system upstream in Slovakia (Somogyi 1999). Following a ruling by the European Court in the Hague in 1997, water should be released into the Danube system below the dam (Kern and Zinke 2000). Bilateral negotiations between the two countries on the timing and volume of releases have been difficult and require information on discharge volumes and the stage, timing and duration of floods needed by the Danube forests. This case highlights the value of understanding and quantifying links between hydrological inputs, geomorphological processes and forest regeneration and growth. It also demonstrates the inherent difficulties in reaching agreement on water allocation where rivers cross international boundaries, a situation which is very common in Europe.

It is clear from these descriptions that integration of flow allocation and river corridor management objectives is very important (Schaefer and Brown 1992). However, in many countries, water budgeting, including determination of abstraction rates in a catchment, flood defence and the setting of conservation priorities are managed by separate government agencies and are hardly co-ordinated (De Jong *et al.* 1994; Peiry *et al.* 1999; Lorenz *et al.* 2001; Reuss 2005). As a result, the needs of extant floodplain forests, let alone a more complex vision of mobile floodplain vegetation mosaics, cannot be fully catered for within current institutional arrangements.

2.5.2 The need for, and opposition to, river dynamics

The biological diversity of floodplains reflects several inter-dependent influences. Physical habitat diversity depends on the topographical character of the floodplain, and the elevation variation, proximity to groundwater, and propensity to flood. The floodplain sedimentology may range from gravels to clays in different floodplain landforms (point bars, natural levées, channel fills, backswamps), and each material type has its own characteristic influences on the ecology (Richards *et al.* 2002). These factors together give rise to spatial variations in vegetation assemblage, in each of which there will also be differences in species diversity, depending on patch size, connectivity, and time since the last significant disturbance. The latter influence also determines the diversity of ages of individuals of given species within these communities. All of these aspects of diversity then depend on the scale (length of reach) over which the floodplain ecology is assessed; clearly the larger the spatial scale of the assessment of diversity, the greater it is likely to appear to be, which makes comparison of diversity extremely difficult, as the scale must be standardised in some suitable way.

Probably the most important factor determining the biological diversity of floodplains is the degree of disturbance they experience. Biological diversity generally demands disturbance at various scales and of various kinds – from individual tree throw, through fire, local animal and human interference (and other forms of biological competition), to geomorphological and climatic changes in dynamic physical environments. Floodplains may be, but are not invariably, examples of such dynamic environments, and the disturbances they experience include flooding and channel change. Rivers are often in a dynamic equilibrium, in which the dimensions of various morphological attributes (channel width, depth, slope, meander wavelength, sinuosity) appear to be steady when averaged over an appropriate time and space scale, but are variable at specific places and over shorter timescales. This requires the hydrology and sediment supply themselves to be stable on average. River bends migrate and may even become cut-offs, but their average wavelengths over a reach may remain roughly constant. This dynamic behaviour is an intrinsic property of many natural rivers, and attempts to prevent it may be self-defeating and expensive (in terms of the maintenance costs required to sustain the control). For example, channelisation by straightening may simply serve to increase gradient and stream energy, and therefore to initiate a period of rapid bed material transport, alternate bar formation, bank erosion, and bend re-creation (e.g. Lewin 1976).

The key concepts that define these influences on floodplain ecology are the flood pulse concept (Junk *et al.* 1989) and its broader version, the flow pulse concept (Tockner *et al.* 2000), and the intermediate disturbance hypothesis (Connell 1978). These indicate that the hydrological regime and the channel dynamics are critical abiotic factors determining the biological diversity of a river and its floodplain. The former (flow variations, which arise during hydrograph events and may in some cases lead to over-bank floods) drive the latter, causing erosion and deposition, and bend migration and channel avulsion. These in turn create new surfaces for colonisation and interrupt vegetation succession, locally re-setting the ecological clock and restoring the seral communities that would otherwise be effaced in a less disturbed successional sequence. Without this disturbance, the floodplain would be occupied by a uniform mature woodland, and its diversity would be lower. In some rivers, of course, this is the natural tendency (indeed, therefore, the reference condition;

see below) if the channel lacks the necessary dynamism. But where channels are naturally dynamic and their floodplains have the mosaic of habitats shown in Figure 2.2, the question of restoration of this process is a challenge.

Ecologically, the question of restoration (especially if indexed by a meaningful reference condition) is not simply answered by creation of habitat. This has pejoratively been referred to as "gardening", and is in principle little different from engineering. The maintenance of created habitat is likely to require continual intervention to sustain an isolated, island feature. Ecologically, what restoration needs is restoration of function and process, and this is more systemic and less controllable. Such processes include the channel dynamics and the flood regime that drives it. However, there are innumerable objections to restoring this dynamic, not least of which are legal constraints related to land ownership. Rivers often form boundaries, between land-holdings, administrative districts and states. When rivers migrate, the bank erosion destroys one person's property and the compensating deposition adds to that of a neighbour (although it is likely to be unproductive for some years). Such gradual change is conventionally taken to imply that the boundary changes position, remaining, for example, along the centre-line or thalweg. However, when a river changes its course by avulsion (for example, through meander cut-off), the legal position differs; the boundary in this case follows the old, abandoned channel alignment. The case of the State of Missouri *versus* the State of Nebraska concerning a meander cut-off on the Missouri River, which formed the state boundary, was heard by the U.S. Supreme Court in 1904, and is a classic example, although similar cases still arise along the same river (Camp 1977). In neither the case of slow migration nor rapid avulsion, however, will both protagonists be happy to accept what the river has done, and the preference will be to prevent the channel change by bank "protection": rip-rap, gabion baskets or concrete mattresses to prevent erosion. These are not, in general, designed to provide the most ecologically sound response to bank erosion.

2.5.3 The need for, and arguments against, reference systems

As hinted above, defining ecological goals in a river restoration project is less easy than defining technical goals for activities like flood management, and is often centred around the provision of physical conditions suitable for one or more target species (Downs *et al.* 2002). River restoration projects are usually carried out within the context of a river system state that is changed in many essential properties from any previous, more "natural" system state. However, ecological goals are often set by reference to some pre-determined historical or contemporary "reference" river system. The goals of improving river ecology under the Water Framework Directive explicitly require definition of such reference conditions, as well as the assessment of those pressures and impacts that have forced the river away from the reference state (WFD CIS 2003). Known relationships between biotic and abiotic parameters can also be used as a reference for refining objectives and the methods adopted to achieve them, as in the WFD assessment of the river "hydromorphology" as a determinant of "high" ecological status (i.e. equivalent to the reference condition). Under the WFD it is not deemed necessary to define the hydromorphology at lower levels of ecological status ("good", "moderate", "poor" or "bad"). However, if reach-scale characteristics of rivers reflect their catchment condition, it is beneficial to determine the status of hydromorphology even when in these lower status classes, as this may give clues about the catchment processes that have caused deterioration in state, and the remedial measures that are then appropriate.

Using reference systems makes definition of goals and objectives in a floodplain restoration project appear to be relatively straightforward but they can also give a misplaced confidence in the predictability of the outcomes. They can be helpful in defining the levels of physical and ecological variability that might be returned to a river system but should not be used as rigid templates against which to evaluate project success. The problems arise because biophysical systems are, as outlined above, generally considered to be in a state of dynamic equilibrium. However, many systems are in reality changing along poorly-documented trajectories, which means that at a practical level it is very difficult to use past or present river systems as templates for the future (Hughes *et al.* in press). In addition, it is well established that every river system is unique in terms of its specific biotic-abiotic relationships. So while similar general relationships can be traced between rivers, specific relationships cannot be reproduced. Hughes *et al.* (2005) categorise several problems with using reference systems, including those considered below.

(a) There are often no suitable contemporary reference systems to use as templates

In some cases, reference systems are chosen which might not be biogeographically matched with the system to be restored, although they are the best available. European floodplain forests are now so rare that it is difficult to find remaining suitable stretches in some biophysical regions to inform restoration projects (Middelkoop *et al.* 2003).

(b) Many different reference conditions will be required

There can be no single reference system. Within a single river system, from its headwaters to its downstream reaches, the character of the river will vary in terms of its discharge regime, bed and bank sedimentology, gradient, cross-section dimensions and shape, bar- and bed-forms, and channel pattern (Richards *et al.* 2002). This continuous variability has to be converted into a discrete typology of river-floodplain assemblages, for each of which a reference state is necessary. There is no agreed typology, although examples have been produced for particular purposes in the USA (Rosgen 1994; Montgomery and Buffington 1997) and Australia (Brierley and Fryirs 2005), and are also being developed in Europe for WFD implementation (WFD CIS 2003; UK TAG 2004). A limitation of these typologies is that they may constrain a river restoration manager into reproducing particular reach types even in the face of changing base-levels and climate (Downs *et al.* 2002).

(c) Catchment parameters have changed since the time of the historic reference systems

Even where rivers are well documented through maps and estate records, it is often not known by how much different parameters have changed (Wohl 2005). Parameters might include flow volume and variability (Wade *et al.* 1998), water quality, land use and run-off patterns and river management practices. In some cases these changes may have been natural, and in all cases any decision on a desirable previous state to mimic is arbitrary (Sterba *et al.* 1997).

(d) Climate change through the Holocene has been considerable and future change is predicted to affect run-off patterns and hydrological regimes

Although it is difficult to produce correlations between climate change and detailed changes in fluvial systems (Patton and Baker 1980), there is nevertheless evidence during the Holocene of phases of stability and high flood activity in many rivers in temperate zones (Starkel 1991). General Circulation Models (GCM) of future climate change are forecasting greater variability than experienced at present with considerable variation in impacts between regions, though at present these models are not consistent or reliable. Coping with increased variability between seasons may well be one of many adjustments that have to be made by river managers and factored into any restoration project.

(e) The presence of alien or non-native species in most river systems has increased

The presence of alien species cannot be discounted during river restoration projects and may need to be included in predicted vegetation communities within project objectives. Changes in climate will inevitably have variable impacts on many native and alien species, for example through impacts on physiological characteristics such as susceptibility to water stress, and on phenology.

(f) In a natural landscape habitats are influenced by their landscape context

A small area within a large mobile habitat mosaic is influenced by its location within the mosaic and the vectors of change experienced there. Restoring that small area without the larger mobile mosaic, which is the norm for site-specific restoration projects, will probably produce communities that are not identical to any communities that formerly occupied that location (the "gardening" referred to above). This can be understood by simply considering the situation with propagules that may no longer be available from previous upstream sources for plant regeneration (see Andersson *et al.* 2000).

These challenges imply that restoration science is inexact and uncertain, and the goals of a restoration project are likely to be contentious. Whether the strategy adopted is to create habitat or to naturalise processes, it will remain difficult to predict the response of a river reach to changes at the catchment scale, and some form of continual intervention (or adaptive management; Schreiber *et al.* 2004) will be necessary. This may lead to problems of maintaining public acceptance, since it will reduce confidence in restoration, especially if the untidy, disorganised character of a natural, wet, riparian woodland does not accord with a sanitised public perception of nature.

2.6 CONCLUSIONS

Benefits, both quantifiable and intangible, will derive from integrated river and floodplain conservation and restoration in Europe, where the present status of floodplains is so severely degraded that there is little understanding or even historical record of wooded floodplain reference conditions. However, the Water Framework Directive provides a policy framework for catchment and river management that will require environmental assessment and actions to improve the status of rivers. Because lateral connectivity with the floodplain is a necessary aspect of a healthy river ecology, this may encourage greater attention to the riparian and broader floodplain ecosystem. It may initially seem difficult to undertake restoration in floodplains simply because of the question of land ownership. Nevertheless, some of the instruments developed in the process of reform of the Common Agricultural Policy may increase the chance that recovery of the natural functions and processes of floodplains can gradually occur.

This will challenge the science required to guide restoration, which remains uncertain, but it will be possible to learn from experience such as that at Regelsbrunner Au on the Donau, and initiatives such as those discussed in this book. Guidance is needed not only on how to restore, but also on what the appearance, functions and processes of naturalised river and floodplain systems would be in European catchments that are heavily impacted by human pressures. The FLOBAR2 project generated a guidance document entitled *The flooded forest: guidance for policy makers and river managers in Europe on the restoration of floodplain forests* to contribute to this learning process.[4]

However, there is also a need for learning at all scales – individual, social, organisational and institutional (Ostrom 1990; Pahl-Wostl 2002) – in order to benefit from the opportunities presented by the Water Framework Directive. This implies that institutions (*sensu lato*) must be an integral part of the research projects that each restoration initiative will represent, rather than simply being assumed to be external to them as directors, sponsors, funders, or responsible bodies. The functions of agencies, their personnel, their policies and their interaction with, and accountability to, the public will all need re-thinking in a world where naturalisation of the environment is on their agenda. But equally, as the case studies in this book show, there are different kinds of restoration, in terms of scale, purpose, range of phenomena to be managed, and range of stakeholders to be engaged. This is fertile ground for working through an inter-disciplinary research agenda with a necessarily pluralist methodology early in the 21st Century.

ACKNOWLEDGEMENTS

We acknowledge the contributions to all our partners in the FLOBAR2 project: Catherine Allain, Nadia Barsoum, Philippe Belleudy, Jacky Girel, Hélène Guilloy-Froget, Adrian Hayes, Rachel Horn, Axel Klaphake, Mats Johansson, Luc Lambs, Gerard Marigo, Jochen Monstadt, Tim Moss, Etienne Muller, Christer Nilsson, Guy Pautou, Jean-Luc Peiry, Stéphane Petit, Stewart Rood, Manuela Siener, Franck Vautier, Thomas Weith, Mark Winfield and Shaojun Xiong.

2.7 REFERENCES

Adams, W. M., Perrow, M. R. and Carpenter, A. (2004) Conservatives and champions: river managers and river restoration discourses in the United Kingdom. *Environment and Planning A* **36**, 1929-1942.

Amoros, C., Bravard J.P., Castella C. *et al.* (1986) *Recherches interdisciplinaires sur les ecosystèmes de la Basse-Plaine de l'Ain (France): potentialités évolutives et gestion.* Documents de Cartographie Écologique 29. Université Scientifique, Technologique et Médicale de Grenoble, CNRS, Grenoble, 166 pp.

Andersson, E., Nilsson, C. and Johansson, M.E. (2000) Plant dispersal in boreal rivers and its relation to the diversity of riparian flora. *Journal of Biogeography* **27**, 1095-1106.

Barsoum, N. (2002) Relative contributions of sexual and asexual regeneration strategies in *Populus nigra* and *Salix alba* during the first years of establishment on a braided gravel bed river. *Evolutionary Ecology* **15**, 255-279.

Brewer, G.D. (1999) The challenges of inter-disciplinarity. *Policy Sciences* **32**, 327-37.

[4] This guidance document is available at: http://www.geog.cam.ac.uk/research/projects/flobar2/reports/final/flobar2.pdf.

Brierley, G.J., and Fryirs, K.A. (2005) *Geomorphology and River Management, Applications of the River Styles Framework*, Blackwell Publishing, Oxford.

Camp, L.S. (1977) Land accretion and avulsion: the battle of Blackbird Bend. *Nebraska Law Review* **56**, 814-835.

Connell, J.H. (1978) Diversity in tropical rain forests and coral reefs. *Science* **199**, 1302-1310

Downs, P. W., Skinner, K. S. and Kondolf, G. M. (2002) Rivers and Streams. In *Handbook of Ecological Restoration, Volume 2 Restoration in Practice* (ed. M.R. Perrow and A.J. Davy), Cambridge University Press, Cambridge, UK, pp.267-296.

Dufour, S. (2005) Contrôles naturels et anthropiques de la structure et de la dynamique des forêts riveraines: exemples de différents hydrosystèmes rhodaniens. PhD, Université Jean Moulin, Lyon, France.

Eden, S., Tunstall, S. M. and Tapsell, S. M. (2000) Translating nature: river restoration as nature-culture. *Environment and Planning D* **18**, 257-274.

European Centre for River Restoration (2006) About the ECRR. http://www.ecrr.org/

European Commission, DG Environment (1999) *Interpretation Manual of European Union Habitats* EUR 15/2. Brussels.

Gibbons, M., Limoges, C., Nowotny, H., Schwartzman, S., Scott, P. and Trow, M. (1994) *The new production of knowledge: the dynamics of science and research in contemporary societies*. Sage Publications, London.

Girel, J., Hughes, F.M.R., Moss, T., Muller, E., Nilsson, C. and Richards, K.S. (2003) A case for floodplain forests? In *The flooded forest: guidance for policy makers and river managers in Europe on the restoration of floodplain forests* (ed. F.M.R. Hughes), FLOBAR2, Department of Geography, University of Cambridge, Cambridge, UK, pp.6-23.

Hager H. and Schume, H. (2001) The floodplain forests along the Austrian Danube. In *The Floodplain Forests in Europe: current situation and perspectives* (eds. E. Klimo and H. Hager), Koninklijke Brill NV, Leiden and European Forest Institute Research Report No. 10, pp.83-100.

Haraszthy, L. (2001) The floodplain forests in Hungary. In *The Floodplain Forests in Europe: current situation and perspectives* (eds. E. Klimo and H. Hager), Koninklijke Brill NV, Leiden and European Forest Institute Research Report No. 1.

Horn, R.P. and Richards, K.S. (in press) Modelling hydraulics in restored floodplain environments. In *Hydroecology and Ecohydrology: Past, Present and Future* (eds. P. Wood, D.M. Hannah and J.P. Sadler), John Wiley & Sons, Ltd, Chichester.

Hughes, F.M.R., and Rood, S. B. (2003) The allocation of river flows for the restoration of woody riparian and floodplain forest ecosystems: a review of approaches and their application in Europe. *Environmental Management* **32**, 12-33.

Hughes, F.M.R., Adams, W.M., Muller, E., Nilsson, C., Richards, K.S., Barsoum, N., Decamps, H., Foussadier, R., Girel, J., Guilloy, H., Hayes, A., Johansson, M., Lambs, L., Pautou, G., Peiry, J.L., Perrow, M., Vautier, F. and Winfield, M. (2001) The importance of different scale processes for the restoration of floodplain woodlands. *Regulated Rivers: Research and Management* **17**, 325-345.

Hughes, F.M.R., Colston, A. and Mountford, J.O. (2005) Restoring riparian ecosystems: the challenge of accommodating variability and designing restoration trajectories. *Ecology and Society* **10**(1), 12
[online] URL: http://www.ecologyandsociety.org/vol10/iss1/art12/

Hughes, F.M.R., Moss, T. and Richards, K.S. (in press) Uncertainty in riparian and floodplain restoration. In *River Restoration: managing the uncertainty in restoring physical habitat* (eds. S. Darby and D. Sears) Wiley, Chichester.

Junk, W.J., Bayley, P.B. and Sparks, R.E. (1989) The flood pulse concept in river-floodplain systems. *Canadian Special Publication of Fisheries and Aquatic Sciences* **106**, 110-127.

Kern, K., and Zinke, A. (2000) Rehabilitation of the Danube in the reach affected by the hydropower system of Gabcikovo. In *Proceedings of the international symposium for living rivers: River rehabilitation of international waterways*, (ed. G. Kovacs), 21 January 2000. Budapest, Hungary, pp.130-139.

Irwin, A. (1995) *Citizen science: a study of people, expertise and sustainable development.* Routledge, London.

Lewin, J. (1976) Initiation of bed forms and measures in coarse-grained sediment. *Geological Society of America Bulletin* **87**, 281-285.

Lorenz, C.M., Gilbert, A. and Cofino, W.P. (2001) Environmental auditing: indicators for transboundary river management. *Environmental Management* **28**, 115-129.

Middelkoop, H., Berg, J.H., van den Schoor, M.M., Alabyan, A.M., Babich, D.B., Shoubin, M.A. and de Krammer, J. (2003) Bio-morphodynamics of the Allier and lower Volga rivers a reference study for river rehabilitation in the Netherlands. In *Lowland river rehabilitation 2003 - An international conference addressing the opportunities and constraints, costs and benefits to rehabilitate natural dynamics, landscapes and biodiversity in large regulated lowland rivers* (eds. A.D. Buijse, R.S.E.W. Leuven, and M. Greijdanus-Klaas), Netherlands Centre for River Studies, Delft, The Netherlands.

Montgomery D.R. and Buffington, J.M. (1997) Channel reach morphology in mountain drainage basins. *Geological Society of America Bulletin* **109**, 596-611.

Moss. T., Klaphake, A., Monstadt, J. and Siener, M. (2003) What prospects for restoration? In *The flooded forest: guidance for policy makers and river managers in Europe on the restoration of floodplain forests* (ed. F.M.R. Hughes), FLOBAR2, Department of Geography, University of Cambridge, Cambridge, UK, pp.71-84.

Naiman, R.J., Bechtold, J.S., Drake, D., Latterell, J.J., O'Keefe, T.C. and Balian, E.A. (2005) Origins, patterns and importance of heterogeneity in riparian systems. In *Ecosystem function in heterogeneous landscapes* (eds. G. Lovell, C.G. Jones, M.G. Turner and K.C. Weathers), Springer-Verlag, New York.

Nilsson, C., Lepori, F., Malmqvist, B., Törnlund, E., Hjerdt, N., Helfield, J.M., Palm, D., Östergren, J., Jansson, R., Brännäs, E. and Lundqvist, H. (2005) Forecasting Environmental Responses to Restoration of Rivers Used as Log Floatways: An Interdisciplinary Challenge. *Ecosystems* **8**(7), 779-800.

Ostrom, E (1990) *Governing the commons: the evolution of institutions for collective action.* New York: Cambridge University Press.

Pahl-Wostl, C. (2002) Towards sustainability in the water sector. The importance of human actors and processes of social learning. *Aquatic Sciences* **64**, 394-411.

Patton, P. C. and Baker, V.R. (1980) Geomorphic responses of central Texas stream channels to catastrophic rainfall and runoff. In *Geomorphology in arid regions* (ed. D.O. Doehring), Allen and Unwin, London, UK, pp.189-217.

Peiry, J.-L., Girel, J., Pautou, G. and Nguyen, T.D. (1999) Planning of the River Isère in France. In *River Design and Environmental Protection in Europe*, (ed. J.M. de Ureña), Universidad de Cantabria, Spain, pp.195-300.

Peterken, G.F. and Hughes, F.M.R. (1995) Restoration of floodplain forests in Britain. *Forestry* **68**(3), 187-202.

Petit, S. (2001) Essai de quantification de la rugosité d'une plaine d'inondation: example de la Station de Chatel-de-Neuvre sur la rivière Allier. Unpublished thesis, University of Clermont Ferrand, France.

Petts, G.E. (1989) Historical analysis of fluvial hydrosystems. In *Historical change of large alluvial rivers: Western Europe* (ed. G.E. Petts), Wiley and Sons, Chichester, pp.1-18.

Richards, K.S., Brasington, J. and Hughes, F.M.R. (2002) Geomorphic Dynamics of Floodplains: Ecological Implications and a Potential Modelling Strategy. *Freshwater Biology* **47**, 1-22.

Rosgen, D.L. (1994) A classification of natural rivers. *Catena* **21**, 169-199.

Rotach, P. (2001) General considerations and basic strategies. In *In situ conservation of Populus nigra* (eds. F. Lefèvre, N. Barsoum, B. Heinze, D. Kajba, P. Rotach, S.M.G. de Vries and J. Turok), EUFORGEN Technical Bulletin, International Plant Genetic Resources Institute, Rome, Italy, pp.8-15.

Reuss, M. (2005) Ecology, planning, and river management in the United States: some historical reflections. *Ecology and Society 10*(1): 34.
[online] URL: http://www.ecologyandsociety.org/vol10/iss1/art34/

Robinson, M.A. (1978) A comparison between the effects of man on the environment of the first gravel terrace and floodplains of the upper Thames valley during the Iron Age and Roman Periods. In *The effect of man on the landscape: The lowland zone*, (eds. S. Limbrey and J.G. Evans), Council for British Archaeology, 21, pp.35-43.

Roux, A.L., J.-P. Bravard, Amoros, C. and Pautou, G. (1989) Ecological Changes of the French Upper Rhone River since 1750. In *Historical Change of large alluvial rivers: Western Europe*, (eds. G.E. Petts, H. Moller and A.L. Roux) Wiley, Chichester, pp.323-352.

Schaefer, J.M., and Brown. M.T. (1992) Designing and protecting river corridors for wildlife. *Rivers* **3**, 14-26.

Scheimer, F., Baumgartner, C. and Tockner, K. (1999) Restoration of floodplain rivers: the Danube restoration project. *Regulated Rivers, Research and Management* **15**, 231-244.

Schreiber, E.S.G., Bearlin, A.R., Nicol, S.J. and Todd, C.R. (2004) Adaptive management: a synthesis of current understanding and effective application. *Ecological Management & Restoration* **5**(3), 177-182.

Somogyi, Z., Szabados, I. and Vesperdi, G. (1999) Growth and health of floodplain forests in the Szigetkoz (N-W Hungary) before and after diversion of the Danube: results of a ten-year monitoring. *Ekologia* **18**, 59-68.

Stanturf, J.A., Schoenholtz, S.H., Schweitzer, C.J. and Shepard, J.P. (2001) Achieving restoration success: myths in bottomland hardwood forests. *Restoration Ecology* **9**(2), 189-200.

Starkel, L. (1991) Long-distance correlation of fluvial events in the temperate zone. In *Temperate palaeohydrology: fluvial processes in the temperate zone during the last 15000 years,* (eds. L. Starkel, K.J. Gregory and J.B. Thornes), Wiley, Chichester, UK, pp.473-495.

Sterba, O., Mekotova, J., Krskova, M., Samsonova, P. and Harper, D. (1997) Floodplain forests and restoration. *Global Ecology and Biogeography Letters* **6**, 331-337.

Tabacchi, E, Planty-Tabacchi, A.M., Salinas, M.J. and Decamps, H. (1996) Landscape structure and diversity in riparian plant communities: A longitudinal comparative study. *Regulated Rivers: Research and Management* **12**, 367-390.

Tockner, K., Malard, F. and Ward, J.V. (2000) An extension of the Flood Pulse Concept. *Hydrological Processes* **14**, 2861-2883.

UK TAG (2004) *Guidance on Typology for Rivers for Scotland, England and Wales*. UK Technical Advisory Group on the Water Framework Directive.
http://www.wfduk.org/tag_guidance/Article_05/Folder.2004-02-16.5312/TAG2003%20WP%202a%20%2803%29

UNEP-WCMC (World Conservation Monitoring Centre) (2000) *European forests and protected areas: gap analysis.* Technical Report, UNEP-WCMC, Cambridge, UK. 27pp.

Van Urk, G. and Smit, H. (1989) The Lower Rhine: Geomorphological Changes. In *Historical change of large alluvial rivers: Western Europe* (eds. G.E. Petts, H. Moller and A.L. Roux), Wiley, Chichester, pp.167-182.

Wade, P.M., Large, A.R.G. and de Waal, L.C. (1998) Rehabilitation of degraded river habitat: an introduction. In *Rehabilitation of rivers: principles and implementation,* (eds. L.C. de Waal, A.R.G. Large and P.M. Wade), Wiley, Chichester, UK, pp.1-10.

WFD CIS: Water Framework Directive, Common Implementation Strategy (2003) *Guidance Document No. 10 Rivers and Lakes – Typology, reference conditions and classification systems.* Common Implementation Strategy for the Water Framework Directive Working Group 2.3 – REFCOND.

Wohl, E. (2005) Virtual Rivers: Understanding Historical Human Impacts on Rivers in the Context of Restoration. *Ecology and Society* **10**(2): 2.
[online] URL: http://www.ecologyandsociety.org/vol10/iss2/art2/

Yon, D. and Tendron G (1981) *Alluvial Forests in Europe.* Nature and the Environment Series, 22, Strasbourg: Council of Europe.

Part B

POLICY CONTEXTS OF FLOODPLAIN RESTORATION

3

Flood risk management and floodplain restoration in Europe: recent policy developments at EU level

Thomas Dworak

3.1 INTRODUCTION AND BACKGROUND

The modification of water courses, the construction of barrage weirs, the lack of water retention due to the expansion of settlement areas, intensive farming, damaged forests in mountainous areas and the correction of small rivers and streams have led to a reduced number of floodplains and the acceleration of flood pulses and thus to an increased number of flood events across Europe. Climate change has also increased awareness of the issue of flooding.

However, flooding is a natural phenomenon with important positive effects for the environment. Floods and floodplain areas perform a range of ecologically valuable functions (cf. Richards and Hughes in Chapter 2 of this volume). For example, floods distribute large amounts of water and suspended sediment over vast areas, restocking land with valuable soil nutrients. Floodplains provide crucial habitats for plants and animals, including many rare and endangered species. They are spawning grounds for fish and act as important migration corridors, especially for water birds but also for other plant and animal communities. Species which live on floodplains usually suffer when water levels remain constant and require a seasonal variation in water levels, consisting of natural flooding and dry periods, to thrive.

Nevertheless floods also have negative consequences and cause considerable economic damage as well as personal losses such as damage to housing (see EEA 2003). The impacts of floods depend very much on how the surrounding land is used. Infrastructure, such as roads and transport networks, as well as industrial plants is likely to be badly affected. The effects on human health are varied and can also be serious (WHO 2002). Some of the environmental consequences of flooding events can also be highly negative, including the destruction of wetland areas resulting in biodiversity loss or the flooding of industrial installations leading to water pollution. From January to December 2002, 15 major floods occurred in Europe in countries such as Austria, the Czech Republic, Germany, Hungary and the Russian Federation. These floods killed approximately 250 people and adversely affected a further one million (EEA 2005).

Against this background, it has become evident that technical solutions have only a limited effect. New approaches to flood management, including the restoration of rivers' natural flood zones in order to reactivate the ability of natural wetlands and floodplains to retain water and alleviate flood impacts, have become indispensable. This chapter analyses current EU policies and how they can be used for the restoration of floodplain areas. It presents a detailed and systematic assessment of the opportunities and challenges arising from water and non-water related EU policies.

3.2 THE DEVELOPMENT OF A EUROPEAN FLOOD DIRECTIVE

The major flooding events in August 2002 killed more than 100 people in Germany, Russia, Austria, Hungary and the Czech Republic, and caused considerable damage. This played a particular role in the process of recognition at EU level that more effort should be devoted to dealing with flooding. In the wake of these events the EC Council of Ministers (Environment), following a French initiative, addressed the issue of precautionary and sustainable flood protection in Luxembourg on 17 October 2002. The following month, in Copenhagen, the European Commission and the Water Directors – who are government officials of the individual EU Member States responsible for water management issues – were requested to move forward on an EU-wide strategy for sustainable flood prevention as a basis for further Council decisions. To this end, a "Core Group on Flood Protection" was established by the Water Directors.

At the meeting in Copenhagen, officials also agreed to hold a high-level meeting on flood protection in Budapest at the end of November 2002, and an international workshop in Germany in early 2003. At this workshop, it was agreed that an ad hoc drafting group should be formed to prepare a "best practice document" (Water Directors 2003). The Group was co-chaired by France and the Netherlands. Its members were Hungary (to provide input from the Budapest initiative on strengthening international co-operation on sustainable regional development), Germany (to provide input from the Bonn workshop on precautionary flood protection) and the European Commission.

This best practice document, which is an update of the 2000 United Nations and Economic Commission for Europe (UN/ECE) guidelines on sustainable flood prevention (United Nations and Economic Commission for Europe 2000), lays out best practice examples of flood prevention, protection and mitigation (Water Directors Core Group on Flood Protection 2003). The document covers only river and flash floods, and does not address marine and tidal floods. It is designed as a living document that requires continuous input and improvement. Its character is strategic rather than technical. It clearly stresses that international efforts should be intensified to restore rivers' natural flood zones in order to reactivate the ability of natural wetlands and floodplains to retain water and alleviate flood impacts.

As a next step, the Water Directors decided at their meeting in Rome on 24-25 November 2003 to promote a concerted EU Action Programme in order to improve flood risk management. The Commission Communication "Flood Risk Management", including the essential features of the proposed EU Action Programme, was introduced on 12 July 2004 (European Commission 2004) and was discussed at the Informal Environment Council under the Dutch presidency, held in Maastricht on 16-18 July 2004. On 14 October 2004, the Environment Council formally invited the Commission to submit a proposal for a "European Action Programme on flood risk management" before mid-2005 (see also Farmer 2005). As a result, and with the support of an impact assessment, a proposal for a Directive on the assessment and management of floods (European Commission 2006) was finally put forward in January 2006

3.3 THE PROPOSAL FOR A FLOODS DIRECTIVE

The objective of the Directive is to reduce and manage the risks which floods pose to human health, the environment, infrastructure and property. Therefore, the Directive addresses all phases of the flood risk management cycle (prevention, protection, preparedness). This approach and the strong link to the Water Framework Directive (see below) strengthen the importance of the restoration of floodplains and wetlands as measures to reduce the likelihood and/or the impact of floods. The Directive does not call for specific measures to be taken, however, and the way in which flood management is handled strongly depends on the national approaches.

The Directive proposes three main steps:

(1) Member States should undertake a preliminary flood risk assessment of their river basins and associated coastal zones. This assessment should include a description of flooding processes and their sensitivity to change, including the role of floodplain areas as a natural retention/buffer of floods and flood conveyance routes now or in the future;

(2) Where real risks of flood damage exist, Member States shall then develop flood risk maps for (a) floods with a high probability (once in every 10 years) (b) floods with a medium probability (once in every 100 years), and (c) floods with a low probability (extreme events). The maps should indicate areas that are at risk of flooding and the damage that could occur;

(3) finally, flood risk management plans must be drawn up for river basins and associated coastal zones. The management plans are to include measures to *reduce the probability* of flooding and to reduce the potential consequences of flooding. These plans are to address all phases of the flood risk management cycle, focusing on prevention, protection and preparedness.

Changes in the economic value of land potentially affected by flooding could ensue from the adoption of the Floods Directive as a result of increased transparency and greater access to information via the flood risk plans. This in turn might lead to changes in behaviour in terms of land use and purchase (including the use of insurance against flooding as well as the terms of the insurance). This can be considered a positive development since it would reflect the actual situation more closely and avoid distortions based on limited information, to the overall benefit of society.

The proposal for the Floods Directive provides an approach for linking the Water Framework Directive (WFD) with flood management (see also sections below). The Floods Directive builds on the main organisational aspects of WFD implementation, requiring a synchronisation of planning cycles and allowing the use of the same management bodies. Flood risk assessment is to be carried out at the river basin district level; this further strengthens this concept of action and allows Member States to integrate their flood plans into the river basin management plans required by the WFD.

Member States are responsible for setting the required level of protection from flooding, and for selecting and implementing measures to achieve this level of protection within a given time schedule. The success of the Directive in managing flood risks using floodplain restoration will thus strongly depend on the commitment of the Member States.

3.4 FUTURE STEPS

At the EU Water Directors meeting held in London in November 2005, the Water Directors requested the Commission to prepare a mandate and work programme for a future Floods Working Group under the Common Implementation Strategy (CIS) of the WFD. At the Water Directors meeting in December 2006 the work programme for such a future Floods Working Group was agreed.[1]

The objective of this working group is to provide an information exchange platform for the European Floods Action Programme and to support the implementation of the (future) Directive on the assessment and management of flood risks. Information sharing, information and data management, the application, testing and validation of methods, the integration of flood risk management in other policy areas, notably with the WFD, and civil protection and land-use management are the central elements of the work programme for 2007 to 2009.

3.5 LINKAGE OF EU FLOODS POLICY TO OTHER POLICY AREAS

The "best practice document" agreed at the 2003 Bonn workshop and several statements from the Water Directors in Budapest and Bonn assert the importance of integrating the issue of flood protection into other policy fields, particularly nature conservation and soil policy, but also transportation, shipping, urban development and emergency management.

Legislation to date has not been integrated, however, and interests and decision-making procedures regarding regional planning, industry, agriculture and

---

[1] The future Floods Directive is formally not part of the WFD implementation process but for practical reasons co-ordination and information exchange will be organised through the CIS.

transportation are still mostly dealt with as separate dossiers. The most important policies influencing floodplain restoration are summarised briefly below.

3.5.1 EU water policy - The Water Framework Directive

Increasing demands by citizens and environmental organisations for cleaner rivers and lakes, groundwater and coastal beaches at the beginning of the 1990s had a significant impact on EU water policy. As a result of these demands, European water policy has undergone an important restructuring process, leading to a holistic approach on how to deal with waters and related aquatic ecosystems. This approach is now represented in the Water Framework Directive (European Directive 2000/60/EC) which is the most substantial piece of EC water legislation to date. It establishes a strategic framework for managing the water environment and a common approach to protecting and setting environmental objectives for all water bodies within the European Community: groundwater and surface waters (defined as rivers, canals, lakes, reservoirs, estuaries and coastal waters up to one mile from the shore) and water-related aquatic systems (such as wetlands).

This makes the Water Framework Directive (WFD) currently the most influential piece of water policy in Europe. Indeed, it shifts the focus of policy-making from addressing problems individually to integrated river basin management, requiring work to take place beyond and across administrative borders. The Directive specifies how objectives must be set, which surface waters must include a chemical quality objective and, for the first time, an ecological quality objective which also covers hydro-morphological aspects.

At the heart of the Directive is the requirement to produce management plans for each river basin setting out how the objectives that have been set for the water bodies within the river basin are to be achieved. The plans must be based on a detailed analysis of the pressures on the water bodies within the river basin and an assessment of their impact. This allows a comprehensive programme of measures to be drawn up that is tailored to the specific circumstances in the river basin and, in particular, targets improvements and monitoring efforts on those water bodies that are most at risk of failing to meet the objectives.

The WFD does not address flood management as such directly but, due to the link between the ecological status and the hydro-morphological structure of a river, the WFD will in many cases require the restoration of river and coastal hydro-morphology. This provides a unique opportunity for the restoration of floodplain areas. The immediate impact of the WFD on flood management will relate to the following situations:

- new interventions are likely to have to undergo a greater degree of environmental appraisal before they can be implemented. In this context dams can only be implemented if the required standard of flood defence cannot be achieved by other measures such as floodplains. In such a case the dams can be justified and the river has to be designated as heavily modified.

- existing interventions will have to be assessed and possibly altered to meet the WFD standards. This could improve the potential for floodplain restoration.

This approach is also clearly mentioned in the CIS policy paper on the WFD and Hydro-morphological Pressures (CIS Drafting Group on Hydro-morphology 2006) which supports the Member States in the implementation of the WFD. It calls for integrated flood risk management focusing on prevention, protection and preparedness (including forecasting). In this framework, *making space for river and coastal flooding* in the areas where the human and economic stakes are relatively low represents a more sustainable way of dealing with floods. The conservation and the restoration of the natural functions of wetlands and floodplains, with their ability to retain floodwaters and reduce the flood pulse, are a key feature of this strategy. This provides important opportunities for synergies with the implementation of the programme of measures. In other words, flood risk management plans should be included in the subsequent update of the River Basins Management Plans under the WFD. Such a combined approach would also avoid unnecessary duplication and reduce administrative burdens and costs.

A further link between the WFD and the proposed Floods Directive is established via the public participation requirements (WFD Art. 14, Floods Directive Art 13(3)). The active involvement of all relevant stakeholders in water management is one of the pivotal elements of the concept of integrated water resources management (IWRM) as it allows for the integration and consideration of the views, needs and interests of water users and those affected by water management planning.

3.5.2 Common agricultural and rural development policies

Agriculture and forestry – covering more than three quarters of the European Union's land – play a key role in determining how land is used and to what extent flood prevention and protection measures other than technical solutions such as dyking can be applied. The success of the restoration and reactivation of natural floodplains will strongly depend on the future development of the agricultural sector and on the co-operation of flood managers and farmers.

Even though flooding is not directly addressed within the regulation defining the Common Agricultural Policy (CAP) (Council Regulation (EC) No 1782/2003) the 2003 revision offers several possibilities for flood prevention especially under the second pillar (rural development). The EU's rural development policy has evolved as part of the historical development of the CAP. It is a response to the various characteristics of Europe's rural areas, which differ in their geographical and landscape features as well as in the challenges they face. These challenges range from the restructuring of the agricultural sector, remoteness, poor service provision and depopulation, to population influx and pressure on the natural environment, particularly in rural areas near urban centres (Bendz 2004).

In recent years rural development has increased in importance in EU agricultural policy. As part of the "Agenda 2000" CAP reform, the introduction in 1999 of the Rural Development Regulation (Council Regulation (EC) No 1257/1999) (RDR) marked an important step in strengthening this policy. The latest RDR (Council Regulation No. 1698/2005) came into force at the beginning of 2007 and sets out rural development measures which can be included in national programmes, providing crucial opportunities for Member States to fund measures related to flood prevention and protection, even though the available measures do not explicitly focus on flooding or floodplain restoration (Dworak *et al.* 2005).

The successful prevention of, and protection from, flooding will strongly depend on the level of co-operation between farmers and flood managers. Farm land that can serve as floodplain areas but is used for intensive agricultural production is often the subject of controversial discussion. The RDR payments linked to nature conservation can be used to compensate farmers for income losses incurred due to the establishment of floodplain areas or the restoration of hydro-morphological structures.

In addition to these indirect uses of RDR measures for flood prevention or awareness raising, the Directive offers further options as it also refers directly to natural disaster prevention actions (Council Regulation No. 1698/2005 Art. 20 b (vi)). These measures can contribute to mitigating the effects of floods by giving aid for restoring agricultural and forestry production damaged by natural disasters (including floods) and at the same time introducing appropriate prevention measures such as floodplains.

Further, the new RDR Leader approach (Council Regulation No. 1698/2005Art. 61 and 62) provides the opportunity to implement local development strategies designed to help rural actors consider the long-term potential of their local region. Encouraging the implementation of integrated, high-quality and original strategies for sustainable development, it has a strong focus on partnerships and networks for experience exchange. This offers a unique chance to create a bottom-up approach in "problem areas".

3.5.3 EU nature conservation policy

Poor planning, wasteful land use and intensive farming methods have resulted over the years in the destruction of many natural habitats, such as wetlands, upon which many wild species depend for their survival. As a response to this the EU has decided to take action. The Sixth Environmental Action Plan (EAP), which sets out the EU's environmental policy agenda until 2012, highlights nature and biodiversity as a top priority, stating that responses must be found to the pressures on biodiversity resulting from human activities.

The European Union has adopted ambitious goals to stop the decline of biodiversity by the year 2010. The legislative tools for this are two directives, the Birds Directive and the Habitats Directive, at the heart of which is a new network of protected areas called Natura 2000. Under Natura 2000 a wide range of animal and plant species, but also habitat types, like wetland, different types of forests, meadows, sand dunes, caves etc., have to be protected by Member States. The protection and restoration of floodplains is an important issue covered by these directives.

The EU Water Framework Directive with its river basin approach for integrated water management requires EU Member States to establish a register of protected areas (WFD Art. 6 and Annex IV.v) including relevant Natura 2000 sites. The Water Framework Directive and the nature conservation directives are thus fully compatible and reinforce each other if implemented properly and fully. Measures taken under the Natura 2000 framework can help reinforce the creation of floodplains lying in catchment areas governed by the WFD.

In 1998, the European Commission adopted a Biodiversity Strategy (European Commission 1998) aimed at preventing and actively addressing the causes of biodiversity losses, and ultimately halting biodiversity loss by 2010. A total of five action plans have been adopted under this Biodiversity Strategy, the latest of which was unveiled by the European Commission in May 2006. The first of these action plans (European Commission 2001) is the most relevant to floodplain restoration as it focuses on wild plants and species and their related ecosystems and habitats. In this context win-win situations can be created as the restoration or creation of wetlands can be used for biodiversity benefit and to meet flood control and water quality objectives. The Action Plan makes clear that Natura 2000 and the conservation of threatened species will not be viable in the long term without a wider terrestrial, freshwater and marine environment favourable to biodiversity. This also strengthens the role of the restoration of floodplain areas, as they are important not only for flood protection but also to provide for biodiversity.

3.5.4 Urban and regional planning

The European Union has competencies in social, economic and environmental policy, and decisions by European institutions in various policy fields have far-reaching consequences for spatial development. However, the European Union has no common agreed spatial planning policy as competency for this policy area rests firmly with the Member States.

The European Commission tries to influence spatial planning, however, through its rural development and regional policies. The DG Regional Policy, as part of the European Commission, distributes money in various ways (e.g. Interreg III, URBAN II) and this funding influences the spatial development of the supported regions and their neighbours (Wiebusch, 2004). One important driver for floodplain restoration projects is the Interreg initiative, which was launched in 1990 with the aim of supporting the balanced development of European territory while promoting transnational co-operation and ensuring that national borders do not constitute a barrier to this development. Three programmes have since been put in place: Interreg I (1991-1993), Interreg II (1994-1999) and Interreg III (2000 to the end of 2006). Interreg IV will cover the period 2007-2013. Cross-border flood mitigation represents one important aspect of the Interreg initiatives. One example of such an initiative is the Sustainable Development of Floodplains (SDF), to be implemented from January 2003 to June 2008. The SDF involves transnational co-operation and interaction between Germany and the Netherlands that deals with flood prevention and nature development along the river Rhine. The SDF project is investing €32 million in relocating dykes, creating new polders, side channels and inlet works, and in nature development. In addition, several Interreg projects dealing with river basin management have been set up. Further research would be needed to assess the extent to which these projects have been linked to floodplain restoration.

Another important European policy in regional planning is the European Spatial Development Perspective (European Commission 1999), which was adopted in 1999 in Potsdam, Germany, by the Ministers for Spatial Planning. It was drawn up by the Member States with the European Commission and is based on the EU aim of achieving balanced and sustainable development that combines and reconciles social and economic aims with a region's environmental and cultural functions. This legally non-binding document represents a policy framework for better co-operation among community sector policies with significant spatial impacts, as well as among Member States, their

---

[2] See http://www.sdfproject.nl/portal/alias__Rainbow/lang__en-US/tabID__3339/ DesktopDefault.aspx, 15.05.2007.

regions and cities. The wise management of natural and cultural heritage is among its policy options and it covers support for the extension of the Natura 2000 network, the preparation of integrated spatial development strategies for environmentally sensitive areas and the protection of soil and reduction of erosion. Most importantly, it encourages the development of strategies at regional and transnational level for risk management in disaster-prone areas and for countering the destruction of cultural landscapes prone to flooding.

3.5.5 Integrated coastal zone management

As estuaries connect the river basin and coastal zone it is important to integrate coastal zone management into flood management. Especially susceptible are low-lying areas such as deltas, which may also have to cope with high river discharges and are therefore highly prone to flooding. These deltas and the related flood-prone areas provide a wide variety of physical and biological functions and perform processes related to fish and wildlife habitats, biological diversity, recreation, flood control, economic resources, aesthetic enjoyment and so on.

Integrated Coastal Zone Management (ICZM) is a multi-disciplinary approach to promoting the sustainable management of coastal zones including river deltas. It covers the full cycle of information collection, planning (in its broadest sense), decision-making, management and monitoring. ICZM uses the informed participation and co-operation of all stakeholders to assess the societal goals in a given coastal area and therefore should also take into account the creation of new, and the protection of existing, floodplain areas.

In 2000, based on the experiences and outputs of its Demonstration Programme, the Commission adopted two documents:

- a Communication on "Integrated Coastal Zone Management: A Strategy for Europe" (European Commission 2000a);
- a proposal for a European Parliament and Council Recommendation concerning the implementation of Integrated Coastal Zone Management in Europe (European Commission 2000b). This Recommendation was adopted by Council and Parliament on 30 May 2002.

The strategy should lead to the improved management of coastal zones and should also improve the implementation of a wide range of EU legislation and policies in coastal zones. Such an improved management based on the WFD principles and considering the good status of watercourses should also include the restoration of floodplain areas.

3.5.6 Funding flood protection

Financial resources will play a major role in effective flood prevention, protection and mitigation measures. Currently there are several EU funding possibilities for sustainable flood prevention, protection and mitigation. With respect to floodplain restoration the most important source of funding seems to come from the European cohesion policy. During the new programming period 2007-2013 a total volume of €308 billion (at 2004 prices) will be available to support investments in order to sustain regional growth agendas and stimulate job creation.

The Community Strategic Guidelines (Council Decision 2006) set out the principles and priorities of cohesion policy and suggest ways in which the European regions can take full advantage of the amount of money that has been made available for national and regional aid programmes over the next seven years. National authorities will use the guidelines as the basis for drafting their national strategic priorities and planning for 2007-2013, the National Strategic Reference Frameworks (NSRFs). According to the guidelines and in line with the renewed Lisbon strategy, programmes co-financed through the cohesion policy should also be used for preserving environmental potential. Direct reference is made to the issue of flooding in two cases:

- strengthening the synergies between environmental protection and growth;
- cross-border and transnational cooperation.

In both cases floodplain restoration is not addressed directly but it can be seen as one potential option for achieving the objectives set out in the guidelines. Such measures would also fit into the European Commission's best practices on flood prevention, protection and mitigation. For example, it is stated there that transnational efforts should be intensified to restore rivers' natural flood zones in order to reactivate the ability of natural wetlands and floodplains to retain water and alleviate flood impacts (European Commission 2003).

3.5.7 EU research policy

The European Commission has funded research into the different aspects of flood risk management for nearly a quarter of a century through successive Framework Programmes addressing the various knowledge gaps in flood management. These gaps include the need for a better forecasting system, early warning and socio-economic issues.

The Sixth Framework Programme (FP6), for instance, sponsored the EU's largest flood research project to date. Launched in March 2004 the five-year FLOODsite project[3] aims to deliver an integrated European flood risk analysis and management methodology. Its seven themes are: risk analysis; risk management (pre-flood measures and flood emergency management); technological integration; pilot applications for river, estuary and coastal sites; training and knowledge uptake (such as guidance for professionals, public information and educational material); networking, review and assessment procedures; and co-ordination and management activities.

The projects FLOBAR 1 and 2[4] were established to address knowledge gaps in present scientific understanding of river-floodplain biological and physical systems. The projects assessed the impacts of different flow regimes on the growth and regeneration of floodplain plant species and the impact of floodplain vegetation on flood retention. They aimed to integrate scientific understanding of interactions between abiotic and biotic factors on floodplains with the decision-making mechanisms involved in water allocation in river catchments.

In addition, the CRUE project[5] was set up in order to complement the work carried out under FLOODsite. It aims to facilitate networking among European national flood risk management research programmes.

3.6 CONCLUSIONS

Flooding is a natural phenomenon and not a new problem. However, in many areas of Europe flood risk appears to have grown. Between 1971 and 1995, more than 150 major floods occurred in Europe (Lammel 2002). Since August 2002, when the worst natural disaster in living memory hit Central and Eastern Europe, floods have been recognised as a major problem in Europe. The main driving forces behind increased flood risk are land sealing, human encroachment into floodplain areas, population growth, urbanisation, increasing settlement, roads, railways and hydraulic engineering measures (EEA 2001).

In the last 15 years, not only because of such catastrophic events, it has been recognised that action is required concerning the way in which rivers are utilised and managed. With respect to flood management, the limitations of technical solutions such as dykes and weirs have become apparent and new approaches, including the restoration of rivers' natural flood zones, are required. In recent years river development has therefore been shifted towards river management, leading to a more holistic approach that considers the various functions of rivers and their

---

3 Cf. http://www.floodsite.net/, 18.05.2007.

4 Cf. http://www.geog.cam.ac.uk/research/projects/flobar2/, 18.05.2007.

5 Cf. http://www.crue-eranet.net/, 18.05.2007.

dependent eco-systems. In addition, it is widely recognised that a key way forward in the management of floodplains is the adoption of a river basin approach.

This holistic approach to river management is largely laid down in the Water Framework Directive. This Directive should be recognised as one of the most far-reaching pieces of environmental legislation ever to emerge from Europe. It shifts water management from a sectoral towards a more integrated approach which considers social, economic and ecological issues. For the first time, water is seen not only from a qualitative point of view; quantitative and hydro-morphological issues are also addressed by the Directive. Thus, the WFD will in many cases require the restoration of river and coastal hydro-morphology which has been adversely impacted by dykes and land drainage schemes in the past. This should result in giving more space back to the rivers.

It should be noted however that this approach is still far from being implemented as no specific action with respect to floodplain restorations is actually required. The forthcoming EU Floods Directive does not require any particular measures with respect to floodplain restoration. In addition, the WFD allows for a broad range of exemptions even though the approach to floodplain restoration as a prevention and protection measure has been strengthened by the Directive's general objective of good water status. Floodplain restoration and protection will therefore continue to rely on the provision of wider mechanisms (Maltby and Blackwell 2005) such as rural development or nature conservation, as these policies are often more important drivers in the development of river basins. Both policies provide a link between floodplain protection as a flood prevention measure and the protection of nature and biodiversity. In addition, several funding opportunities exist that support such a restoration process by providing additional funding to Member State expenditure.

However, although there are various opportunities under EU policy for protecting and restoring floodplains, the success will strongly depend on national implementation plans and the extent to which Member States use existing opportunities in the policy framework. Even if flooding is high on the political agenda, the application of non-technical measures such as the restoration of floodplain areas cannot be guaranteed under the existing policy framework. Other interests such as agricultural production and housing are often considered more important and therefore in many cases the restoration of floodplain areas will remain a challenge.

3.7 REFERENCES

Bendz, K. (2004) EU-25 Agricultural Situation Rural Development in the European Union 2004, Global Agriculture Information Network (GAIN). GAIN Report Number: E34095, 11/30/2004 (http://www.fas.usda.gov/gainfiles/200411/146118181.pdf, 21.05.2007).

Council Decision of 6 October 2006 on Community strategic guidelines on cohesion (2006/702/EC).

Council Directive 92/43/EEC of 21 May 1992 on the conservation of natural habitats and of wild fauna and flora. Official Journal L 206 , 22/07/1992 P. 0007 – 0050.

Council Regulation (EC) 1257/1999 of 17 May 1999 on support for rural development from the European Agricultural Guidance and Guarantee Fund (EAGGF) and amending and repealing certain Regulations, OJ L 160, 26. 06.1999, p. 80.

Council Regulation (EC) 2012/2002 establishing the European Union Solidarity Fund, OJ L 311, 14. 11.2002, p. 3.

Council Regulation (EC) 1782/2003 of 29 September 2003 establishing common rules for direct support schemes under the common agricultural policy and establishing certain support schemes for farmers and amending Regulations (EEC) No 2019/93, (EC) No 1452/2001, (EC) No 1453/2001, (EC) No 1454/2001, (EC) 1868/94, (EC) No 1251/1999, (EC) No 1254/1999, (EC) No 1673/2000, (EEC) No 2358/71 and (EC) No 2529/2001. OJ No. L270, 21.10.2003, p. 1-69.

Council of the European Union (2005): Council Regulation (EC) No 1698/2005 of 20 September 2005 on support for rural development by the European Agricultural Fund for Rural Development (EAFRD).

Council Regulation (EC) No 1084/2006 of 11 July 2006 establishing a Cohesion Fund and repealing Regulation (EC) No 1164/94.

Directive (2000/60/EC) of the European Parliament and of the Council establishing a framework for Community action in the field of water policy, OJ L327, 22.12.2000, p.1.

Dworak, T., Karaczun, Z., Herbke, N., Schlegel, S. and Landgrebe-Trinkunaite, R. (2005) *Deliverable 3 - WFD and Agriculture linkages at the EU level.* Final Report about Rural Development Programmes. Ecologic/Warsaw Agricultural University, Berlin/Warschau, commissioned by European Commission, DG Research.

EEA (2001) *Sustainable Water Use in Europe, Part 3: Extreme Hydrological Events: floods and droughts.* Environment issue report number 21, Copenhagen 2001, pp.17-20.

EEA (2003) *Mapping the impacts of recent natural disasters and technological accidents in Europe.* Environmental issue report no. 35 for a list of recent European floods and their consequences. Copenhagen. (http://reports.eea.europa.eu/environmental_issue_ report_2004_35/en, 21.05.2007)

EEA (2005) *EEA Briefing 01/2005, Climate change and river flooding in Europe,* http://reports.eea.europa.eu/briefing_2005_1/en/briefing_1_2005.pdf.

European Commission (1998) *Communication of the European Commission to the Council and to the Parliament on a European Community Biodiversity Strategy* (COM (98)42).

European Commission (1999) *ESDP. European Spatial Development Perspective. Towards Balanced and Sustainable Development of the Territory of the European Union.* Agreed at the Informal Council of Ministers responsible for Spatial Planning in Potsdam. May 1999 (http://ec.europa.eu/regional_policy/sources/docoffic/official/ reports/pdf/sum_en.pdf, 21.05.2007).

European Commission (2000a) *Communication from the Commission to the Council and the European Parliament on "Integrated Coastal Zone Management: A Strategy for Europe"* (COM/00/547 of 17 Sept. 2000).

European Commission (2000b) *A proposal for a European Parliament and Council Recommendation concerning the implementation of Integrated Coastal Zone Management in Europe* (COM/00/545 of 8 Sept. 2000).

European Commission (2001) *Communication from the Commission to the Council and the European Parliament - Biodiversity Action Plan for the Conservation of Natural Resources* (COM/2001/0162 final).

European Commission (2003) *Best practices on flood prevention, protection and mitigation* (http://ec.europa.eu/environment/water/flood_risk/pdf/flooding_bestpractice.pdf, 21.05.2007).

European Commission (2004) *Communication from the Commission to the Council, the European Parliament, the European economic and social committee and the committee for the regions. Flood risk management, Flood prevention, protection and mitigation,* COM(2004) 472 final.

European Commission (2006) *Proposal for a Directive of the European Parliament and of the Council on the assessment and management of floods.* COM(2006) 15 final.

European Commission CIS Drafting Group on Hydro-morphology (2006) *WFD and Hydro-morphological pressures. Focus on hydropower, navigation and flood defence activities – Recommendations for better policy integration.* Draft, 25 July 2006.

Farmer, A. (2005) A European Union Directive on flood management. *Journal of Water Law* **16**, 85-89.

Lammel, B. (2002) *Managing floods in Europe: The answers already exist. More intelligent river basin management using wetlands can alleviate future flooding events.* WWF Background Briefing Paper 26, September 2002.

Malty, E. and Blackwell, M.S.A. (2005) Managing riverine environments in the context of new water policy in Europe. *International Journal of River Basin Management* **3**(2), 133-141.

Regulation (EC) No 1080/2006 of the European Parliament and of the Council of 5 July 2006 on the European Regional Development Fund and repealing Regulation (EC) No 1783/1999.

Regulation (EC) No 1081/2006 of the European Parliament and of the Council of 5 July 2006 on the European Social Fund and repealing Regulation (EC) No 1784/1999.

United Nations and Economic Commission for Europe (UN/ECE) (2000) *Guidelines for sustainable flood prevention.* The Hague, Netherlands, 23-25 March 2000.

Water Directors (2003) *Initiative on flood prevention at the working-level meeting on flooding on 5/6 February 2003 in Bonn* (http://forum.europa.eu.int/Members/irc/env/wfd/library?l=/framework_directive/floods_programme&vm=detailed&sb=Title, 21.05.2007).

Water Directors Core Group on Flood Protection (2003) *Best practices on flood prevention, protection and mitigation* (http://www.verkeerenwaterstaat.nl/object/?lc=uk&tb=Object&id=122, 21.05.2007).

WHO (2002) *Factsheet 05/02: Flooding: Health effects and preventative measures.*

Wiebusch, M. (2004) *On the road to a European spatial planning system?* (http://www. ceu-ectp.org/inc/cgi/dd/dd20041125.pdf, 21.05.2007).

4

Policy innovations in the aftermath of a disaster: contexts of floodplain restoration in Germany

Jochen Monstadt and Timothy Moss

4.1 INTRODUCTION

As in most European countries, German rivers and floodplains have been subject to extensive modification by human action. River flows have been controlled by 'hard engineering' flood defence methods like the construction of dams, dykes and water retention basins, channels have been straightened and deepened for navigation, rivers have been dammed for electricity supply, and floodplains have been embanked and drained for agriculture, settlements and industry (cf. Adams *et al.* 2004: 1931). However, a number of severe floods hit Germany in the 1990s (e.g. Rhine 1993 and 1995, Oder 1997, Danube 1999), the increasing frequency and magnitude of which served as a reminder that

 Restoring Floodplains in Europe: policy contexts and project experiences edited by Timothy Moss and Jochen Monstadt. ISBN: 1843390906. Published by IWA Publishing, London, UK.

flood protection is a goal that can rarely be achieved by the regulation of rivers (cf. Lange and Garrelts 2007). Moreover, the loss of valuable and species-rich riparian and aquatic ecosystems and river landscapes became an important issue in the discourse not only on nature conservation but also on soft tourism and sustainable rural development. River managers and flood defence engineers began to acknowledge the significant environmental impact of channel modifications and the limited protection afforded by technical flood defence.

Accordingly, a number of reforms in federal water, spatial planning and nature conservation legislation have been adopted in Germany since the mid-1990s addressing alternative methods of flood prevention and river management. In addition to this more favourable legal framework for floodplain restoration a number of operational concepts, funding programmes and pilot projects, as well as research studies, have been introduced or conducted in recent years which relate to some extent, at least, to floodplain restoration (BBR 1998). Furthermore, the planning authorities in some German states (Länder) began to tighten restrictions on the use of land, developing regulations at regional and state planning levels, which in turn have been implemented at the municipal level (Petrow 2006: 721). However, owing to strong federalism, national activities were rarely launched and regulations in land-use planning and flood-risk management remained different from state to state. Policy initiatives in floodplain restoration were, in the late 1990s, limited to a few frontrunners among the Länder; integrated river basin approaches to precautionary flood management and floodplain restoration were very rare. Whilst new federal policies opened windows for restoring floodplains, very few schemes for relocating dykes or widening the floodplains were actually put into practice.

In the summer of 2002 the German Federal Government announced a major policy change. Heavy rains in Central Europe had led to some of the worst flooding Germany had ever witnessed. The Elbe flooding affected the Czech city of Prague and the city of Dresden and many villages in Germany. Twenty-one people lost their lives due to the floods in Germany, and the direct damage to property amounted to €9.2 bn.[1] Faced with a concurrent election campaign in Germany, the government gave high priority to the issue of flood management as a means of demonstrating political effectiveness. In response to the disaster, flood management practices came under increased scrutiny. The Elbe flooding led to intense public discussion of the inadequacies of the current flood management model, as both traditional river engineering works – changing the course, cross-section and regime of rivers – and intensified land use in river basins – urbanisation in floodplains, sealing of large areas in the basin and losses of retention capacity – were seen to have increased the flood damage

[1] BMU (2002) (http://www.bmu.de/english/news/pressrelease030205.php, 19.02.2003).

potential (Lange and Garrelts 2007; Petrow *et al.* 2006). Furthermore, the floods raised public awareness of global warming, which is likely to increase the occurrence of extreme weather conditions and flooding events.

This chapter explores the altered policy context for restoring floodplains in Germany in the aftermath of the 2002 floods. The aim is to analyse the new institutional opportunities for, and restrictions to, restoring floodplains. The study is based on the analysis of policy documents and legislation of the national government and exemplary regional governments, and the relevant secondary literature. The main issues raised in this chapter were explored in qualitative interviews with experts from national and regional administrations and research institutes (see list under References). The analysis starts with a selection of relevant policy statements on floodplain restoration. Then the institutional context of floodplain restoration in different policy fields is examined, as well as funding programmes and modes of cooperation and consultation across policy fields. The chapter concludes by reflecting on the principal drivers of, and constraints to, floodplain restoration.

4.2 POLICY STATEMENTS ON FLOODPLAINS AND THEIR RESTORATION

In Germany, there exists no national policy or strategy document relating specifically and exclusively to floodplains and their restoration (cf. The Ramsar Convention 1999: 2). Nevertheless, references to protecting and restoring floodplains can be found in a wide range of relevant federal and state legislation relating to water management, spatial planning and nature conservation, particularly in laws revised since the mid-1990s. Above all the "5-Point Programme" of the Federal Government to improve flood protection, designed to give space back to rivers, systematically addresses the restoration of floodplains (cf. Section 4.3.1). Moreover, extensive and detailed references to floodplain restoration can be found in policy statements and operational guidelines produced by a number of trans-national and national water management and spatial planning organisations. References to floodplain restoration in recent policy documents reflect an increasing interest in preventive forms of flood protection and river restoration and, in broader terms, a more holistic, catchment-based approach to river management.

Table 4.1 presents an overview of the references (explicit and implicit) to floodplain restoration in the principal laws and policy documents in Germany. The survey comprised primarily the federal acts on water management (WHG), regional planning (ROG), building and nature conservation (BNatSchG), selected state acts for all these issues, policy statements by leading river management and spatial planning organisations (LAWA, MKRO, IKSR, IKSE) and selected state programmes on river restoration, floodplain restoration and flood protection.

Table 4.1 Policy statements relating to floodplain restoration in Germany. [Text in square brackets is the authors' own].

Document	Policy statement
Federal Water Act (WHG)	Art.31b para. 4: "In floodplains… no new building zones may be identified in building plans". Art.31b para. 6: "The functioning of floodplains [i.e. areas in which a flood event is statistically expected once every 100 years, J.M.] …as retention areas is to be maintained … Former floodplains which are suitable for use as retention areas should be restored as far as possible…" Art.31d para. 1: "In flood protection plans [which the *Länder* have to draw up regularly, J.M.] measures are to be included for the maintenance or restoration of retention areas, for their flooding and draining in accordance with optimal flood outflow conditions for river basins, for the relocation of dykes, for the maintenance or restoration of water meadows and for the retention of rainwater".
State water acts	As a consequence of the Federal Flood Control Act, several reforms of state water acts were instigated in 2006, no longer referring only to the protection of existing floodplains but also to their restoration. For example the State Water Act in Saxony-Anhalt states: "The functioning of floodplains as natural retention areas is to be maintained… Areas which were previously flooded and which are suitable to be used as retention areas should as far as possible be restored as retention areas" (WG LSA, Art.97 para. 1).
Federal Regional Planning Act (ROG)	Art.2 para. 2 No. 8: "Preventive flood protection is to be achieved… above all by securing or restoring floodplains, retention areas and areas in danger of flooding".
State regional planning acts	Only a limited number of state regional planning acts include planning provisions for securing new floodplains (e.g. Hesse, Northrhine-Westphalia, Rhineland-Palatinate) One such objective from the Regional Planning Act of Brandenburg is: "Reserved areas for... floodplains... are to be kept free of uses which could infringe upon the interests of water resource management" (Art. 4 BbgLPlG).
Federal Building Code (BauGB)	Art.5 para. (4a): "Designated floodplains as specified by… the Federal Water Act should be listed. Not yet designated floodplains… and areas in danger of flooding as specified in the… Federal Water Act should be recorded in land use plans [development plans]."
Federal Nature Conservation Act (BNatSchG)	No explicit reference to floodplains, but implicitly e.g. in Art.2 para. 1 (protecting biotopes) and Art.20 c (protection of water meadows and riparian woodlands).

Document	Policy statement
State nature conservation acts	Similar references to BNatSchG, though more specific, e.g. Art.1 para. 2 of the Nature Conservation Act of Brandenburg BbgNatSchG (protecting and restoring natural rivers).
State landscape planning programmes	E.g. Landscape Programme of Brandenburg (MUNR 1998: 12): "The maintenance and development of the functioning of connecting watercourses requires above all... measures for the restoration of natural-like water meadows (extensive land use, expansion of the overflow areas, resettlement of floodplain woodland, retention of backwaters)."
Inter-ministerial Conference on Regional Planning (MKRO)	MKRO resolution on flood protection (MKRO 1996): "In order to meet the requirements of preventive flood protection… the protection and expansion of floodplains should be ensured."
Inter-state Working Group on Water (LAWA)	Guidelines for Flood Protection for the Future (LAWA 1995): Principle 1: Retain water: including restoration of floodplains (p.20) On floodplain woodland restoration: "A natural succession of riparian forests not only serves to retain flood waters but is also important for nature conservation, because riparian forests are today among the rarest and most endangered types of biotope." (p.10)
International Commission for the Protection of the Rhine (IKSR)	Flood Protection on the Rhine. Taking Stock. (IKSR 1997): "Future flood protection must be based... on the retention of water on surrounding land and water meadows through the restoration of floodplains..." (p.8). Rhine 2020. Programme for the sustainable development of the river Rhine (IKSR 2001: 13): The objective formulated is "reactivating at least 300 sq. km. of floodplains by 2005 and 1,000 sq. km. by 2020".
International Commission for the Protection of the Elbe (IKSE)	Strategy for Flood Protection in the Elbe River Basin (IKSE 1998: 5): "It is above all necessary to maintain the natural floodplains and to use them in a suitable fashion, and also to restore former floodplains as far as possible." Plan of Action for Flood Protection on the Elbe (IKSE 2003: 14): "Within this framework [of sustainable flood protection, J.M.] designated floodplains are absolutely necessary, since only in this way can these areas be guaranteed in the long term for water drainage or retention and thus contribute to a reduction in flooding risk."

4.3 PRINCIPAL INSTRUMENTS AND POLICIES RELEVANT TO FLOODPLAIN RESTORATION

The institutional regimes for floodplain restoration in Germany are influenced to a great extent by the principles of federalism and subsidiarity. The regulatory activities of the Federal Government in regional and landscape planning, nature conservation, flood protection and water management have long been confined to the setting of a legal framework offering general policy guidelines, defining procedural minimum standards etc. ("Rahmengesetzgebung"). The "Länder", the state governments, had a large amount of legislative latitude in putting this framework legislation into a concrete legal form and in implementing it in cooperation with municipalities. Only since the reform of federalism in 2006 have the responsibilities for regional and landscape planning, nature conservation, flood protection and water management been transferred from framework legislation to concurrent legislation. The Federation thus now has full legislative competence in these fields. As compensation – many experts call it unwarranted overcompensation (cf. Koch and Krohn 2006) – the states were given the right to create diverging legislation in several areas, and thus to permit far-reaching exceptions from national environmental regulations. Accordingly, the Länder have extensive responsibilities in the legislation and implementation of policies relevant to floodplain restoration in each state. Together with the far-reaching responsibilities of municipal self-government in the fields of land-use planning as well as water supply and sewage management, the German floodplain restoration regime is highly complex, fragmented and decentralised.

Despite this decentralised regime structure, the regulatory resources of the federal policy level should not be underestimated. Even though the responsibilities for national floodplain restoration policies are divided among several ministries, the Federal Ministry for Environment, Nature Conservation and Nuclear Safety (BMU), the Federal Ministry for Transport, Building and Urban Development (BMVBS) and their subordinate agencies are the main actors in floodplain restoration at the federal level. Particularly since the end of the 1990s they have launched several policy initiatives in spatial planning, nature conservation, water management and flood protection promoting the restoration of floodplains (Heiland 2003; Petrow *et al.* 2006). In 1998, a reform of the Federal Spatial Planning Act (Raumordnungsgesetz – ROG) promoted the designation of floodplain restoration areas. In addition, the Federal Water Act (Wasserhaushaltsgesetz – WHG) as well as the Federal Building Code (Baugesetzbuch – BauGB) include rules and regulations for precautionary flood protection. Furthermore, the reform of the Federal Nature Conservation Act (Bundesnaturschutzgesetz – BNatSchG) in 2002 targeted the restoration of floodplains. Finally, the Act to Improve Preventive Flood Control (Hochwasserschutzgesetz – HWSchG) in 2005 reformed the WHG, the ROG and the BauGB with regard to flood prevention issues.

However, particularly owing to the recent reform of federalism, the Federal Government acts only as a *primus inter pares* in legislating and regulating floodplain restoration. Each of the 16 German Länder has ample scope to define the legislative framework for long-term land use, such as the protection and extension of retention areas as well as the protection and management of floodplains (Böhm *et al.* 2004). Accordingly, there are 16 state Water Resources Acts, Regional Planning Acts and Nature Conservation Acts, and on this basis 16 state development plans and various regional plans, landscape plans, flood prevention plans etc. Within the states, water management powers are in most cases split up between three levels of water administration, the functions and territorial reach of which differ from state to state (for details see Kahlenborn and Kraemer 1999). In order to harmonise water legislation and to evaluate and develop water management policies, the supreme water authorities of the Länder cooperate in the Working Group of the Federal States on Water Problems (LAWA). Additionally, the Länder are the main actors in the international committees for the protection of the rivers Rhine, Elbe, Danube and Oder and in many river basin committees for smaller rivers. These committees have gained in importance in developing river basin management plans in line with the Water Framework Directive and flood prevention plans following the reform of the Federal Water Act (cf. Section 4.5).

Owing to the German constitutional principle of municipal self-government and the comparatively high level of municipal budgetary autonomy, the Federation and the Länder are required to allow the municipalities flexibility in their decision-making unless there are binding mandatory regulations which have to be followed throughout the regional and municipal planning processes. Local authorities develop statutory plans for the regulation of communal interests within their territory. With the help of landscape plans, preparatory zoning plans and binding development plans, the municipal authorities define local land use. They thus have far reaching influence on the restriction of building permits and the designation of nature conservation, retention and floodplain restoration areas. Moreover, they have extensive responsibilities in averting flood hazards and in protecting the local population from floods. Therefore, the municipalities play a key role in flood management and in initiating and implementing land-use changes for flood prevention.

Table 4.2 illustrates the scope of functions and the relevant legislation of the corresponding administrative level.

Table 4.2: Administrative levels and regulations for floodplain restoration in Germany. (Source: Based on Petrow et al. 2006: 726).

Administrative level	**Scope of functions**	**Land-use planning**	**Nature conservation**	**Water and flood management**
European Union	Overall principles and regulations	European Spatial Development Perspective (ESDP) (not binding)	Birds Directive and Habitat Directive (binding)	Water Framework Directive (WFD) (binding); Flood Risk Management Directive
Federal (national) level	Principles, aims, general and obligatory regulations	Federal Spatial Planning Act (ROG); Federal Building Code (BauGB)	Federal Nature Conservation Act (BNatSchG)	Federal Water Act (WHG)
State level (16 German Länder)	Comprehensive planning for every state	16 state Planning Acts and corresponding state development plans	16 state Nature Conservation Acts and corresponding landscape programmes	16 state Water Acts; river basin management plans and flood prevention plans
Regional level	Comprehensive planning for every region	Regional development plans	Regional landscape framework plans	River basin management plans
Municipal level	Mandatory regulations	(Preparatory) zoning plans and (binding) development plans for every municipality	Local landscape plans and green structures plans	Local water and flood protection plans

4.3.1 Flood protection and water management

For centuries, river management in Germany, as in many other countries, has been powerfully shaped by the drive to control floods by ‘hard engineering’ methods. However, the serious floods on the river Rhine in 1993 and 1995 led to shifts in thinking towards flood risk management and floodplain restoration within a wider community. As a consequence, the environment ministers of the German Länder instructed the Working Group on Water of the German Federal States (LAWA) to develop guidelines for precautionary flood management, which were drawn up in November 1995. These guidelines depicted a progressive model of flood protection addressing, among other things, spatial planning issues and the designation of floodplain areas, but they were never implemented by the Länder (Lange and Garrelts 2007). However, due to the flood events and the recommendations of the LAWA, the planned reform of the Federal Water Act (WHG) of 1996 was extended to advance floodplain restoration and other flood prevention measures. The principal tool of the WHG relating to floodplain restoration is the requirement for floodplains to be granted a legally protected status as defined by the law. In earlier versions of the WHG this related only to areas which were already regularly flooded; the restoration of former floodplains was not addressed. Article 32 of the 1996 version sets downs aims and procedures for granting floodplains legal status, which include provisions for the restoration of former floodplains. Article 32 paragraph 1 requires the states to introduce regulations to "maintain and improve the ecological structures of water courses and their floodplains" and to retain and restore natural retention areas. Article 31 paragraph 5 states specifically that planning permission for the alteration of rivers is to be refused "if destruction of natural retention areas, in particular floodplain woodlands, is to be expected".

However, the reform did not meet initial expectations and a policy change towards floodplain restoration could not clearly be identified (Böhm *et al.* 1999). Since the states are responsible for water and flood management issues and most of the state water acts were not substantially revised, the older, more restrictive references to the protection of existing floodplains largely remained, although some state water acts included policy objectives on the need for more natural forms of water retention (e.g. Art. 3a WasserG BadW).[2] The German Advisory Council on the Environment declared in its Environmental Report of 2000 that the instruments of flood protection provided by Federal legislation are adequate (SRU 2000: 303). However, it charged the Länder with not applying effectively the legal opportunities provided by the WHG for flood prevention. The Council criticised in particular the implementation deficit at the regional and local levels in designating flooding zones or flood retention areas. Instead of reforming the Federal Water Law, the Council recommended better application of the instruments provided by the water and regional planning acts of the Länder and the Federal Building Code and called for intensified trans-boundary cooperation between the Länder on flood management (ibid.). The Federal Office for Building and Regional Planning has also criticised the serious bureaucratic limitations to the designation of floodplains by the Länder and municipalities in accordance with the WHG (BBR 1998: 24-25). Granting a floodplain legal status, which is binding for regional and land-use planning, is a

[2] Some state water acts specified the procedures for determining floodplains by law, the activities prohibited on these floodplains and the activities which require permission (e.g. Arts. 100 & 101 BbgWasserG, Arts. 96-98 WasserG Sachsen-Anhalt, Arts. 112-114 WasserG NRW, Arts. 77-79 WasserG BadW).

very time-consuming, costly and complex administrative procedure. As a result, only a small proportion of floodplains along Germany's major rivers have been accorded legal status under the WHG. Many of the legally protected floodplains are founded on Prussian legislation dating back to the 19th Century.

Following the implementation of the EU Water Framework Directive (WFD) the WHG was reformed in 2002, strengthening the integrated approach to river basin management. Even though this reform forces the Länder, regions and municipalities to cooperate more in the interest of river basin management, neither the WFD nor the German reform of the WHG address specifically the advancement of flood prevention and floodplain restoration.

It was in the aftermath of the flood disaster on the River Elbe in 2002, shortly before the federal elections, that the Federal Government initiated a sudden policy shift in flood protection. The government immediately announced a five-point programme[3] to be implemented in cooperation with the Länder and neighbouring countries that was based mainly on the policy guidelines of the LAWA of 1995 (Lange and Garrelts 2007). The main objectives were:

- to give space back to the rivers by preserving and restoring floodplains and their forests, by revitalising straightened and bank-impounded tributaries and by conserving meadows as natural floodplains. In order to protect existing settlements, natural floodplains should be kept free of construction and as far as possible floodplains should be restored, e.g. by relocating dykes, and serve as natural floodplains;
- to retain precipitation throughout the entire river basin in a decentralised way in order to reduce the flow rate;
- to control urban development more effectively in order to minimise the flood damage potential. The remaining floodplains must be preserved;
- to improve the forecasting and early warning systems and to link them internationally (Deutsche Bundesregierung 2002).

Based on the five-point programme, the German Federal Ministry for the Environment developed a Flood Control Act, which aims at improving precautionary flood protection (for details cf. Jekel 2005). The bill – entitled The Act to Improve Preventive Flood Control (Hochwasserschutzgesetz – HWSchG) passed the German cabinet in May 2005. The act is a so-called article

[3] The "Five-Point Programme of the Federal Government: Working steps towards an improvement of flood protection", announced at a river conference held in Berlin on 15 September 2002 emphasises five programmatic tasks: (1.) coordinated flood management programmes of the Länder, (2.) interstate/catchment-based action plans, (3.) support of European cooperation, (4.) review of structural river interventions and (5.) immediate flood protection measures (cf. Deutsche Bundesregierung 2002).

law, requiring amendments to other federal laws relevant to flood protection in general and preventative measures in particular. Amendments have to be made to the Federal Water Act (WHG), the Federal Building Code (BauGB), the Federal Regional Planning Act (ROG), the Federal Waterways Act (WaStrG) and the Act on the German Meteorological Service (DWDG). Changes in the Federal Water Act comprise, among other things, further regulations for the Länder to enforce the creation of retention areas, the relocation of dykes, the conservation and restoration of floodplain and flood-prone zones and alluvial meadows (WHG Arts. 31b-31d). Accordingly, within a period of 5 to 7 years the Länder have to identify and designate flood-prone zones (defined as the enlarged floodplain area that will be affected if technical flood protection fails) and floodplain zones (defined as the area affected by the 100-year flood). For the latter, regulations ensure the identification and the appropriate conservation status of these areas within the next 5 to 7 years. The floodplain areas and flood-prone areas have to be designated in local zoning and development plans and the planning of new buildings is restricted (WHG Art. 31b para. 4; cf. Section 4.3.2 below). Another policy innovation towards effective flood prevention and river management is the new federal framework for catchment-based flood protection plans, which have to be adjusted to international standards (Art. 31d). These plans also incorporate measures to deal with a 200-year flood.

In sum, the Flood Control Act and the corresponding reform of the WHG constitute various amendments promoting flood risk management and floodplain restoration at the federal level. However, the crucial point will be its adoption into water legislation at the level of the Länder and its implementation at the regional and local level. Especially since the reform of federalism in 2006, the Länder have extended rights to diverge from Federal water legislation und thus to create far-reaching exceptions to national flood regulations. On the other hand, the recently proposed Flood Risk Management Directive of the EU (cf. Dworak in Chapter 3 of this volume) could further promote a uniform approach for catchment-wide flood risk management. It could exert further pressure on the Länder to adopt and implement policy innovations and strengthen the role of the international commissions for large river systems.

4.3.2 Regional and local land-use planning

Regional planning legislation at the federal and state levels refers specifically to the need to protect and restore floodplains in the context of the spatial implications of preventive flood protection (see Table 4.1). As with water management, the reform of federal legislation on regional planning in 1998 created new opportunities for floodplain restoration which were substantiated and specified by the recent reform in 2005 following the Flood Control Act. Since the reform of the Federal Regional Planning Act (Raumordnungsgesetz –

ROG) of 1998, flood protection has been defined as a spatial planning issue. Precautionary flood protection became one of the main principles of Federal planning legislation (Art. 2 para. 2 ROG) and has to be implemented into all subsequent directives at all administrative levels. The law explicitly aims at the protection and restoration of floodplains, retention areas and flood-prone areas (Jekel 2005: 399). The 1998 reform of the ROG introduced new categories of planning zones to secure natural areas for flood protection. These categories – so-called priority and precautionary zones according to Article 7 paragraph 4 of the ROG – are seen by some commentators as being a more effective instrument for restoring floodplains than granting them legal status according to the WHG, being less complex to introduce and yet, in the case of the former, binding for subordinate plans (Böhm *et al.* 1999: 51; BBR 1998). These designated planning zones for floodplains may extend beyond those set by the WHG if they are in accordance with spatial development objectives and have been weighed up against other land-use interests. The 1999 State Development Plan for Saxony-Anhalt, for instance, establishes priority zones for flood protection areas behind dykes which could be restored as floodplains in the future.

As with water management legislation, it is as yet unclear what impact these new instruments will have on state regional policy and practice concerning floodplain restoration. Particularly under the new conditions of the reform of federalism, the Länder were given extensive rights to diverge from Federal regional planning legislation (Koch and Krohn 2006: 678). Even today, state regional planning laws, like those for water management, differ considerably from one another and mostly pre-date the ROG. Consequently there is little reference to the new planning zones and their application to floodplains. Spatial planning provisions for actively restoring floodplains exist in only a few states (e.g. Hesse, Northrhine-Westphalia, Rhineland-Palatinate, Saxony-Anhalt); references are more commonly directed at protecting existing floodplains (see e.g. in Table 4.1) (Böhm *et al.* 1999: 104). Where the new planning zones have been introduced into regional planning documents, there is often no reference to how or why they differ from the floodplains with legal status set according to the WHG. Here, again, the absence of reliable data and conflicts of user interests have contributed to weakening the instruments for floodplain restoration.

Below the level of regional planning, land-use planning also considerably affects floodplains. The Federal Building Code (BauGB), which was also reformed by the Flood Control Act in 2005, contains a number of instruments which have the potential, at least, to promote floodplain restoration. Via zoning plans (Flächennutzungspläne) and binding development plans (Bebauungspläne) the municipalities define the most important conditions for the loss, protection or restoration of floodplains. Since the reform in 2005 floodplain areas and flood-prone areas have to be adopted into the existing and new zoning and development plans (Arts. 5 para. 4a and 9 para. 6a BauGB) in order to give the public the necessary information about existing flood risks, which can be taken into account for site selection and building design at an early planning stage (Jekel 2005: 398; Petrow *et al.* 2006: 727f.). According to the new Article 246a BauGB, zoning plans are required to show floodplains legally protected according to Articles 31b and 31c WHG, and to identify further areas to be maintained for flood prevention requirements (Jekel 2005: 398). In addition they may identify areas to compensate environmentally for interventions on other sites according to the regulations on compensatory measures for interventions (Eingriffs-/Ausgleichsregelungen), which could apply for instance to potential new floodplains (Art. 1a para. 3 BauGB).

Even before the reform of the BauGB the development plan could keep land free from building for specific purposes (Art. 9 para. 1 no. 10 BauGB) and specify uses of land for nature conservation, agriculture etc. In practice, however, it was very rare for local authorities to take the initiative in extending water retention areas according to the BauGB. On the contrary, planning permission was occasionally granted on floodplains at odds with state regulations and plans. In future, federal law prohibits the planning of new housing areas and buildings in floodplain areas. However, exceptions are still possible if nine conditions are all met in the individual case (e.g. no alternative for human settlement in the community; flood-adapted building is ensured; no negative consequences for upstream or downstream communities). The effectiveness of the reform of the BauGB will thus depend considerably on the application of these exceptions for building permits by local zoning and development plans.

4.3.3 Nature conservation

After a controversial debate lasting more than fifteen years and several failed reform attempts, the Federal Nature Conservation Act of 1976 was amended in April 2002. The reform has the potential to strengthen nature conservation by defining the contribution of agriculture to nature conservation, strengthening the legal position of nature conservation associations, legally supporting habitat networks and consolidating the role of landscape planning. This, however, is dependent on statutory implementation and substantiation by the nature conservation legislation of the Länder.

While the former Federal Nature Conservation Act did not make any explicit reference to the protection or restoration of floodplains, the reformed law partially addresses those topics. Nature conservation measures relevant to floodplain restoration are founded primarily in the basic principles and objectives of nature conservation set down in Article 2 paragraph 1, BNatSchG.

The protection, development and restoration of natural or nearly natural waterways and their riparian zones is defined as one of the general principles of nature conservation and landscape management (Art. 2 para. 4 BNatSchG). As a principle of "good technical practice" for agriculture, it is now illegal to convert grassland on floodplains (Art. 5). Article 30, BNatSchG, lists water meadows and riparian woodland, natural or nearly natural ox-bows and regularly flooded areas amongst the biotopes to be accorded legal protection under the law. Additionally, surface water and riparian zones are to be protected and developed in order to fulfil their networking function for habitats (Art. 31). The principles and instruments of nature conservation have to be substantiated in state nature conservation legislation relating to specific biotopes (e.g. Art. 1 para. 2 BbgNatSchG).

The principal instruments of nature conservation in Germany comprise landscape planning, protected areas, the conservation of species and habitats and compensatory measures, all of which can to some extent contribute to floodplain restoration. Landscape planning, as the major spatial and precautionary planning instrument for nature conservation and landscape development under state responsibility, assesses and evaluates the objects of landscape and nature protection and development in spatial contexts and defines principles and priority objectives (SRU 2006: 12). It is conducted conventionally according to a three-tier structure, with state landscape programmes (Landschaftsrahmenprogramme) at the level of the Länder, landscape framework plans at regional or district level (Landschaftsrahmenpläne) and landscape plans at the level of municipalities (see Table 4.2). State landscape programmes often make detailed reference to floodplains and their restoration, as for instance that in Saxony-Anhalt relating to a programme of dyke relocation to increase the flooding of floodplains, and in Brandenburg to develop floodplains (see Table 4.1). Municipal landscape plans indicate the current and targeted status of nature and the landscape. The plans can include technical information on nature conservation of relevance to floodplain restoration, e.g. the retention capacity of municipal land, the status of the floodplains and their forests, the role of the municipality in flood defence for downstream communities or specific solutions in urban and natural areas to protect, develop or restore floodplains (cf. Riedel 2000: 56).

Another instrument of relevance to floodplain restoration is the designation of protected areas. The reformed law makes it possible to identify protected areas not only according to their present ecological value, as was the case previously, but also regarding their future development and their ecological potential. As a result, river ecosystems and floodplains which have been transformed by human interventions but have a high ecological potential can be defined as protected areas in order to develop and restore them ecologically. Amongst the protected areas envisaged by the Federal Nature Conservation Act, nature conservation zones (Art. 23 BNatSchG) are useful instruments for protecting, developing and restoring biotopes such as floodplains. Several of these nature conservation zones are located on the Elbe and Rhine as well as other rivers. National parks (Art. 24) are more relevant for larger areas relevant to flood protection. One of the 13 German national parks was founded in 1995 in order to conserve the river ecosystems and floodplains (and their woodlands) in the lower Oder valley. This trans-boundary conservation area covers about 9,500 ha in Poland and Germany. Floodplain restoration projects can also benefit from the existence or creation of other protected areas, such as biosphere reserves (Art. 25), e.g. the biosphere reserve on the River Elbe, but also landscape protection areas (Art. 26), nature parks (Art. 27) and Natura 2000 sites (Art. 32).

Beyond these restrictive, regulatory instruments, the reform of the Federal Conservation Act reinforces opportunities for promoting floodplain restoration via financial incentives (Art. 8 BNatSchG). Farmers who enter into a contractual agreement with nature conservation and agricultural authorities can receive compensatory payments for the additional costs of farming their floodplains more extensively (known as contractual nature conservation). The 5-year limit to these contracts in the past has, however, made it difficult to maintain nature conservation standards in the long term.

A further instrument of some importance to floodplain restoration is compensation for the impairment of the performance and functions of ecosystems or landscapes on a site affected by physical interventions by means of nature conservation measures elsewhere (Arts. 18-21 BNatSchG) (BBR 1998: 17). There are several instances of floodplains having been restored by land developers in conjunction with the granting of planning permission for a major development, such as a new railway line or industrial estate on a nearby site (see Kruse in Chapter 7 of this volume).

In sum, the reform of the BNatSchG considerably amended the conditions for floodplain restoration. However, these innovations in federal nature conservation policy have recently been undermined by the reform of federalism in Germany. The Länder now have far-reaching rights to diverge from national legislation. The German Advisory Council on the Environment is critical that key elements of nature conservation legislation have been put at the disposition of the Länder, such as the duty to build up a European and national biotope network, landscape planning, compensatory regulations, the definition of types of protected areas, environmental monitoring and partially also the legal position of nature conservation associations (SRU 2006: 11-14; see also Koch and Krohn 2006).

4.3.4 Agriculture and forestry

Federal and state laws on forestry and agriculture are less directly relevant to the restoration of floodplains and their woodlands. The Federal Forestry Act (Bundeswaldgesetz – BWaldG), designed to regulate a balance between individual and general interests, includes the preservation of forests and woodland in respect to their natural functions as a key objective. State forestry laws are more specific, for instance prohibiting uses of woodland which limit its natural functions. Within the scope of the Federal Agriculture Act (Landwirtschaftsgesetz – LwG) there are three main instruments relevant to floodplain restoration (cf. Böhm *et al.* 1999: 89-92; Frerichs *et al.* 2003: 103f). One of the most important instruments is the funding by the Joint Task of Improving Agricultural Structures and Coastal Protection (see Section 4.4 below). Another instrument is agricultural planning (Agrarstrukturelle (Vor-/Entwicklungs-)Planung), which develops concepts and guidelines for future agricultural land use. It has the power to limit agricultural uses on, for instance, floodplains. Moreover, land consolidation procedures enable the exchange of land parcels and more extensive land use in retention areas and floodplains. In practice, however, fields are usually only designated as serving the purposes of flood prevention and nature conservation when they are no longer viable for agricultural production (Böhm *et al.* 1999: 133). On the whole, it is too early to assess the consequences of the ongoing restructuring of agriculture for floodplain restoration. Although potentially significant and very supportive, the outcome is currently too open-ended.

4.4 FUNDING PROGRAMMES AND ECONOMIC INSTRUMENTS

The principal economic instruments for floodplain restoration in Germany are funding programmes (for an overview, BBR 1998: 65-66). These draw on sources from EU, federal and/or state governments, as well as NGOs, and come primarily from broader sectoral programmes for nature conservation, agricultural restructuring and flood protection. Foremost amongst EU programmes are INTERREG programmes for preventive flood protection measures, such as the IRMA (Rhine-Maas) programme of €135 million and the Odraregio programme, which were both in response to the flood disasters on the rivers Rhine (1993 and 1995) and Oder (1997). Additionally, the LIFE programme for nature conservation has, for instance, contributed to several floodplain restoration projects along the Elbe river (see Monstadt in Chapter 10 of this volume). Agricultural restructuring funds form an important financial resource for floodplain restoration. The EU offers subsidies to farmers for measures which can contribute to floodplain restoration, such as transforming arable land into grazing land, funding new forestry plantations or leaving fields fallow. The Federal Government contributes 60% of the costs for maintaining waterways in a 'natural' state or extensive forms of agriculture within its own restructuring fund (Joint Task of Improving Agricultural Structures and Coastal Protection). In this context, land consolidation procedures are subsidised, promoting the transition from arable to grazing land or ecological river development. Between 1998 and 2001 the Federal Government contributed some €124 million to flood protection measures by the Länder, especially via this programme. Since 2002 the emphasis of this programme has changed to give greater priority to sustainable flood management, including floodplain restoration, rather than technical flood defence measures (BMU 2002: 5).

The funding programmes of the Federal Government and the Länder are sometimes used to finance state programmes for river and floodplain restoration, such as the Nahe-Programme in Rhineland-Palatinate, the Riverbank Programme in Brandenburg, the Programme for the Restoration of Watercourses in Schleswig-Holstein, or the large protection area on the River Elbe. It is still rare, however, for such programmes to be funded by more than one federal state. Many measures are delayed, or even terminated, as a result of disagreements over the distribution of the financial burden between states and between funding sources. In addition to state sources, the importance of NGO funding, in particular from WWF Deutschland and EURONATUR, should be emphasised at this point. From whichever source, the funds are commonly used to either buy or lease land identified as important for floodplain restoration or to pay subsidies to landowners and farmers in return for their observance of certain cultivation requirements (for instance via contractual nature conservation, see Section 4.3.3 above). Some funding also goes into research, for instance via the Elbe-Ökologie programme of the Federal Research Ministry.

4.5 FORMS OF COOPERATION, CONSULTATION AND EDUCATION

It is important to distinguish between two types of cooperation and consultation: that required by law in the context of planning procedures and compensation contracts on the one hand and voluntary forms of cooperation on the other. In general, obligatory consultation procedures are relatively far-reaching in Germany. Representatives of the public interest (Träger öffentlicher Belange), primarily the principal organised stakeholders, are engaged in detailed consultation procedures for all major plans, whether spatial or sectoral, including regional planning procedures (Raumordnungsverfahren), planning appraisal procedures (Planfeststellungsverfahren) or environmental impact assessments (Umweltverträglichkeitsprüfung). This applies also to planning procedures for establishing a floodplain as a protected area. Since the adoption of the Federal Flood Control Act, public participation is also obligatory for the designation of floodplain areas in order to raise the public awareness of flood risks (Art. 31b para. 2 no. 5 WHG, see Jekel 2005: 396). Beyond these legal requirements, however, consultation (in particular with the general public) is underdeveloped (cf. The Ramsar Convention 1999: 17). Coordination and cooperation between state water authorities in Germany is rather limited and is particularly weak between water authorities and other water-relevant policy-making or planning bodies.

In order to overcome some of the problems of coordinating river management issues across state (and national) boundaries, working groups or commissions have been established between up- and downstream actors along the major rivers. These comprise national working groups of water authorities along, for instance, the Elbe, the Weser and the Rhine, and the International Commissions for the Protection of the Rhine (ICPR), Elbe (ICPE) and Oder (ICPO). Whilst the focus of several of these organisations was initially on water quality problems, recent flooding events have contributed to placing issues of preventive flood protection, including floodplain restoration, onto the agenda (IKSR 1995; 1997; 2001; IKSE 1998; 2003; 2006, see Table 4.1). The Action Plan for flood protection on the Rhine, for example, envisages the restoration (over the period 1998-2030) of some 1000 km^2 of floodplains in the whole Rhine basin and 160 km^2 along the Rhine itself, at a cost of €2.03 billion and €1.45 billion respectively (IKSR 1995). To take the case of the Elbe River, the ICPE specifically notes in its flood protection strategy that the removal of floodplain woodland in the past was a mistake. It calls for the retention and, where possible, restoration of natural retention areas in the catchment area, the restoration of former floodplains via dyke relocation, the use of floodplains primarily with extensive grazing and, where possible, the development of floodplain woodland (IKSE 1998: 5-6). Moreover, the ICPE recommends giving space back to the river Elbe by implementing 15 dyke relocation projects and restoring more than 2,600 km^2 of floodplain along the German stretch of the river (IKSE 2003: 24f.; cf. Monstadt in Chapter 10 of this volume). It should be emphasised, however, that these river commissions have an advisory function only; their decisions are not binding on the participating countries or states.

Cooperation between stakeholders of floodplain protection and restoration beyond the water authorities is generally limited in Germany, but certain forms of partnership have been formalised or partially formalised in the context of river or floodplain restoration programmes funded by individual states. In these instances, working groups are established to coordinate and plan activities as well as to resolve conflicts. The ICPE has recently announced that it will intensify its cooperation with NGOs from nature conservation, industry and commerce and agriculture. Greater participation of NGOs in plenary meetings and working groups will partially be funded by the ICPE. The floodplain programme of North Rhine-Westphalia is coordinated by working groups comprising representatives from water and nature conservation authorities and NGOs (MULR 1990). The floodplain programme for the River Ems, working on a smaller scale, has both a core working group to resolve conflicts (with equal representation from public water management, ecology and farming agencies) and an extended working group also including interest groups and local authorities, responsible for planning tasks (Staatliches Umweltamt Münster 1999). Planning and implementation of the Naheprogramm in Rhineland-Palatinate is coordinated by a working group comprising state and local authorities, nature conservation associations and farming representatives (MUF and MWVLW 1999). Such working groups are accredited with gaining support for floodplain restoration measures in the community. However, they would appear to have been less successful in overcoming sectoral differences between the principal water management, nature conservation and land-use authorities (Bunzel-Drüke 1995). In terms of awareness-raising, the nature conservation academies of several states offer seminars, guided tours and exhibitions for both professionals and the general public, including the topics of wetland and floodplain restoration.

4.6 PROGRAMMES AND PROJECTS TARGETED AT FLOODPLAIN RESTORATION

The most extensive and successful forms of floodplain restoration have been achieved in the framework of strategic and funding programmes geared towards preventive flood protection, river restoration or, specifically, floodplain (woodland) restoration. Several of the most significant programmes, initiated by international river commissions or individual German states, have been mentioned above. Out of such programmes, site-based floodplain restoration projects emerge (for detailed case studies see Kruse and Monstadt in Chapters 7 and 10 of this volume). Recently, interest in the restoration of floodplain woodland has also been growing, with projects for woodland development in Brandenburg, Saxony-Anhalt, Lower Saxony, Mecklenburg-Pomerania, Baden-Württemberg and Rhineland-Palatinate (see Interview 4). Familiar problems of implementation encountered in the course of these programmes and projects relate to land availability and use, funding and cross-sectoral cooperation. Existing and potential floodplains often have a high value for agricultural production and other uses, which can be incompatible with floodplain restoration. Funding for land acquisition or compensation payments often has to come from a variety of sources, each with their own specifications and limitations. Joint responsibility and implementation by several state authorities or agencies is very rare, making an integrated strategy on floodplain restoration difficult.

Since 2002, Länder that were not affected by the August flood have also intensified their flood prevention efforts by hazard mapping and by developing new concepts and plans for flood prevention. These are catchment-wide concepts, which will include the delineation of flood hazards on a municipal scale as well as recommendations on how to implement prevention measures (Petrow *et al.* 2006: 727). Saxony has developed flood protection concepts for all catchments, especially for tributary rivers which were severely damaged by flash floods in 2002 (ibid.). Despite various programmes and guidelines aiming at improved flood prevention and floodplain restoration, the implementation of projects is still rare and remains uncertain.

4.7 CONCLUSIONS: DRIVERS AND CONSTRAINTS OF FLOODPLAIN RESTORATION

The analysis of policy and practice with regard to floodplains and their restoration in Germany reveals a marked discrepancy between favourable policy at the federal level and difficulties of implementing restoration projects at regional and local levels.

Recent reforms to federal legislation in the key fields of water management, nature conservation and spatial planning have all created a number of potentially powerful drivers for floodplain restoration, reflecting widespread recognition in policy-making circles of the need for a more strategic, integrated approach to flood risk management and the value of floodplains to this end (see Section 4.3.1). In particular, regulatory and planning instruments have been strengthened so as to protect existing and restore currently defended floodplains for flood defence and biodiversity purposes (see Section 4.3.2), for instance creating openings for designating areas for future floodplain restoration. The new Flood Control Act in particular has introduced a set of innovative guidelines and legislation for effective flood management and floodplain restoration.

It remains to be seen, however, to what extent the Länder will use the constitutional rights they have been given by the recent reform of federalism. Even if this reform might not automatically lead to a progressive dismantling of regulatory standards (a "race to the bottom") in nature conservation, flood prevention and regional planning legislation, it is very probable that initiatives of river basin management and catchment-wide flood risk management will continue to suffer from heterogeneous legislation and policies from the Länder. Moreover, it is uncertain what impact these legal reforms will have, since their effective implementation depends on backing by the laws and authorities of the Länder and significant changes in the land-use and development planning practices of local authorities. In the field of spatial planning we can observe how relatively favourable drivers at the federal level are confronted by the contradictory priorities of many local authorities to use floodplains for urban development purposes. Even if planning permission in floodplain areas may generally not be granted since the recent reform of the BauGB, it is quite likely that particularly those communities which are not themselves affected by flood damage will try to apply the exceptions defined by the legal reform or will not be motivated to designate floodplain areas. At the same time, the farming community is still generally sceptical of the benefits of floodplain restoration. For these reasons some of the largest floodplain restoration projects have only been made possible through land acquisition by NGOs, state agencies or other interested groups, or by land exchange in compensation for development elsewhere (see the case studies of Kruse and Monstadt in Chapters 7 and 10 of this volume).

Although some Länder have undertaken ambitious initiatives to restore floodplains and to institutionalise flood prevention, coordination between them on the one hand and between communities on the other is still weak. Efficient cross-boundary strategies are still inadequate and uniform concepts are largely missing (Petrow *et al.* 2006: 727). One important reason for this territorial fragmentation in Germany lies in the absence of any powerful river basin management institutions. Water management, including flood defence, is the responsibility of the individual Länder and, largely as a consequence, there has in the past been only limited trans-boundary cooperation across whole catchments. The EU Water Framework Directive is already proving a powerful instrument for creating awareness in water policy circles of the benefits (and difficulties) associated with river basin management. Although it is difficult as yet to assess the impact of the WFD, the recently proposed Flood Risk Management Directive and the reforms by the Federal Flood Control Act are likely to strengthen the growing appreciation of a catchment-based approach to flood risk management and floodplain restoration.

In conclusion, the flood disasters and the growing awareness of the value of floodplains for flood prevention, nature conservation, rural development etc. have led to policy innovations at the national level which have opened up new opportunities for the preservation and restoration of floodplains. However, it is as yet unclear how far the German Länder will diverge from federal legislation in water management, regional planning and nature conservation and how far they will, in collaboration with the municipalities, use their implementation resources to initiate new floodplain restoration projects.

4.8 REFERENCES

Adams, W. M., Perrow, M. R. and Carpenter, A. (2004) Conservatives and champions: river managers and the river restoration discourse in the United Kingdom. In: *Environment and Planning A* **36**(11), 1929-1942.

BBR (Bundesamt für Bauwesen und Raumordnung – Federal Office for Building and Regional Planning) (1998) *Leitfibel vorbeugender Hochwasserschutz*. BBR. Werkstatt: Praxis No. 6.

BMU (Bundesministerium für Umwelt, Naturschutz und Reaktorsicherheit – Federal Ministry for Environment, Nature Conservation and Nuclear Safety) (2002) *Perspektiven für eine ökologisch ausgerichtete Hochwasservorsorge*. Progress report of the German Federal Cabinet, 17 April 2002,
Berlin (http://www.bmu.de/download/dateien/hochwasser_bericht.pdf, 29.04.2002).

Böhm, H. R., Heiland, P., Dapp, K. and Mengel, A. (1999) *Anforderungen des vorsorgenden Hochwasserschutzes an Raumordnung, Landes-/Regionalplanung, Stadtplanung und die Umweltfachplanungen – Empfehlungen für die Weiterentwicklung*. UBA-Texte 45/99, Umweltbundesamt: Berlin.

Böhm, H. R., Haupter, B., Heiland, P. and Dapp, K. (2004) Implementation of flood risk management measures into spatial plans and policies. *River Research and Applications* (Special Issue "Towards Sustainable Flood Risk Management in the Rhine and Meuse River Basins") **20**(2) 255-267.

Bunzel-Drüke, M. (1995) Auenschutz in NRW. In *Seminarbericht Auenschutz in NRW* (Vol. 13) Naturschutzzentrum NRW (ed.), pp. 46-51.

Deutsche Bundesregierung (2002) *5-Punkte Programm der Bundesregierung: Arbeitsschritte zur Verbesserung des vorbeugenden Hochwasserschutzes*. Berlin, 15 September 2002.

Dister, E. (2002) *Leben am Fluß 2050. Final Report contributing to the IRMA-project "Multifunktionale Entwicklung von Überschwemmungsauen"* (http://www.irma-lfw-rp.de/Irma_text/03fachbeitrag_irma1.zip, 20.12.2006)

Förster, S., Kneis, D., Gocht, M. and Bronstert, A. (2005) Flood risk reduction by the use of retention areas at the Elbe River. *International Journal of River Basin Management* **3**(1), 21-29.

Frerichs, S., Hatzfeld, F. Hinzen, A., Kurz, S., Lau, P. and Simon, A. (2003) *Sichern und Wiederherstellen von Hochwasserrückhalteflächen*. UBA-Texte 34/03. Umweltbundesamt: Berlin.

Heiland P. 2002. *Vorsorgender Hochwasserschutz durch Raumplanung, interregionale Kooperation und ökonomischen Lastenausgleich*. Ph.D. Thesis Technische Universität Darmstadt, Schriftenreihe WAR 143: Darmstadt.

IKSE – Internationale Kommission zum Schutz der Elbe (1998) *Strategie zum Hochwasserschutz im Einzugsgebiet der Elbe*. IKSE: Magdeburg.

IKSE (2003) *Aktionsplan Hochwasserschutz Elbe*. Magdeburg, 24.10.2003.

IKSE (2006) *Erster Bericht über die Erfüllung des "Aktionsplans Hochwasserschutz Elbe" im Zeitraum 2003 bis 2005*. Magdeburg.

IKSR – Internationale Kommission zum Schutz des Rheins (1995) *Grundlagen und Strategie zum Aktionsplan Hochwasser*. IKSR: Koblenz.

IKSR (1997) *Hochwasserschutz am Rhein. Bestandsaufnahme*. IKSR: Koblenz.

IKSR (2001) *Rhein 2020 – Programm zur nachhaltigen Entwicklung des Rheins*. IKSR: Koblenz.

Jekel, H. (2005) Das Gesetz zur Verbesserung des vorbeugenden Hochwasserschutzes. *Zeitschrift für Umweltrecht* **9**, 393-400.

Länderarbeitsgemeinschaft Wasser (1995) *Leitlinien für einen zukunftsweisenden Hochwasserschutz.* LAWA o.O.

Kahlenborn, W. and Kraemer, R. A. (1999) *Nachhaltige Wasserwirtschaft in Deutschland.* Springer Verlag: Berlin et al.

Koch, H.-J. and Krohn, S. (2006) Umwelt in schlechter Verfassung? Der Umweltschutz nach der Föderalismusreform. *Natur und Recht* **11**, 673-680.

Lange, H. and Garrelts, H. (2007) *Risk management at the science-policy interface. Two contrasting cases in the field of flood protection in Germany.* Unpublished and revised paper based on the presentation on the Conference "Governance for Sustainable Development. Steering in Contexts of Ambivalence, Uncertainty and Distributed Control", Berlin, 06.02.2006, University of Bremen.

MKRO (Ministerkonferenz für Raumordnung) (1995) *Beiträge räumlicher Planungen zum vorbeugenden Hochwasserschutz.* Entschließung vom 8. März 1995.

MKRO (Ministerkonferenz für Raumordnung) (1996) *Grundsätze und Ziele der Raumordnung und Landesplanung zu einem grenzübergreifenden vorbeugenden Hochwasserschutz an Fließgewässern.* Entschließung der MKRO vom 29. März 1996.

MUF/MWVLW (Ministerium für Umwelt und Forsten des Landes Rheinland-Pfalz und Ministerium für Wirtschaft, Verkehr, Landwirtschaft und Weinbau des Landes Rheinland-Pfalz) (1999) *Das Naheprogramm. Bilanz 1994-1999.* Mainz.

MUNR (Ministerium für Umwelt, Naturschutz und Raumordnung des Landes Brandenburg) 1998: *Materialien zum Landschaftsprogramm Brandenburg.* Potsdam.

MURL (Minister für Umwelt, Raumordnung und Landwirtschaft des Landes Nordrhein-Westfalen) (1990) *Gewässerauenprogramm.* MURL, Düsseldorf.

Petrow, T., Thieken, A. H., Kreibich, H., Merz, B. and Bahlburg, C.H. (2006) Improvements on Flood Alleviation in Germany: Lessons Learned from the Elbe Flood in August 2002. *Environmental Management* **38**(5), 717-732.

Purps, J. (1999) Auenwaldentwicklung an der Unteren Mittelelbe und Perspektiven des Auwaldes an der Elbe und an weiteren Flüssen. In *Auenregeneration durch Deichrückverlegung* (ed. Fachtagung Elbe) Wittenberge 04.-07.05.1999

Riedl, U. (2000) Was können Landschaftsplanung, Naturschutz und Landschaftspflege zum vorbeugenden Hochwasserschutz beitragen? In *Vorbeugender Hochwasserschutz auf kommunaler Ebene.* Report of the Workshop on 13th and 14th December 2000 (ed. Umweltbundesamt und Institut für ökologische Raumentwicklung) Dresden, pp. 55-69.

Rühl, C. (2000) Rechtliche Vorgaben und Instrumente der kommunalen Hochwasservorsorge. In *Vorbeugender Hochwasserschutz auf kommunaler Ebene.* Report of the Workshop on 13th and 14th December 2000 (ed. Umweltbundesamt und Institut für ökologische Raumentwicklung) Dresden, pp. 14-23.

SRU (Deutscher Rat von Sachverständigen für Umweltfragen – German Advisory Council on the Environment) (2000) *Umweltgutachten 2000: Schritte ins nächste Jahrtausend.* Stuttgart, Poeschel.

SRU (2004) *Umweltgutachten 2004. Umweltpolitische Handlungsfähigkeit sichern.* Bundestagsdrucksache 15/3600, Berlin.

SRU (2006) *Der Umweltschutz in der Föderalismusreform.* Stellungnahme Nr. 6, Berlin.

Staatliches Umweltamt Münster (1999) *Gewässerauenprogramm Ems. Ems-Auen-Schutz konzept.* Staatliches Umweltamt Münster. Berichte und Informationen Band 4.

The Ramsar Convention on Wetlands (1999) *National Report of Germany for COP7.* Gland.

Volkers, S. (1998) Gewässer- und Auenentwicklungskonzept der Ilm. In *Fließgewässer Schutz und Entwicklung* (ed. Alfred Toepfer, Akademie für Naturschutz) NNA Berichte **11**(1), 63-70.
Weichselgartner, J. (2003) Nach der Elbeflut 2002: Und danach? In *GAIA – Ecological Perspectives for Science and Society* **12**(4), 245-248.
Zaugg, M. (2003) Mehr Raum den Fließgewässern – Der Weg zu einem nachhaltigen Hochwasserschutz. In *GAIA – Ecological Perspectives for Science and Society* **12**(3), 201-207.

LAWS

Federal Building Code (Baugesetzbuch - BauGB) of 23 June 1960, last amended on 5 September 2006, Federal Law Gazette I, pp. 2098ff.
Federal Flood Control Act (Gesetz zur Verbesserung des vorbeugenden Hochwasserschutzes Hochwasserschutzgesetz (HWSchG)) of 3 May 2005, Federal Law Gazette I of 9 May 2005, pp. 1224ff.
Federal Nature Conservation Act (Gesetz über Naturschutz und Landschaftspflege – Bundesnaturschutzgesetz (BNatSchG)) of 25 March 2002, last amended on 21 June 2005, Federal Law Gazette I, pp. 2833ff.
Federal Regional Planning Act (Raumordnungsgesetz – ROG) of 18 August 1997, last amended on 9 December 2006, Federal Law Gazette I, pp. 2833ff.
Federal Water Act (Gesetz zur Ordnung des Wasserhaushalts – Wasserhaushaltsgesetz (WHG)) of 27 July 1957, last amended on 25 June 2005.
State Nature Conservation Act of Brandenburg – (Gesetz über den Naturschutz und die Landschaftspflege im Land Brandenburg – Brandenburgisches Naturschutzgesetz (BbgNatSchG)) of 26 May 2004.
State Regional Planning Act of Brandenburg (Landesplanungsgesetz und Vorschaltgesetz zum Landesentwicklungsprogramm für das Land Brandenburg – Brandenburgisches Landesplanungsgesetz (BbgLPlG)) of 12 December 2002, last amended on 28 June 2006, GVBl. I No. 1, pp. 96ff.
State Water Act of Saxony-Anhalt (Wassergesetz für das Land Sachsen-Anhalt – WG LSA of 12 April 2006, GVBl. LSA 2006, pp. 248ff.

INTERVIEWS

Interview 1: Professional environmentalist at the Green League (Grüne Liga), Berlin, 2 February 2002.
Interview 2: Officer at the State Agency for Environmental Protection, Saxony-Anhalt, Magdeburg, 27 February 2002.
Interview 3: Director of the Elbe River Landscape Biosphere Reserve, Dessau, 27 February 2002.
Interview 4: Environmentalist at the Floodplains Institute of the World Wide Fund for Nature (WWF-Auen-Institut), Rastatt, 28 February 2002
Interview 5: Head of Division for General, Fundamental and International Aspects of Water Management at the Federal Ministry for the Environment, Nature Conservation and Nuclear Safety (BMU), Berlin, 1 March 2002.

Interview 6: Officer of the State Ministry for Environment, Forestry and Consumer Protection, Rhineland-Palatinate, 4 March 2002.
Interview 7: Officer of the International Commission for the Protection of the Elbe, Magdeburg, 5 February 2003.
Interview 8: Officer of the Water and Shipping Directorate East, Magdeburg, 5 February 2003.
Interview 9: Two conservation officers at the Federal Agency for Nature Conservation, Bonn, 29 January 2003.

5

Floodplain restoration in the context of river basin management: policy development and implementation in France

Jochen Monstadt

5.1 INTRODUCTION

Conditions for policies promoting the restoration of floodplains and flood prevention are generally favourable in France as the national territory was subdivided into six river basins in the 1960s in order to promote solidarity among the users of the same fluvial system. However, although it has been possible to develop and implement relevant policies more efficiently at a catchment level, the relevance of these river basin institutions in the restoration of river environments and in flood prevention has in fact remained rather weak up until the present. As in most other European

© 2008 IWA Publishing. *Restoring Floodplains in Europe: policy contexts and project experiences* edited by Timothy Moss and Jochen Monstadt. ISBN: 1843390906. Published by IWA Publishing, London, UK.

countries, in France policy or strategy documents relating specifically to floodplains and their restoration exist neither at the national nor at the catchment level. However, following severe flood events in the early 1990s and studies demonstrating an alarming degradation of wetlands and river ecosystems, a moderate policy shift in the management of floodplains has been initiated by the national government and in individual river basins. Several legislative reforms and policy documents relating to water management, flood prevention, spatial planning, nature conservation and agricultural policies refer to the protection and restoration of floodplains. In particular the legal reforms of water management in 1992 and 2006, the reform of flood risk management in 1995 and 2003, the reform of regional planning in 1999, the national action plan for wetlands of 1996, several funding programmes and the reinforcement of contractual nature conservation have created new openings for floodplain restoration and innovative forms of flood risk management. In addition to national policies, more detailed references to floodplain restoration can be found in the water management plans, in local river contracts and in risk prevention plans. Moreover, public or public-private agencies and nature conservation associations promote integrated regional development and floodplain restoration projects by means of operational guidelines and incremental approaches.

The following section contains a policy analysis of the context for restoring floodplains in France. The aim is to scrutinise the new drivers for, and the persistent constraints to, restoring floodplains under the current conditions. The study is based on an extensive analysis of policy documents and legislation of the national government and of exemplary regional governments, and the relevant secondary literature. Apart from this policy analysis, which was mainly carried out in the years 2000 to 2002 and updated in winter 2006/2007, the main issues raised in this chapter were explored in qualitative interviews with experts from national and regional administrations and research institutes (see list under References). The analysis starts with a selection of relevant policy statements on floodplain restoration. The regulation and planning of floodplain restoration in different policy fields is then examined, as well as funding programmes and modes of cooperation and consultation across policy fields. The chapter concludes by recapitulating the main drivers of, and constraints to, floodplain restoration.

5.2 POLICY STATEMENTS ON FLOODPLAINS AND THEIR RESTORATION

In France, specific references to floodplains and their restoration are to be found in a wide range of relevant federal legislation relating to water management, spatial planning and nature conservation, particularly in laws revised since the mid-1990s, e.g. the Environmental Code, which includes legislation on water management, risk prevention and nature conservation. Moreover, various planning guidance documents, operational guidelines and policy statements refer to the restoration of floodplains, e.g. a national action plan for wetlands, water management and risk prevention plans. These are published by the statutory bodies responsible for flood protection, water management, nature conservation and spatial planning, notably national ministries, water agencies, and regional water and regional planning authorities. References to floodplain restoration in recent policy documents reflect an increasing interest in preventive forms of flood protection, in river restoration and in catchment-based approaches to flood management.

Table 5.1 presents an overview of the references (explicit and implicit) to floodplain restoration in the principal laws and policy documents in France.

Table 5.1: Policy statements relating to floodplain restoration in France.

Document	Policy statement
Environmental Code	*Water and aquatic environments* EC Art. L211-1 General regime and management of the resource: "Balanced management implies – among other things – ensuring the protection of aquatic ecosystems, sites and wetlands (wetlands are defined as land, developed or not, usually flooded with fresh, salt or brackish water in a permanent or temporary way; the vegetation, if existent, is dominated by hygrophile plants during at least part of the year)". *Water management plans (SAGE)* EC Art. L 211-7: The water management plans at the catchment or sub-basin level aim *inter alia* at "the protection and restoration of sites, aquatic ecosystems and wetlands as well as bordering wooded formations". *National Water Solidarity Fund:* EC Art. L213-8: The water solidarity fees should be spent *inter alia* on "investments relating to the restoration of rivers and floodplains/flooding zones" and "subsidies for investments relating to the restoration of rivers and floodplains/flooding zones, ...the conservation and restoration of wetlands". *Risk Prevention Plans (PPP):* EC Art. L562-8: "In the lower-lying parts of the river basins and other flooding zones, Risk Prevention Plans ...define, if needed, the technical prohibitions and prescriptions to respect, in order to ensure free water run-off and the conservation, restoration or extension of floodplains and flooding zones." The priority orientations of the multi-annual intervention programme 2007–2012 of the water agencies are the following (Art. 83 of the water legislation reform in 2006): "7. to guide and privilege activities in order to preserve, restore, maintain and enhance the management of aquatic environments and wetlands". 9. to contribute to flood regulation by the extension of the flood retention capacity of natural floodplains..."

Document	Policy statement
Law project comprising the reform of water policy (adopted by the National Assembly, but not implemented)	*Water and aquatic environments; General regime and resource management (Art. 28):* Obligations of public interest can be defined at the request of the State, the territorial bodies or their groupings located at rivers aiming at one or several of the following objectives: 1. To allow periodic flooding in zones called "flood retention zones" in order to limit floods in urbanised areas located downstream; 2. To allow the natural displacement of the river bed in zones called "zones of mobility of a river", in order to preserve or to restore its essential hydrological, geomorphological and ecological character; 3. To confine or to prohibit some agricultural practices in the zones near the river called "zones of protection", in order to preserve the water quality. *Fees for the modification of water regimes (Art. 41, para. 5)* "Fees for the modification of water regimes are due from every person whose installations, works or activities affect 1. the derivation of the whole or a part of a river…; 2. the presence of a continuous obstacle joining two banks of a river…; 3. the storage of the whole or a part of the volume running in a river....; 4. the restitution of water volumes for hydraulic power use…; 5. the sealing of land…; 6. the reduction of the surface of floodplains/flooding zones".
Collective service schemes (Schémas de services collectives)	*Collective service scheme for natural and rural areas* (DATAR 2002): "The general objective of spatial planning in large river valleys is to preserve their multifunctionality and to preserve their natural zones in particular. Accordingly, the building of certain large infrastructures and equipment will have to be reconsidered". "Wetlands, crucial for good water management in quantity and quality, for biodiversity and landscapes, constitute, in addition, multipurpose spaces useful for flood control, offering facilities for original products and various amenities, walking, fishing, hunting and recreation. In order to preserve the wetlands, it is necessary to organise a concerted action at the local level between all the actors to publish a joint management plan aiming at guaranteeing the sustainability of their ecological functions."
MATE (Ministère de l'Aménagement du Territoire et de l'Environnement, Direction de l'Eau et Direction de la Nature et Paysages) 1996	*Functions and values of alluvial wetlands:* "Alluvial wetlands, like floodplains ("champ d'expansion des crucs") regulate the river flows by reducing high waters and by compensating minimum flows by the retention of water in critical periods. They also play a part in the purification of water and some of them in the replenishment of groundwater. They play a significant role with respect to the conservation of biodiversity. …As natural basins, they prevent risks of flooding by decreasing the intensity of rising water levels. The vegetation on riverbanks contributes to the prevention of erosion. …The riverbanks through their heritage, landscape and cultural value also support tourism and recreation" (MATE 1996: 46).

Document	Policy statement
	Conservation strategy: "Problems of conservation are due to the functional isolation of natural river environments caused by hydraulic modifications and disturbances of river dynamics. Indeed, the management of alluvial zones must be conducted at the basin level because of functional interdependencies of the upstream and downstream environments. Thus, the restoration of ecological performance requires taking water and flow dynamics into account" (MATE 1996: 46).
Circulaire interministérielle relative à la prévention des inondations et à la gestion des zones inondables du 24.01.1994[1]	The circular relating to flood prevention and the management of floodplains specifies the objectives to be achieved: - to exclude human settlements in the most dangerous zones where the security of people cannot be completely guaranteed; - to preserve the capacities of flow and of flood expansion…; - to conserve the balance of environments dependent on small floods and the quality of the landscapes.
Master plan for water management, SDAGE, for the basin of Loire-Bretagne (1996)	*Conservation and restoration of the natural character and ecological quality of floodplains as a tool of flood prevention:* A common policy should be implemented for "conserving or restoring the natural character and the ecological and landscape quality of floodplains preserving their ecological functions and diversity" (p. 28). *Conservation and restoration of wetlands:* A fundamental principle should be the "conservation of the natural system of quantitative and qualitative regulation of the water resource (natural flood retention, self-purification, restitution of high quality water for periods with low water)" (pp. 40-43). It is recommended to statutory services and other promoters of the SDAGE that they "guarantee wetland restoration programmes if required" (p. 40/43). Furthermore, financial resources from the EU LIFE Nature programme are to be spent primarily on wetlands indicated by EU directives or the Ramsar Convention (p. 43).
Agricultural Act	Art. 1: "Agricultural policy takes into account the economic, environmental and social functions of agriculture and participates in regional planning for sustainable development. ...In particular agricultural policy considers the specific regional context, especially …the limited number of wetlands the characteristics of which require implementation of a specific policy…"

1 Cf. http://aida.ineris.fr/textes/circulaires/text4188.htm, 08.03.2007.

5.3 PRINCIPAL INSTRUMENTS AND POLICIES RELEVANT TO FLOODPLAIN RESTORATION

France is a centralised democracy with, traditionally, two sovereign entities: the central State, representing the nation, and the municipalities, representing – in contrast to Anglo-Saxon countries – the assembly of local citizens (Barraqué 1998: 88). However, decentralisation reforms in the 1960s, 1980s and in 2003 have partially counter-balanced central government power to the benefit of local and regional levels of government. One important element of this decentralisation, the regions (régions), was created in the 1960s as an intermediate level of government between the central State and the departments (départements) to promote economic policies and planning. Despite the efforts at decentralisation in recent decades and the constitutional reform of 2003 on the decentralised organisation of the State, the central State still has a constant and far-reaching influence on the policy-making process of the regions, departments and municipalities. The presence of the central State is not only manifested in the capital, Paris, but also in the existence of administrative bodies subordinate to the central administration throughout France. The central State's influence is stabilised by its directly supervised administration bodies at the regional level (dispersed administrations) and the delegated representatives of the central State in the regions and departments (prefects). This dispersed national network is in charge, under the regional prefect's authority, of implementing locally the policies defined at national level. This far-reaching influence of the central State at the regional level interacts with the weak position – relative to Germany – of the municipalities and the stronger supervision of municipal operations by the central administration. For one thing, the municipalities are on average the smallest in Europe, for another they are traditionally financially weaker than their counterparts in most other European countries.

Accordingly, the institutional regimes for floodplain restoration in France are influenced to a great extent by the centralised structure of government and administration. This centralised regime structure is also characteristic in the specific areas of water management, flood protection, regional planning and nature conservation. However, where these policy domains relevant to the management of floodplains are concerned, both decentralised and centralised bodies (as well as their departmental and regional representatives) are involved (cf. Sangaré and Larrue 2004: 187f.)

Central government and its representative bodies at the regional and departmental levels exercise regulatory power. At the central level, the competencies for floodplain restoration are mainly shared among the following three ministries and their subordinate bodies: the Ministry of the Environment and Spatial Planning, the Ministry of Public Works and the Ministry of Agriculture (for details see Barraqué 1998). An inter-ministerial Committee, chaired by the Minister for the Environment, is responsible for coordinating national water policy. Within the State administration, a special organisation has been set up to implement water law (cf. Sangaré and Larrue 2004: 188): six special prefects have been appointed to coordinate State water policy at the level of the six main water basins. They are assisted by six special administrative structures (water basin administration – DIREN de bassin).

The regional (26) and departmental (100) councils are also involved in policies relevant to floodplain restoration. They usually subsidise local authorities to equip their territories with modern sewage plants and to deliver good quality water, as well as to implement measures for protecting the population from floods and to set up hydraulic works for agricultural purposes. Departmental councils are also able to protect wetlands by using special taxes to purchase sensitive areas. They also provide local authorities with technical advice and funds to improve the maintenance of river banks (Sangaré and Larrue 2004: 188). However, in the case of water policy, the regions and departments have only limited power and they in fact compete with the six water agencies (cf. Section 5.3.1).

At the local level, the mayors of about 36,000 municipalities are responsible for the supply of drinking water, for wastewater treatment and for flood protection. In many cases the municipalities form inter-municipal bodies, which deliver these services collectively, finance waterworks[2] and develop various common projects.

5.3.1 Water management

The regulatory framework of French water policy is defined by the *Water Act of 1964,* reformed in 1992 and 2006, which was integrated into the Environmental Code in 2000. Since the water reform of 1964, French territory has been subdivided into six river basins (Artois-Picardie, Seine-Normandie, Loire-Bretagne, Adour-Garonne, Rhin-Meuse, Rhone-Méditerranée) in order to promote solidarity among users of the same fluvial system by legally linking them together and by establishing financial solidarity among water users in each hydrographic system. The Water Act created six river basin agencies as administrative public organisations endowed with civil status and financial autonomy and aiming at a balance between water resources conservation and social requirements. These agencies collect levies paid by water polluters and

---

[2] However, most communes and inter-communal bodies have delegated their responsibilities in water supply and sewage management to private companies. Here, the two largest utility companies supply more than 60% of the population with drinking water and collect the wastewater of about 40% of the population (Sangaré and Larrue 2004).

water consumers and distribute subsidies to protect water resources. Their tasks traditionally range from securing the supply and quality of drinking water and treating water for domestic use to reducing water pollution from point sources, especially from industrial polluters. More recently the need to reduce pollution from diffuse sources, especially agricultural nitrogen emissions, to reduce water consumption as a whole and to preserve aquatic ecosystems was put on the agenda and defined as the task of the water agencies. The activities of the water agencies are supervised by river basin committees consisting of 60 to 100 local councillors, manufacturing and agricultural users, associations and government representatives. These "water parliaments" are consulted on the rates and bases of water charges levied for water abstractions and discharges and on the priorities for the agencies' 5-year action programmes. However, as the meetings of the committees are rare and effective supervision is limited by the high costs of coordination among the members, the role of the committee is usually to approve the decisions prepared by the water agencies.

In January 1992, the national water legislation was revised to restore the quality of rivers and watercourses and to reinforce the role of local authorities, particularly in maintaining waterways. The Water Act underlines that water is part of the common national heritage, and that water resources must be protected, enhanced and developed as a matter of public interest. The reform gave a legal definition to wetlands[3] and established the principles of well-balanced management between the aquatic environment, wetlands, landscape protection and the economic use of water resources. Even though several regulations involve an integrated approach towards the conservation and sustainable management of aquatic environments, the act primarily focuses on water quality management with the introduction of the new water planning tools, the master plan for water management (schéma directeur d'aménagement et de gestion des eaux – SDAGE) and the water management plans (schéma d'aménagement et de gestion des eaux – SAGE).

Each of the obligatory *master plans for water management (SDAGEs)* defines for one of the six river basins the fundamental policy position for water resource management for the next fifteen years. SDAGEs take into account the main programmes decided by public communities and define, in a general manner, objectives for water quantity and quality as well as the developments and improvements to be undertaken to attain them. They define the spatial scope of the sub-basins corresponding to hydrographical units. The master plans aim at guaranteeing sustainable water management, taking aquatic environments into account. Several SDAGEs promote ecological objectives like the conservation or

3 Wetlands are defined as land, developed or not, often flooded with fresh, salt or brackish water in a permanent or temporary way. The vegetation, if existent, is dominated by hygrophile plants during at least part of the year (EC art. L211-1: I 1).

restoration of floodplains, major aquifers and other extraordinary aquatic environments, and the opening up of watercourses for migratory fish. Even the concept of "free space for river courses" ("espace de liberté"[4]), developed by the Environmental Ministry in 1990, has been integrated into some SDAGEs, e.g. for the Rhone-Mediterranean-Corsica and Loire-Brittany basins (cf. IFEN 2000: 79).

After the adoption of the SDAGE by the Basin Committee it has to be approved by the administrative authorities. Administrative decisions in the field of water management and land-use planning have to be made compatible with the guidelines and the provisions of the SDAGE, or at least not contradict them. However, in most cases the SDAGE cannot prevent urban planning decisions with negative impacts on aquatic systems (Sohnle 2001: 182).

Complementary to the SDAGE, the Water Act introduced *water management plans* (*SAGE*) at the level of sub-basins corresponding to hydrographic units or aquifer systems. In contrast to the obligatory SDAGE they are optional and depend on the will of the local authorities located in the sub-basins. Within the framework of the SDAGE they define the general objectives for the use, development and protection of water resources and aquatic environments, as well as the preservation of wetlands. Once the plan has been approved, decisions by administrative authorities concerning water, which are applicable to the area covered by the plan must be consistent, or made consistent, with the plan.[5] The preparation, revision and monitoring of the SAGEs fall under the responsibility of local water commissions (Commissions locale de l'eau), comprising users and State and local authorities. The hierarchy in the planning process (SDAGE/SAGE) is to a large extent inspired by urban planning procedures. Because of the local or catchment scale, the nature of the SAGEs is more binding than the SDAGEs, with objectives and measures relating directly to specific spatial contexts and defining concrete measures for wetlands preservation and floodplain restoration. Some Water Management Plans have already delineated the "liberty space" of rivers (including dead arms and alluvial forests), for which different restrictions on use have been adopted and backed up by a judicial framework (Dupont 2000). Despite this potential, the water management plans have so far not turned out to be an effective instrument for floodplain restoration, as they are far from covering all the relevant rivers. In 2001 only three SAGEs had been adopted and about 57 projects were still in the process of development (DATAR 2002: 43).

---

[4] The UNESCO defined this concept as follows: "Space of the major river bed in which lateral translations allow for a mobilisation of sediments as well as an optimal functioning of aquatic and terrestrial environments" (Author's translation of French text found at http://webworld.unesco.org/water/ihp/db/glossary/glu/frdic/dicespac.htm, 17.03.2007).

[5] Office International de l'Eau, http://www.oieau.fr/anglais/gest_eau/part_a.htm, 10.04.2007.

Complementary to the SAGE, *river contracts*, introduced in 1981 and reformed with the Water Act in 1992, promote the restoration and valuation of aquatic environments on a contractual basis. While they traditionally served as an instrument for developing quality objectives for a river, river contracts have been extended in the direction of an integrated approach to the management of water and aquatic environments. Besides the improvement of water quality they promote the restoration and conservation of riverbanks, flood prevention and the revaluation of river environments. Since the reform of water policy in 1992 they have served increasingly as the principal operational translation of the SAGE.[6]

Even if the French water planning regime – according to statements by the Ministry for the Environment[7] – has been progressing from water management to an integrated management of aquatic environments, neither the SDAGE nor the SAGE have been able to contribute effectively to the development of river environments, and more specifically to the protection and restoration of wetlands and floodplains. Despite the objectives relating to the conservation and restoration of aquatic environments and river courses formulated in the SDAGEs, the weak statutory authority of the plans limits their effectiveness (Sohnle 2001: 182). The water management plans and their respective financial and institutional resources for implementation focus above all on the water quality and not on the development of aquatic environments. This is also true for the water agencies, the most powerful financing and coordinating institutions of water management plans at the basin and sub-basin level, which are rarely involved in the conservation or restoration of aquatic environments or in flood defence (cf. Section 5.4). The elaborate French water regime has thus not achieved the integrated management of aquatic environments incorporating flood risks and the conservation and restoration of aquatic environments.

Responding to the European Water Framework Directive (WFD) and various criticisms of the French system, an ambitious *water policy reform project* was launched by the former Green environmental minister Dominique Voynet in 1999 and was approved at a plenary meeting of the French National Assembly in January 2002. After the elections in 2002, the new government did not pass the reform but proposed a new national debate among different actors on the reform of water policy in order to define the main criteria for a future water law in line with the WFD. After a three-year debate, a legislative reform, the "*Water and Aquatic Environment Act*", was finally passed in December 2006. Without challenging the original laws on water management, this new law was primarily set up to equip France with the means to meet the requirements of the WFD by

6 Cf. Ministry of Environment and Spatial Planning 2002 (http://www.environnement.gouv.fr/actua/cominfos/dosdir/DIREAU/contrivi.htm, 17.04.2002).

7 Ministry of Environment and Spatial Planning 2002 (http://www.environnement.gouv.fr/ministere/sdage.htm, 01.04.2002).

2015, to prepare for the impact of climate change on aquatic environments and to ensure more transparency and municipal control in the supply of water and wastewater services. Its main objectives are:

- to transpose the WFD into national law by defining objectives for the good ecological status of water and aquatic environments;
- to extend the polluter/user pays principle to more sectors influencing water quality or regimes, so that agricultural and industrial emissions and water use, as well as water use for electricity production, are subject to higher levies;
- to submit the programmes of, and the fees levied by, the water agencies and their programmes to the control of the Basin Committees;
- to reinforce the transparency of public services and to shorten the length of concession and delegation contracts with the water/wastewater utilities;
- to reinforce the role of the water police.

Following the earlier reform project, one of the main legislative innovations in the field of floodplain restoration is to extend the intervention programmes of the water agencies to the management and restoration of aquatic environments, especially watercourses and wetlands, and to flood regulation *inter alia* by extending the flood retention capacity of natural floodplains. This is to be achieved primarily by promoting and financing projects in the restoration of aquatic environments and by levying fees for the construction of significant obstacles to watercourses (cf. Section 5.4).

However, the adopted Water Act lags far behind earlier reform initiatives that aimed at levying fees for the modification of water flows and water regimes by various land uses (cf. Section 5.4) and at introducing three types of zoning categories. Within the *flood retention zones*, landowners and users were to have been prohibited from all activities likely to impair the good functioning and conservation of openings permitting the flooding of those zones. It had been planned to authorise the prefect to prohibit all activities likely to create an obstacle to the natural course of the river in the *zones of mobility* (dams, embankments and other constructions) or to prescribe modifications for the displacement of the river, in order to preserve or to restore its essential hydrological, geomorphological and ecological characteristics. Moreover, the Water Act had aimed to prohibit some agricultural practices in the *zones of protection* near the river.

To conclude, recent reforms have opened new windows of opportunity for floodplain restoration. However, the reform was unable to meet earlier expectations. This is not only true for the concessions to the agricultural lobbyists concerning the lowering of the financial levies for diffuse pollution by nitrates and pesticides and for the modification of water regimes. Even though ambitious targets for floodplain restoration are defined, the way in which they are to be implemented by the water agencies and local and regional authorities in water and land-use planning remain rather vague and efficient instruments and financing mechanisms are largely missing.

5.3.2 Flood risk management

In France flood risks are considered the primary natural risk in terms of the extent of damage and the number (8,000) of towns and villages concerned, the size of floodable areas (more than 20,000 km^2) and the population residing in these areas (4.5 m).[8] Though certain human activities have been able to contribute to reducing the local intensity of floods, in practice the extent of extreme floods could not be modified appreciably (MATE 2000: 2). The reasons for this are the prevailing land-use patterns (particularly the intensified urbanisation and sealing of land in the alluvial valleys), the canalisation of rivers and adjustment of river beds, the lack of river maintenance or unchanged maintenance practices and the removal of wetlands and flood retention areas (ibid.: 2-4). As a result the damage caused by floods has increased significantly over recent decades (DATAR 2002: 7; 12). Between 1991 and 2000 flood damage in France amounted to approx. €750 million per annum (DATAR 2002: 62).

As a basic instrument of urban planning and local risk management, *local land-use plans* (Plans d'Occupation des Sols), which define general rules for local land use, were created in 1967 (in 2003 they were replaced by "plans local d'urbanisme"). Some areas susceptible to major floods or other major natural hazards were defined as out-of-bounds for any building activities (Michel-Kerjan 2001: 2). Responding to the major flood events in 1981 and the fact that most natural hazards, like floods, droughts or avalanches, were excluded from French insurance policies, a new system based on national solidarity, private insurance and an unlimited state-guaranteed public reinsurance was created. Moreover, the legal reform of 1982 required the creation and implementation by the State of natural risk exposure plans (PER) in order to determine risk zones.

Due to the increasing economic damage caused by natural catastrophes in France, especially the flood events of summer 1992, the French government reinforced its risk management policy with both regulatory measures and funds. A multi-annual programme for protection from natural risks endowed with several billion francs was introduced (MATE 2000: 4). An inter-ministerial circular between the Ministries of the Interior, Infrastructure and the Environment concerning flood prevention and the management of floodplains

[8] Ministry of Ecology and Sustainable Development 2007 (http://www.ecologie.gouv.fr/-Flooding-.html, 13.03.2007).

was passed at the beginning of 1994 specifying the ambitious policy objectives for flood prevention (cf. Table 5.1, Section 5.2).

One year later the *Act on the Reinforcement of Environmental Protection*, the *Barnier Law*, was adopted in February 1995, establishing the main policy instruments for risk prevention. It was reformed by the Act on Technological and Natural Risk Prevention and Reparation of Damages in 2003 (for details cf. OECD 2006: 21-29). The Barnier Law simplified the different statutory provisions for controlling land-use planning in risk-prone areas and replaced them with a single instrument, the *Foreseeable Natural Risk Prevention Plans* (Plans de prévention des risques naturels prévisibles – PPR), which cover all natural risks, including floods (the most frequent), land movements, earthquakes, forest fires, avalanches, storms and cyclones. PPRs are drawn up by the State. After a public hearing and consultation with local authorities, the PPR is approved in decrees issued by the Prefect of the department. They are subsequently annexed to the local land-use plans, thus making compliance a statutory requirement for granting planning permission, which is a responsibility of the mayors.

The PPRs' objective is to map the risk-prone areas and to define the rules for urban planning and construction. They aim at banning all new construction work in the most exposed areas and regulating construction activities in less exposed areas (IFEN 2000: 141). For example, people living in the floodable areas are supposed to make flood prevention investments to the tune of a minimum of 10% of their house value in 10 years (Barraqué 2003: 3). The PPRs contain a full range of preventive measures in areas exposed to flood risk, ranging from conducting technical studies and risk analyses and providing information for the population, to defining regulations, e.g. by blocking building permission already granted by the mayors. Until 2003, however, these plans did not cover activities that were located upstream from exposed areas and that contributed to aggravating the flood risk, for example by increasing run-off and soil erosion. This deficiency was partially corrected by the reform of 2003 which enabled the central government and local authorities to establish easements in the public interest in order to retain floodwaters or manage changes in a watercourse. It also made it possible for municipalities to expropriate land in cases involving a serious threat to human life (OECD 2006: 25f.).

In order to finance any compensation for measures decided under PPRs, the Barnier Law established a *Fund for the Prevention of Large-Scale Natural Risks* (Fonds de prévention des risques naturels majeur). This fund was set up to finance expropriations in major risk zones, as well as the expenses for the demolition of buildings. Recently the financial resources of the fund have been doubled in order to finance the development of PPR, especially the costly risk analyses.

The dispersion of risk prevention responsibilities prompted the French government to establish a Flood Prevention Action Programme in 2002

(*Programmes d'Actions de Prevention des Inondations* – PAPI), which enabled the various central government and local authority bodies to cooperate on comprehensive basin-wide projects (OECD 2006: 26f.). The purpose of the programme "is to promote an integrated approach to prevention and mitigation focusing simultaneously on the upstream regulation of the rate of flow of watercourses, protection of housing, reduction of vulnerability at a more relevant level than that of administrative units" (ibid.). Since the initial call for projects, 42 PAPIs were adopted for the 2004-2008 period, covering nearly one quarter of French territory, at a total cost of €500 million (ibid.).

Although the restrictions on land use in river basins created by the PPRs and PAPIs are valuable preconditions for the restoration of floodplains and a preventive approach to flood risk management, they were unable to provide sufficient incentives for local authorities to invest in preventive measures (OECD 2006: 34f.; Michel-Kerjan 2001: 9f.). Ten years after the law came into force the 5,000th PPR was approved in 2005 (Ministère de l'Ecologie et du Développement Durable 2006a). This can only be considered a starting-point for efficient risk prevention in view of the estimated 20,000 communities susceptible to damage by natural events (Michel-Kerjan 2001: 9f.) One of the main barriers is that risk analyses are very costly, so that individuals cannot afford them and only municipalities can ask for funding for these studies from the risk prevention fund (ibid.). Besides, in many cases the interests of municipalities restrict the definition and implementation of PPRs. Municipalities are not urged to publicly disclose the level of risk of their land. Municipalities and landowners in risk areas often keep the results of risk analyses confidential, as the definition of municipal territory as a risk-prone area might reduce the number of potential inhabitants and firms wishing to settle there and thus the real price of the land itself (ibid: 10). Moreover, local authorities may be tempted to underestimate the size of the zones because they can expect to obtain support from a national solidarity fund in case of floods (Barraqué 2003: 1). Consequently, urbanisation in flood-prone areas has in certain cases even continued to expand over the last decade (OECD 2006: 35). Furthermore, compliance with PPRs and PISAs is hampered by the lack of adequate incentive mechanisms. The effective implementation of risk prevention is based on the central government's monitoring capacity, which may be questioned in many cases (ibid.).

It needs to be evaluated how far the implementation of PAPIs can indeed overcome the lack of effective basin-oriented planning or a compensation approach to flood risk management, and how far they can minimise the transaction costs of solving conflicts between upstream and downstream residents. However, the PPRs and the PAPIs focus on regulating construction activities in flood-prone areas and on traditional solutions to flood analysis, management and alert and do not systematically promote the restoration of flood retention areas. Their contribution to floodplain restoration thus remains suboptimal.

5.3.3 Regional planning

The main principles and objectives of France's spatial planning policy were developed by the central government more than 45 years ago to meet the challenges in post-war France, to strike a better balance of population, industry, culture etc. between France's regions and to cope with the predominance of Paris. However, various reforms, particularly the initiatives towards decentralisation, have since changed the institutional framework of spatial planning. One of the main reforms was the *Framework Act on Spatial Planning and Regional Development* of *1995* (Loi d'Orientation pour l'Aménagement et le Développement du Territoire – LOADT), which was the first comprehensive regional planning act integrating a number of individual regulations. It reflects the implications of the French reforms towards decentralisation for spatial planning. Even though the national State retains a major role in spatial planning and regional development policies, the trend towards decentralisation was formalised by introducing the concept of *Pays*[9] and by strengthening contractual approaches to regional development in a dialogue process with local and regional stakeholders contributing to development projects (cf. Marcou 1996).

Four years later the *Framework Act on Spatial Planning and Sustainable Regional Development* of 1999 (Loi d'Orientation pour l'Aménagement et le Développement durable du Territoire – LOADDT) reformed the institutional framework of regional planning, defining the concept of sustainable development as its general objective. The LOADDT takes the concept of the *Pays* as the key territorial units of future regional and local development policy significantly further than its predecessor. The establishment of these *Pays* and the promotion of regional projects by planning contracts are under the joint responsibility of spatial planning, a regional development conference and the regional prefect (the State representative in the region). The law seeks to place territorial development policy within a wider agenda that takes on board notions of sustainability and the reform of the European Structural Funds (Buller 2000: 17). The promotion of regional projects like floodplain restoration and flood prevention beyond administrative boundaries is thus facilitated.

---

[9] The *Pays* are territories that do not necessarily conform to existing administrative units: they "can be defined as a culturally homogenous rural area defined by local actors and conveying a distinct and identifiable sense of geographic, cultural, economic and social cohesion around which a local development project is constructed. The *Pays* are not new administrative nor political structures, but are territorial and community based planning and development project areas" (Buller 2000: 17).

For promoting sustainable regional development the LOADDT introduces eight collective service schemes.[10] One of them, the *collective services scheme for rural and natural areas* ("Schéma de services collectifs des espaces naturels et ruraux") focuses on the sustainable development of rural and natural areas. It is the first time that a strategic spatial plan explicitly applies to the management of natural environments. Among other things, this strategic plan mentions the flood risks in France and their enormous economic costs. The scheme takes up the ambitious propositions of an Interministerial Circular between the Ministries of the Interior, Infrastructure and the Environment on flood prevention and the management of floodplains to exclude human settlements in the most dangerous zones, to preserve the capacities of flow and flood expansion and to conserve regularly flooded river environments (cf. Sections 5.2 and 5.3.2).

Inter alia the scheme devises strategies to preserve the multiple functions of major river valleys and to restore wetlands (cf. DATAR 2002: 67-69; 79-81). In order to preserve the multifunctionality of river valleys spatial planning has to promote ecologically compatible agriculture in the river valleys even if it is difficult to organise interaction between local planning and water management. The scheme requires the restoration of land in river valleys, afforestation, the maintenance of the permeability of agricultural land and the valuation of zones reserved for the expansion of floods (DATAR 2002: 68). As a complementary strategy it is regarded as essential to reduce the vulnerability of flood risks in urbanised river corridors as well as to protect and restore the natural flooding zones by encouraging the territorial bodies to coordinate their PPR (ibid. 69).

The wetlands – regarded as crucial for sustainable water management, for biodiversity and landscapes – constitute, in addition, multipurpose functions useful for flood control, offering facilities for original products and various recreational amenities. In order to preserve the wetlands the scheme calls for concerted action and a joint management plan at the local level by the relevant stakeholders to protect the various functions of wetlands (DATAR 2002: 79-81).

The service plans are to be implemented by regional planning documents ("Schémas Régionaux d'Aménagement et de Développement du Territoire" – SRADT), which must be compatible with the new public service documents. The SRADT define guidelines for regional planning which have to be specified in regional development plans (plans régionaux), in planning contracts between the State and the regions, and by five-yearly financial planning acts. As the protection and restoration of natural environments had not proved to be a priority in regional planning in the past, the 1999 spatial planning legislation introduced a new source of public funding, the *Funds for the Management of*

[10] These sectoral policy guidelines concern the fields of higher education and research, culture, health, information and communication, public transport, energy, natural and rural areas, and sports. The schemes were adopted by decree in April 2002.

Natural Environments ("Fonds de Gestion des Milieux Naturels"). The available funding of the Scheme for Collective Services for Rural and Natural Areas amounts to €10.6 m per annum (Buller 2000: 36).

Besides the collective service schemes, regional planning programmes and administrative decisions in water-related areas have to be compatible with the SDAGEs, developed at the level of river basins (cf. Section 5.3.1). Being binding for the public administration they constitute a (constraining) framework for land-use planning and have to be "taken into account" for all land uses. If they are infringed, legal action can be taken against the administration concerned. The same applies to the PPR, set up by the prefect of the department, which are subsequently annexed to the municipal land-use plans (cf. Section 5.3.2).

5.3.4 Nature conservation

In France the conservation of natural environments, landscapes and biological diversity, the maintenance of a biological balance and the protection of natural resources against degradation are defined by the *Nature Conservation Act* of 1976, which has since been integrated into the Environmental Code (Books III "Natural spaces" and IV "Fauna and flora"). This Act designates the conservation of nature as being of common interest ("intérêt general") and environmental impact assessments are established as prerequisites for all activities that may generate relevant risks for the integrity of natural environments (Peiry et al. 1999: 246). The act introduces various regulations for the protection of flora and fauna, hunting, freshwater fishing and protected areas. The most relevant instruments for the conservation and restoration of river ecosystems and floodplains follow below (for an overview cf. Stoll 2001).

Of major significance are *nature reserves* (réserves naturelles), a conservation category which was established under the Nature Conservation Act in 1976. Nature reserves protect parts of territories with animal or plant species or biotopes of special importance or under threat of extinction. Nature reserves are a strict conservation instrument for natural environments prohibiting various land uses. Their management is delegated to nature conservation associations (60%), public agencies (établissement publics) (19%) and local authorities (21%) (IFEN 2000: 55). In France 156 nature reserves were designated by 2007, including various reserves for parts of rivers or river ecosystems.[11] For example, the majority of the 25 French sites ratified by the Convention on Wetlands of International Importance (known as the Ramsar Convention) include nature

[11] French Ministry of Ecology and Sustainable Development (http://www.ecologie.gouv. fr/-Reserves-naturelles-.html, 27.03.2007).

reserves (cf. Ministère de l'Ecologie et du Développement Durable 2006b). The managers of these protected areas work together within a network of river nature reserves. This cooperation allows for the exchange of scientific, technical and legal expertise in favour of ecological river management.

The *voluntary nature reserves* (réserves naturelles volontaires) were also introduced by the Nature Conservation Act of 1976 and – like the nature reserves – promote the protection of exceptional animal or plant species or biotopes threatened with extinction or of specific ecological value. In contrast to the nature reserves private landowners can request classification on a voluntary basis. In 2002 about 30 voluntary nature reserves protected more than 1000 ha of wetlands.[12]

The *national parks* (parcs nationaux), established by legislation in 1960, consist of large territories of national ecological importance, which include a strictly protected nucleus and a buffer zone on the periphery serving sustainable rural development. In none of the nine national parks in France do river ecosystems or wetlands of major importance benefit, however.

The *regional nature parks* (parcs naturels régionaux) aim at protecting fragile environments of rich natural and cultural importance. Their objective is to contribute to sustainable land use and to promote economic, social and cultural development by the appropriate management of nature and landscapes. In 2002 wetlands of major importance were protected and developed in 32 of these territories.[13]

Whereas most of the nature conservation regulations have been integrated into the Environmental Code, the *prefectoral decree for biotope conservation* (Arrêté préfectoral de protection de biotope - APPB) is part of the Rural Code. This conservation area is designated by the prefect in order to conserve the habitats of protected species or any natural formation which is heavily exploited. In 2002 it comprised 81 wetlands (17,040 ha) and approximately 60 nature reserves.[14]

In addition to these instruments of protected zones, compensatory and contractual instruments have gained in importance. *Policies of land acquisition* and *of management conventions* (conventions de gestion) have been developed extensively in France to implement nature conservation objectives. These policies are particularly carried out by regional nature conservationists (conservatoires régionaux d'espaces naturels – CREN). The CREN have thus become key actors in regional and local nature conservation projects. They are

---

[12] French Ministry of Environment and Spatial Planning 2002 (http://www.environnement.gouv.fr/actua/cominfos/dosdir/zonhum3.htm, 18.04.2002).

[13] French Ministry of Environment and Spatial Planning 2002 (http://www.environnement.gouv.fr/actua/cominfos/dosdir/zonhum3.htm, 17.04.2002).

[14] French Ministry of Environment and Spatial Planning 2002 (http://www.environnement.gouv.fr/actua/cominfos/dosdir/zonhum3.htm, 17.04.2002)

responsible for the acquisition and leasing of land and for arranging territorial contracts with farmers (cf. Section 5.3.5) who voluntarily agree to the sustainable cultivation of their land. Today nearly every region has its own CREN (Stoll 2001: 86).

Under the French legal implementation of the European *Habitats Directive* (cf. Dworak in Chapter 3 of this volume) the protection of the relevant habitats can be ensured by the regulatory or contractual instruments described above. Alluvial wetlands include several important types of habitats classified in the annex to the directive, such as standing and running water bodies, wetlands and wet meadows, alluvial forests and thickets etc. (cf. Annex I of the Directive). Once designated "special zones of conservation", the wetlands will be integrated into the European network Natura 2000.[15]

In order to reinforce and coordinate the application of the instruments of zoning, contracting and compensation, a *national action plan for wetlands* was set up by the Ministry for the Environment (cf. MATE 1996). This action plan was a response to the alarming degradation of wetlands that was made apparent by a study on the status of wetlands and national policy for wetlands protection (Ximenes 2001) and to the Ramsar Convention (see above), which was ratified by the French government after 25 years delay in 1996. As a consequence, it was decided to found a National Observatory of Wetlands (Observatoire National des Zones Humides – ONZH) as part of the French Institute for the Environment for the evaluation and orientation of public policies with a direct or indirect impact on these particularly threatened environments. The following policy priorities were formulated (Dupont 2000):

- to inventorise and monitor the wetlands;
- to launch a national research programme on wetlands;
- to ensure coherence between different public policies;
- to restore wetlands.

The National Observatory developed a list of 84 wetlands of major national importance in which alluvial valleys play a key role (cf. Ministère de l'Ecologie et du Développement Durable 2006b).

To stimulate the implementation of European nature conservation policy, specifically the Birds and Habitats Directives, the European Union dedicated a budget to co-finance national conservation programmes. Among them, various projects relating to wetlands conservation and restoration were financed within the

---

[15] The European Habitats Directive was implemented into French law in April 2001 after considerable delay and following several proceedings before the European Court.

programme *LIFE Nature*, e.g. the project "Loire nature" and projects on the restoration of alluvial valleys in the north and east of France (cf. Section 5.6).[16]

5.3.5 Agriculture and forestry

The intensified use of land by agriculture and forestry has been one of the main origins of the loss of floodplains in recent centuries in France and elsewhere. Agricultural and forestry policies are thus of great relevance for floodplain restoration.

Notwithstanding the rather conventional orientation of agricultural practice in France and the enormous subsidies for industrial agriculture, the French government took an important step with the legislative reform of agricultural policy in the *Framework Act on Agriculture* (Loi d'orientation agricole) of 1999. The act defines sustainable land use as the general model for agricultural policy and promotes the equal consideration of the economic, ecological and social functions of agriculture. Among other things, agricultural policy has to be directed at "the conservation of natural resources and biodiversity and the maintenance of landscapes" (Art. 1). The reform emphasises the relevance of agriculture for environmental protection. It is directly connected to the legislative reform of regional planning in 1999 and specifies the agricultural targets defined by the LOADTT (cf. Section 5.3.3).

The ecological objectives of the law are implemented by *rural development agencies* (Sociétés d'aménagement foncier et d'établissement rural), which are explicitly required to contribute to environmental conservation. Together with the regional conservationists of natural environments (cf. Section 5.3.4) they arrange and implement the *territorial contracts of exploitation* (Contrats Territoriaux d'Exploitation – CTE) integrating environmental, employment and rural development issues. These contracts are the main vehicle by which farmers can define, and receive support for, their multifunctional roles. Their aim is to respond to the objectives of the common rural development policy whilst creating coherence between France's agricultural and regional planning policies. The CTEs establish a contractual agreement between the State (represented by the prefect) and the farmers for a period of five years. The CTEs can be suitable instruments to encourage changes in land use in floodplains by paying compensation for limitations to cultivation or for conservation measures.

A further legislative reform of French agricultural policy, the *Framework Act on Agriculture* of 2006, is a response to the fundamental reform of the Common Agricultural Policy in 2003. It is designed to encourage the entrepreneurial initiative of farmers by structuring the relationships between the production,

[16] Cf. Institut Francais de l'Environnement (http://www.ifen.fr/zoneshumides/pages/medd_zh_Manche.htm, 8.4.2007).

processing and marketing of agricultural goods. Among other things, the reform introduces new credit opportunities and economic incentives for organic agriculture and for biofuel, but it has no major impact on floodplain management. The *Framework Act on Forestry* also does not take special consideration of alluvial or floodplain forests. However, it defines a special conservation status for forests, which are recognised as being necessary "for the maintenance of soils in the mountains, for defence against avalanches, erosion and intrusion of water and sands", which could be applied to floodplain forests.

It is a common strategy of French agri-environmental policy to promote the greening of agricultural production and land use with contractual agreements between the landowners and public agencies aiming at the conservation or sustainable development of natural environments. However, these agri-environmental instruments work chiefly on a voluntary basis. In many cases, the economic incentives are counteracted by the prevailing system of subsidies and economic incentives in European and French agricultural policy. Contractual agreements have thus primarily been drawn up for marginal sites, whereas agricultural cultivation in productive areas has even been intensified.

5.4 ECONOMIC INSTRUMENTS AND FUNDING PROGRAMMES

The Water Act of 1992 defines water as part of the common national heritage that cannot be managed on the basis of market forces alone. In order to maintain this common national heritage several economic instruments have been introduced.

With the Water Act of 1964 a financial mechanism was introduced to create financial solidarity between water users and a relatively far-reaching application of the polluter/user pays principle. The water agencies were authorised to levy charges on all users who affect water quality or quantity and water regimes. For industrial users the charges are calculated by using various parameters appropriate to each branch and the amount and consistency of pollutants. For domestic users they are calculated for each community according to both permanent and seasonal population. The rates for calculating the levies are defined by each water agency and must be agreed upon by the respective river basin committee. The agencies' budget and their investment in water treatment, (drinking) water supply and public health are entirely funded by the revenue from these charges.

Even though France has long-running experience in applying the polluter/user pays principle in water management, this economic instrument has so far been mostly irrelevant for the restoration of floodplains or nature conservation. In the first three years of the seventh programme, the water agencies invested only approximately €230 million in the conservation and restoration of natural environments (milieux naturels), but approximately €5.7 billion in water quality management and approximately €900 million in drinking water supply (OIEau 2000). However, under the recent reform of the water law in 2006, the intervention programmes of the water agencies will be extended to the management and restoration of aquatic environments, especially water courses and wetlands, and to flood regulation (cf. Section 5.3.1). At the same time, the water agencies were authorised to levy fees for the erection of significant obstacles to the watercourse, e.g. for constructions disturbing solids transport and the movement of aquatic organisms.

The earlier legislative reform initiative had been much more ambitious, but it was not adopted after the change of government in 2002. According to this reform, the water agencies were to levy fees for the modification of water flows and regimes. It was planned that constructions that reduce the natural water retention capacity and thus increase flood risks (e.g. large urban development and transport infrastructure projects sealing significant surfaces) or that narrow river courses (e.g. dams, barrages) would be subject to fees (cf. Direction de l'Eau 2002: 13f.).[17]

As this reform of water policy has not been implemented, measures in the field of floodplain restoration are still predominantly dependent on public funding programmes at the EU and national level, often co-financed by regional and local communities. Foremost amongst EU programmes are INTERREG programmes for preventive flood protection measures, such as the IRMA (Rhine-Maas) programme and the LIFE programme for nature conservation, which has contributed to several floodplain restoration and wetland projects in France, especially along the Loire river. "Loire Nature" was one of the most important LIFE funding programmes, comprising investments of about €8 million in nature conservation between 1993 and 1998 (EU: 50%; France: 25%; others: 25%). Between 2002 and 2006 the budget of approximately €15 million was financed by the Ministry of Ecology and Sustainable Development (and its regional departments for the Loire basin), the Loire Brittany Water Board, a public institution that formerly promoted large dams (Etablissement Public Loire), and the local authorities of eight regions in the Loire Basin (Loire Nature 2007).

The Act on the reinforcement of environmental protection has established a mechanism to expropriate the areas exposed to natural risks threatening human life and has created a fund for financing the measures. This *risk prevention fund*

[17] Even if the reform had been implemented, it can be doubted whether the foreseen fees would have been effective for floodplain restoration or flood prevention, as they were very low: e.g. €150/ha. for sealing land, €15/ha. for the reduction of the surface of flooding zones, and for obstacles between €90 and €150 per metre (Direction de l'Eau 2002: 14).

(fonds de prévention des risques naturels majeurs) is supported by a levy of 2.5% on insurance premiums for natural catastrophes (about €15 million per annum).

5.5 FORMS OF COOPERATION, CONSULTATION AND EDUCATION

Coordination in the field of floodplain restoration is, on the one hand, stipulated by regulatory and planning instruments in different policy fields (cf. Section 5.3); on the other hand various committees provide for coordination and consultation. Following the tradition of French environmental policy the organised stakeholders participate in these committees. However, due to the number of interest groups involved in the committees and their heterogeneous and often divergent interests, the influence of the committees is limited (Sohnle 2001).

At the national level, the *national water committee* has to be consulted on the development of water legislation, important regional projects and common problems of river basin committees or water agencies. In each of the six basins *river basin committees* are responsible for the coordination of water planning and the monitoring of the water agencies (cf. Section 5.3.1). The water legislation reform of 2006 diminished the representation of the central State in these committees in favour of local governments. Since then, only 20% of the committee's members are representatives of the central State and its regional bodies (previously one third), 30% represent different users such as industrial and agricultural groups, angling and environmental associations, and a minimum of 50% are representatives of local governments. At the level of river catchments or sub-basins, the local stakeholders participate in local water commissions which participate in the planning process of the SAGEs. The water commissions were established under the Water Act of 1992 and consist of public representatives of the regional and local territories (at least 50%), of user groups (at least 25%) and of the State. The Water Act of 1992 also strengthened the role of the general public by establishing procedural minimum standards for public participation and by publicising the drafts on water management. These can then be revised in order to take objections or proposals into consideration prior to final approval by the State (Sohnle 2001: 182f.). To ensure the coherency of the State's regional activities the respective prefect of the basin coordinates State water policing and management policy.

In many cases the implementation of water policy is delegated to specialised intermediary organisations, which manage regional or local projects and coordinate the various stakeholder interests at the project level. For example EPALA (Etablissement Public d'Amenagement de la Loire et ses Affluents) has been established for flood protection by the river communities on the Loire river and its tributaries and is – together with the French State and the water agency – responsible for the interregional programme "Loire Grandeur Nature". Complementary to this, the environmental group "Espace Naturel de France" (ENF) manages nature conservation projects and coordinates the implementation of the programme *LIFE Nature* (cf. Section 5.6). Other examples are the "Nature Midi-Pyrénées", which manages a nature conservation project in the flooding zone of the confluence of the Ariège and the Garonne rivers, and the "Institution Interdépartementale des Barrages-Réservoirs" (IIBRBS), which manages a restoration project in La Bassée on the Seine river (cf. Hagemeier and Klaphake in Chapter 11 of this volume).

Besides regulatory and planning instruments, contractual agreements are particularly important in France. They aim at integrating economic interests (e.g. agriculture, water management, nature conservation, employment) with the ecological management of natural resources. There are a huge variety of contractual agreements, such as contractual forms of nature conservation, e.g. regional nature parks, land use contracts (Contrats Territoriaux d'Exploitation), river contracts between local elected representatives, the river residents and the users, 'agglomeration' contracts between local communities and the water agencies on measures concerning drinking water and purification, and departmental contracts between the General Council and the water agencies for financing agricultural measures to protect drinking water (Basin Seine-Normandie).

5.6 PROGRAMMES AND PROJECTS TARGETED AT FLOODPLAIN RESTORATION

In France, key programmes and projects targeted at floodplain restoration concentrate on the Loire and Rhine rivers, whereas programmes on other French rivers are implemented on a much smaller scale and are less far-reaching. It is only recently that restoration activities were started on the rivers Sarre/Moselle, Meuse, Seine, Garonne etc. (cf. Hagemeier and Klaphake in Chapter 11 of this volume).

Different *LIFE Nature* programmes, amongst them the programme "Loire nature" and those of the alluvial valleys in the north and east of France financed partly by the European Union, relate specifically to alluvial wetlands.[18] In addition to this, an action plan for the Loire river was set up in January 1994 by the French government. The *plan Loire grandeur nature* covers the whole Loire river and its tributaries from the source to the estuary. The plan comprises decisions relating to the prevention of floods by technical flood defence, the

[18] French Ministry of Environment and Spatial Planning 2002 (http://www.environnement.gouv.fr/actua/cominfos/dosdir/zonhum11.htm, 15.4.2002).

management of floodplains, the management of low water levels, the conservation of the estuary of the Loire, the protection of migrating fish and the restoration and management of the natural heritage by the creation of two natural reserves. The plan is mainly financed by the State, the EPALA (Etablissement public d'aménagement de la Loire et de ses affluents), a public agency for the management of the Loire river and its tributaries, and the water agency Loire-Brittany (cf. for details Loire Nature 2007).

Since 1950 the "International Commission for the Protection of the Rhine against Pollution" (ICPR) coordinates river management along the Rhine. Although focussed on pollution control during the early decades, its activities today additionally comprise the conservation, restoration and integrated management of semi-natural zones of the (upper) Rhine River. As a consequence of major flood events the ICPR commissioned the group "*Action Plan on Flood Defence*" in February 1995 to draft an action plan for the Rhine and its catchment area under the joint responsibility of the Member States. In January 1998 an "Action Plan on Flood Defence", involving expenditure of up to €12 billion within the next twenty years was adopted (Conference of Rhine Ministers 1998). This plan and the programme "*Programme on the sustainable development of the Rhine – Rhine 2020*" (ibid. 2001) define activities to provide for the security of the population located downstream and to conserve and restore natural environments on the river Rhine. The programme Rhine 2020 promotes the restoration of the former network of habitats typical of the Rhine (habitat connectivity) and the ecological functionality (up- and downstream migration) of the Rhine. Apart from technical measures, floods are to be prevented by increasing the water retention capacity in a number of ways: by renaturalising streams, reactivating flooding zones, promoting extensive agriculture, initiating measures aimed at nature enhancement and afforestation and by spatial planning introducing and promoting land uses adapted to the risk of flooding (for details cf. Conference of Rhine Ministers 2001: 9f.)

As far as spatial planning is concerned the ministers responsible in France, Germany, the Netherlands, Belgium and Luxembourg set up a transnational working group "Spatial planning and preventive flood protection Rhine/Meuse" in March 1995. The European Union supported these activities within the European Union initiative INTERREG II C. The operational programme developed on this basis under the name of IRMA (INTERREG-Rhine-Meuse-Activities) contributed to the implementation of specific flood prevention measures along the Rhine and the Meuse in the years 1997 to 2003.

Various research programmes are concerned with floodplain restoration. A national research programme on protected environments (*Programme de recherche espaces protégés*) was initiated by the French government between 1999 and 2003. The programme evaluated the effectiveness of networks of protected areas for the conservation of biological diversity and identified priorities for enhancing biodiversity in France. Furthermore, the restoration of river systems (e.g. the Seine, the Rhone, the Rhine, the Garonne rivers) and their attached wetlands profited from basic and applied research programmes within the framework of an interdisciplinary research programme on the environment (PIREN). The results achieved are used to guide the integrated management, conservation and restoration of these very dynamic mediums.

5.7 CONCLUSIONS: DRIVERS AND CONSTRAINTS OF FLOODPLAIN RESTORATION

The management of rivers and aquatic environments in France has on the one hand been dominated for a long time by the regulation of water quality and water quantity. At a comparatively early stage, beginning in the 1960s, water management was organised in a new and progressive way based on the territories of river basins and catchments. On the other hand, the management of river environments was shaped by traditional flood defence policies aiming at controlling river flows and flood risks by dams and weirs. In addition to water management and flood defence, nature conservation policies were aimed at the protection of the most valuable or endangered species and aquatic biotopes by the means of designating protected areas. The problem was that flood defence and nature conservation policies were predominantly organised institutionally and spatially decoupled from the river basin and from catchment institutions of water policy at the level of the municipalities, departments and regions. Despite a progressive system of catchment-oriented water management, the French system rarely supported the integrated management of river environments or the restoration of floodplains.

However, since the 1990s a remarkable policy shift towards the integrated management of water, river environments and flood risks has been initiated by policy reforms in different policy fields and the need for a more strategic, integrated approach to value the multiple functions of floodplains has been widely recognised.

In the field of *water policy*, floodplain restoration is driven by a growing concern for water flow regimes and by reforms of the water planning regime in the 1990s. Since the Water Framework Directive is inspired by the tradition of French river basin management, innovations in floodplain restoration arising from its implementation into national law are rather insignificant. Nevertheless by reforming water legislation, the activities of the water agencies and water planning authorities were extended towards the management of aquatic environments and the regulation of floods. However, the abandonment of the fee system for the modification of water regimes and for the sealing of land makes it rather unlikely that the water agencies will be capable of efficiently meeting their new tasks and actively promoting catchment initiatives in floodplain restoration.

Another key driver in floodplain restoration is growing flood risk awareness as a consequence of flood events in the 1990s and of climate change. Since the limits of technical flood prevention have become apparent, and since such measures have not proven to be particularly cost-efficient, especially in rural areas, interest in the flood retention capacity of floodplains has grown as a new approach to *flood risk management*. Although risk prevention planning has opened new windows of opportunity for floodplain restoration, the persistence of technical solutions is still high since the beneficiaries are economically powerful and the environmental administration – as the main promoter of floodplain restoration – is weak at the regional level and has no representation at the departmental level.

In the field of *spatial planning* the legal reforms of 1995 and 1999 and the introduction of the collective service schemes for rural and natural areas have facilitated the promotion of floodplain restoration. Together with the introduction of risk prevention plans, the protection of floodplains from development was substantially reinforced and projects were initiated on some rivers. However, the interest of the existing land users of floodplains, the compensation of flood damage by the national solidarity system and the interest of river communities in the economic development of their territory undermine national policy priorities in local land-use planning on floodplains.

In addition, the restoration of floodplains is driven by *nature conservation* policies, in particular by the designation of protected areas. A growing interest in biodiversity targets and in natural floodplain forests can also be identified in the case of some rivers, like the Loire or Allier, and a national action plan for wetlands has also inspired the restoration of many alluvial wetlands. Despite these positive developments, the majority of the projects are essentially site-based, and are focused primarily on the protection of intact and ecologically valuable floodplains and not on the restoration of floodplains transformed by human land and water uses.

In the field of *agricultural policies* too, reforms have opened some windows of opportunity for floodplain restoration. The political pressure for more extensive forms of agriculture and for a reform of the subsidy and support systems of the Common Agricultural Policy and the consolidation of agri-environmental instruments in France has been growing. Even if these reforms have the potential to promote floodplain restoration, the compensation payments for agri-environmental measures do not yet stimulate a major change in farming practices. On the contrary, they are even partially counteracted by the prevailing subsidy and incentives systems of the European and French agricultural markets.

Recapitulating these developments, recent reforms have the potential to create new openings and incentives for floodplain restoration. However, although substantial recasting of issues and institutional structures in several policy fields is underway, the protection and restoration of floodplains is still a pending issue for many water and flood managers, land-use planners and users of floodplains. Thus it is vital that further policy reforms are implemented in the different policy fields which define strict regulatory instruments and economic incentives for land use and water management in floodplains, and that raise awareness of new approaches to flood prevention and the management of river environments. In addition to concerted reforms within the relevant policy fields and within regional and local contexts, floodplain restoration requires coordinated action across territorial borders and across sectoral mindsets. The promotion of catchment-oriented private or public-private organisations such as nature conservation associations or agencies in flood management could advance flexible mechanisms to overcome the fragmented character of political-administrative territories.

5.8 REFERENCES

Barraqué, B. (1999) Prévention des inondations en Europe: hydraulique, assurances ou solidarité? *La Houille Blanche* **3**, 71-78.

Barraqué, B. (2000) Assessing the efficiency of economic instruments: Reforming the French Agences de l'Eau. In *Marked-based instruments for environmental management* (eds. M. Skou Andresen and R.-U. Sprenger), Edward Elgar Publ., Cheltenham, pp.215-230.

Barraqué, B. (2003) *From force majeure to solidarity and planning: French recent changes.* Paper presented at the Workshop "Precautionary Flood Protection in Europe", 5-6 February 2003, Bonn.

Buller, H. (2000) *Actors, Institutions and Attitudes to Rural Development: The French National Report.* Research Report to the World Wide Fund for Nature and the Statutory Countryside Agencies of Great Britain, Department of Geography, University of Paris.

Conference of Rhine Ministers (1998) *Action Plan on Flood Defence.* 12th Conference on Jan. 12th 1998, Rotterdam. http://www.iksr.org/hw/icpr/11uk.htm, accessed on 17.04.2002.

Conference of Rhine Ministers (2001) *Rhine 2020. Programme on the sustainable development of the Rhine.* 29 January 2001. Strasbourg.

DATAR (Délégation à l'Aménagement du Territoire et à l'Action Régionale) (2002) *Schéma de services collectifs espaces naturels et ruraux.* Décret no. 2002-560, 18.04.2002, Paris.

Direction de l'Eau (Ministère de l'Aménagement du Territoire et de l'Environnement) (2002) *Le projet de loi portant reforme de la politique de l'eau: une plus grande transparence, plus de solidarité et d'équité, une meilleure efficacité environnementale.* January 2002.

Dupont, P. (2000) Integrated river basin management in France: The role of wetlands in meeting the objectives of the 1992 water law. In *Proceedings of the WWF seminar "The role of wetlands in river basin management", 9 and 10 November 2000*, Brussels. http://www.panda.org/europe/ freshwater/seminars/sem2/dupont.html, accessed on 08.04.2002.

Gouvernement Français (2002) *Propositions pour une Stratégie Nationale de Développement Durable.* Contribution du Gouvernement Français, March 2002, Paris.

IFEN (Institut Français de l'Environnement – Délegation à l'Aménagement du Territoire et à l'Action Régionale) (2000) *Aménagement du territoire et environnement : politiques et indicateurs.* Orléans.

Loire Nature (2007) *Principaux résultats Loire Nature 2002-2006: Pour une gestion durable d'un fleuve et de ses affluents.* Orléans.

Michel-Kerjan, E. (2001) *Insurance against Natural Disasters: Do the French Have the Answer? Strengths and Limitations.* Cahier n° 2001-007, Ecole Polytechnique, Centre National de la Recherche Scientifique, Laboratoire d'Econometrie, Paris.

Marcou, G. (1996) Das neue französische Raumordnungsrahmengesetz vom 4. Februar und seine Auswirkungen auf das Raumplanungssystem in Frankreich. In *Neues Raumordnungsgesetz in Frankreich. Das französische Raumplanungssystem zwischen tiefgreifenden Reformen und Kontinuität* (ed. Akademie für Raumforschung und Landesplanung), ARL-Arbeitsmaterial No. 225. Hanover, pp.10-19.

MATE (Ministère de l'Aménagement du Territoire et de l'Environnement, Direction de l'Eau et Direction de la Nature et Paysages) (1996) *Entre Terre et Eau. Agir pour les zones humides.* Paris.

MATE (Ministère de l'Aménagement du Territoire et de l'Environnement, Direction de l'Eau) (2000) *La politique Française de prévention des inondations.* 24.08.2000, Paris. http://www.oieau.fr/riob/ag2000/JLR-462u.htm, accessed on 09.04.2002.

Ministère des Affaires Etrangères (2006) *Spatial Planning and Sustainable Development in France.* Paris.

Ministère de l'Ecologie et du Développement Durable (2006a) *10 ans de PPR: quelles perspectives pour la prise en compte des risques naturels dans l'aménagement?* Paris.

Ministère de l'Ecologie et du Développement Durable (2006b) *Wetlands of international importance.* Report referring to the Ramsar Convention. Paris.

Ministre de l'Intérieur et de l'Aménagement du Territoire, de l'Équipement et du Tourisme & Ministre de l'Environnement (1994) *Circulaire du 24 janvier 1994 relative à la prévention des inondations et à la gestion des zones inondables.* Paris.

OECD (Organisation for Economic Cooperation and Development) (2006) *France: policies for preventing and compensating flood related damage.* OECD Studies in risk management, Paris.

OIEau (Office International de l'Eau) (2000) *Les programmes quinquennaux des Agences de l'eau.* http://cartel.oieau.fr/guide/g016.htm, accessed on 05.04.2007.

Peiry, J.-L., Girel, J., Pautou, G. and Nguyen, T.D. (1999) Planning of the river Isère in France. A centenarian evolution. In *River design and environmental protection in Europe* (ed. J. M. de Urena), Universidad de Cantabria, pp.195-299.

Sangaré, I. and Larrue, C. (2004) The evolution of the water regime in France. In *The Evolution of National Water Regimes in Europe. Transitions in Water Rights and Water Policies* (eds. Kissling-Näf, I. and Kuks, S.), Kluwer Academics, Dordrecht, pp.187-234.

Sohnle, J. (2001) Gewässerpolitik und Wasserrecht in Frankreich. In *Integrierte Gewässerpolitik in Europa, Gewässerschutz, Wassernutzung, Lebensraumschutz* (eds. T. Bruha and H.-J. Koch), Nomos, Baden-Baden, pp.165-190.

Stoll, T. (2001) *Schutzgebiete als Instrument der Naturschutzpolitik in Deutschland und Frankreich.* Diplomarbeit im Studiengang Landschaftsplanung, Technische Universität Berlin, Fachbereich VII. Berlin.

Ramsar Convention on Wetlands (1999) *Rapport national de la France pour la COP 7 Ramsar.* http://www.ramsar.org/cop7_nr_france.htm, accessed on 13.04.2002.

Ximenes, M.-C. (2001) *A national wetland monitoring programme in France*. Presentation at the workshop on "Wetland inventory and assessment", RAMSAR conference, 16.10.2001. http://www.ramsar.org/mtg_reg_europe2001_3ximenes_e. doc, accessed on 08.04.02.

LAWS

Act no. 82-600 of 13 July 1982 on the compensation of victims of natural disasters: Loi n° 82-600 du 13 juillet 1982 relative à l'indemnisation des victimes de catastrophes naturelles.

Act no. 87-565 of 22 July 1987 on the organisation of civil protection, protection against forest fires and the prevention of large-scale risks: Loi n° 87-565 du 22 juillet 1987 relative à l'organisation de la sécurité civile, à la protection de la forêt contre l'incendie et à la prévention des risques majeurs.

Act no. 93-24 of the 8 January 1993 on the protection and development of landscapes: Loi n° 93-24 du 8 janvier 1993 sur la protection et la mise en valeur des paysages.

Act no. 95-101 of 2 February 1995 on the reinforcement of environmental protection: Loi n° 95-101 du 2 février 1995 relative au renforcement de la protection de l'environnement.

Act no. 2003-699 of 30 July 2003 on the prevention of technological and natural risks and the repair of damage: Loi n°2003-699 du 30 juillet 2003 relative à la prévention des risques technologiques et naturels et à la réparation des dommages.

Environmental Code – Book II, Title I "Water and aquatic environment", Book III "Natural spaces", Book IV "Fauna and flora" and Book V "Prevention of pollution, risks and nuisances" (Code de l'environnement – Livre II, Titre 1 "Eau et milieux aquatiques", Livre III "Espaces naturels", Livre IV "Faune et flore" et Livre V "Prévention des pollutions, des risques et des nuisances").

Framework Act no. 95-115 of 4 February 1995 on spatial planning and regional development: Loi n°95-115 du 4 février 1995 d'orientation pour l'aménagement et le développement du territoire.

Framework Act no. 99-533 of 25 June 1999 on spatial planning and sustainable regional development: Loi n°99-533 du 25 juin 1999 d'orientation pour l'aménagement et le développement durable du territoire.

Framework Act no. 99-574 of 9 July 1999 on agriculture: Loi n° 99-574 du 9 juillet 1999 d'orientation agricole.

Framework Act no. 2001-602 of 9 July 2001 on forests: Loi n° 2001-602 du 9 juillet 2001 d'orientation sur la forêt.

Framework Act no. 2006-11 of 5 January 2006 on agriculture: LOI n° 2006-11 du 5 janvier 2006 d'orientation agricole.

Law project comprising the reform of water policy adopted by the National Assembly 10 January 2002, no. 172 (2001-2002), but not implemented: Projet de loi portant réforme de la politique de l'eau, adopté par l'Assemblée nationale, 10 janvier 2002, n° 172 (2001-2002), mais pas entré en vigueur.

INTERVIEWS

1. Researcher at the National School of Civil Engineering on spatial dimensions of water management, Paris, 2 July 2001.
2. Officer at the Water Directorate, Subdivision on water conservation and management at the Ministry on Spatial Planning and Environment, 2 July 2001.

3. Research Director at LATTS, the Laboratory on technology, territories and societies, Paris, 3 July 2001.
4. Officer at the French Delegation for Territorial Development and Regional Action (DATAR), Paris, 4 July 2001.
5. Officer at the French Water Academy, Paris, 4 July 2001.
6. Officer at the Water Directorate, Ministry on Spatial Planning and Environment, 5 July 2001.
7. Officer at the Directorate for Prevention of Pollution and Risks, Ministry on Spatial Planning and Environment, Paris, 5 July 2001.
8. Officer at the Water Agency Loire-Brittany, Orléans, 9 July 2001.
9. Officer at the Regional Directorate for the Environment for the Basin Loire-Brittany, Orléans, 9 July 2001.
10. Two conservation officers at the nature conservation agency "Espace naturels de France", Orléans, 10 July 2001.
11. Officer at the Regional Directorate for the Environment for the Basin Adour Garonne, Orléans, 11 July 2001.
12. Two conservation officers at the nature conservation agency "Nature Midi-Pyrénées", Toulouse, 12 July 2001.
13. Officer at the Water Agency Adour Garonne, Toulouse, 13 July 2001.
14. Two project managers at the regional development agency "Syndicat Mixte d'Etudes et D'Aménagement de la Garonne, SMEAG", Toulouse, 13 July 2001.

6

Policy shifts and delivery gaps: contexts of floodplain restoration in England and Wales

Timothy Moss

6.1 INTRODUCTION

The policy context of floodplain restoration in England and Wales has changed significantly in the wake of major flooding events in 1998, 2000 and – most recently – in July 2007. The 2000 floods were – in the words of the Deputy Prime Minister John Prescott – a "wake-up call for everyone".[1] Since then the view has been gaining ground in policy-making circles that past practices of

[1] Environment Agency (no date): http://www.environment- agency.gov.uk/aboutus/ 512398/289428/656297/?version=1&lang=_e&lang=_e. Accessed on 04.12.2006.

 Restoring Floodplains in Europe: policy contexts and project experiences edited by Timothy Moss and Jochen Monstadt. ISBN: 1843390906. Published by IWA Publishing, London, UK.

flood defence are no longer tenable or affordable, prompting a reappraisal of flood management priorities (Ledoux *et al.* 2005: 129). The principal statutory agencies responsible for, or with a vested interest in, flood management have since 2000 produced numerous policy reviews, position statements and action plans calling for a more strategic, integrated approach to managing floods across a whole catchment (Morris *et al.* 2005; Ledoux *et al.* 2005). This ongoing debate has generated growing interest in the capability of floodplains to alleviate flooding and thereby contribute to the new government strategy for flood risk management.

If the recent flooding events provided a short-term trigger for policy reappraisal, policy interest in floodplains and their restoration needs to be seen as part of a longer-term trend emerging from the 'greening' of the political agenda in the UK since the late 1980s (Adams *et al.* 2004: 1934; Defra 2005). Over the past two decades, nature conservation bodies in the UK have been arguing the case for the recreation of wetlands and initiating a small number of demonstration projects of river and floodplain restoration. As early as 1993 the Government's Strategy on Flood and Coastal Defence called for a strategic approach to river management which reflected the interrelationship between watercourses, land use and developments within a catchment and spawned a number of new planning instruments (see below). This policy initiative is just one expression of what Adams *et al.* identify as a new discourse of river and (increasingly) floodplain restoration which, from the late 1990s onwards, has been challenging – though not supplanting – the long-established flood defence discourse (2004: 1940). Government policy to promote biodiversity and debates over the future of farming and land management are creating additional openings for floodplain restoration (Morris *et al.* 2005). So also is the growing concern over climate change and the anticipated increase in the scale, severity and frequency of flooding events. Given the huge cost of constructing and maintaining flood defence structures to provide adequate protection from flooding, the Government is keen to explore alternative, less cost-intensive means of (precautionary) flood protection. As one cost-benefit analysis of flood protection concluded: "it would appear to be in the public interest to redirect funding, both from agricultural support and flood defence for agriculture, into flood storage and wetland creation" (Morris *et al.* 2002: viii; see also Morris *et al.* 2005; Swash 1998).

Whilst the level of interest in restoring floodplains has grown markedly in national policy and scientific circles over the last decade, very few schemes for setting back flood defences or widening the floodplain have actually been put into practice. As Leroux *et al.* recently noted: "Whilst managed realignment has been a topical issue for many years, it has not been widely implemented to date" (2005: 130). Most restoration projects in the UK are limited to the river channel; few have sought to rehabilitate ecosystems in the floodplain (Adams and Perrow

1999: 90). In a survey published in 2002 Halcrow *et al.* identified 31 schemes for the managed realignment of rivers in the UK at either the planning, design or implementation stage, some of which were targeting environmental enhancements as well as flood defence benefits (2002). They concluded, however, that implementation was proving problematic. Translating into practice the recent policy initiatives for the integrated management of floodplains would appear to be a particularly challenging task.

This chapter explores the policy context of restoring floodplains in England and Wales today.[2] The purpose is to analyse what new opportunities are emerging for restoring floodplains, but also what institutional constraints are confronting efforts to implement policy objectives. The review is based on the analysis of policy documents of the responsible national agencies and the relevant secondary literature. The key issues raised in this desk-top analysis were explored in greater depth in face-to-face interviews with 11 national experts from most of these organisations (see list under References). Early findings from the research were discussed at a UK stakeholder workshop held on 16-17 May 2002. The chapter begins with selected policy statements on floodplains and their restoration, then analyses the relevant instruments and organisations from the fields of flood defence and water management, land-use planning, nature conservation, agriculture and forestry, as well as modes of funding and coordination across policy fields. It concludes by briefly reviewing existing schemes in practice and summarising the principal drivers of, and constraints to, floodplain restoration.

6.2 POLICY STATEMENTS ON FLOODPLAINS AND THEIR RESTORATION

In England and Wales specific references to floodplains and their restoration are to be found not in the relevant legislation – the Water Resources Act of 1991, the Land Drainage Act of 1991, the Environment Act of 1995 and the Town and Country Planning Act of 1990 – but in policy statements and planning guidance documents. Table 6.1 lists the most specific references to floodplains made in policy documents by the statutory bodies responsible for flood protection, water management, nature conservation and spatial planning, notably the Department of the Environment, Food and Rural Affairs (Defra), the Environment Agency, and Natural England (formerly English Nature), cited in chronological order.

---

[2] The author would like to thank Ross Beveridge, Manuela Siener and Thomas Weith for their assistance in researching this chapter.

Table 6.1: Policy statements relating to floodplain restoration in England and Wales.

Document	Policy statement
Ministry of Agriculture, Fisheries and Food and Welsh Office (1993): Strategy for Flood and Coastal Defence	"Measures to avoid flooding or alleviate the damage it causes within a river catchment are best taken in the context of the interrelationship between watercourses, land use and developments within that catchment" (p.21)
Environment Agency (1997): Policy and Practice for the Protection of Floodplains	The Environment Agency aims to "secure and, where necessary, restore the effectiveness of floodplains for flood defence and environmental purposes" (p.3) Sub-policy FD-P6: "The Environment Agency will seek to retain and where necessary restore the effective flood flow conveyance and flood water storage capacities of floodplains" "At some locations, redevelopment schemes may provide an opportunity for the restoration of the natural floodplain or at least the removal of obstructions to flood flow routes" (p.14)
English Nature (2001): Action for Floodplain Diversity	"English Nature is seeking to promote strategies for floodplains which develop a vision for the catchment and which include restoration objectives to deliver Biodiversity Action Plan targets" (p.8)
Department of Transport, Local Government and the Regions (DTLR) (2001): Planning Policy Guidance Note 25: Development and Flood Risk	Includes "recognising that flood plains and washlands have a natural role as a form of flood defence as well as providing important wildlife habitats and contributing to landscape value" (Pt.8) "Soft engineering techniques such as creating, preserving and enhancing natural flood meadows and washlands or salt marshes and mud flats can be of great value in attenuating flooding as well as contributing to biodiversity" (Pt.11) "Local plans should also identify sites where …restoration of flood plains could contribute to more sustainable flood management, as well as to amenity, landscape and biodiversity objectives" (Pt.53)
Environment Agency (2003):"Effective Flood Management: Position Statement"	"Effective policies can steer inappropriate development away from areas at a high risk of flooding, create storage for flood water and also, by allowing the re-establishment of natural floodplains, optimise benefits to wildlife." The Environment Agency calls for the "designation of strategic flood plain storage areas, with appropriate mechanisms in place, to encourage landowners to participate in allowing their land to be used for this purpose."
Department of Environment, Food and Rural Affairs (Defra) (2005): Making Space for Water	"As part of the wider portfolio of responses to flood and coastal erosion risk, we will promote the environmental pillar of sustainable development by making greater use of rural land use solutions such as the creation of wetlands and washlands, and managed realignments of coasts and rivers." (p.9)

6.3 PRINCIPAL INSTRUMENTS AND POLICIES RELEVANT TO FLOODPLAIN RESTORATION

The location of references to floodplains and their restoration in strategy documents, rather than in legislation, is indicative of policy formulation and implementation in the United Kingdom (on the following, Lowe and Ward 1998; Héritier *et al.* 1996: 77-102; Smith 2001: 157-166). Environmental policy in Britain is characterised by a legal system with a strong emphasis on procedural – rather than substantive – regulation and a political system dominated by central government and parliament. There is a strong tradition of informal regulation through consensus rather than coercion. Negotiation and compromise between interested parties in particular contexts have conventionally taken precedence over the imposition of environmental standards. Public policy is formulated substantially by central government departments and powerful public – or quasi-public – national agencies. These agencies possess extensive discretionary powers to specify and implement policy, operating via their own regional and local offices. Local and – especially – regional government is comparatively weak. Since the 1990s the environmental policy process in Britain has been opening up to a wider range of stakeholders and more institutionalised modes of consultation. However, cross-sectoral integration – for instance between flood defence, land-use planning and nature conservation – remains under-developed in a political system where, in comparison with many EU Member States, sectoral hierarchies are strong and local or regional accountability is weak.

6.3.1 Flood defence and water management

Organisational responsibilities

The organisational responsibilities for flood defence and water management in England and Wales are highly complex (on the following, Ledoux *et al.* 2005). They are set out in the Water Resources Act of 1991, the Land Drainage Act of 1991 and the Environment Act of 1995. Setting the policy framework, providing strategic guidance and administering legislation is the responsibility of the Department for Environment, Food and Rural Affairs (Defra) in England, and the National Assembly for Wales in Wales. **Defra**, created by the second Labour Government in 2001, comprises the former Ministry of Agriculture, Fisheries and Food (MAFF) and the environmental divisions of the former Department of the Environment, Transport and the Regions (DETR) and includes – of relevance to floodplain restoration – a flood management division and a water resources division. In the field of flood defence, Defra provides strategic and policy guidance, approves proposed measures, administers grant-aid for capital flood defence schemes, funds research and development, and disseminates best practice. Flood defence measures are implemented by the so-called "operating authorities": the Environment Agency, flood defence committees, internal drainage boards and local authorities.

The **Environment Agency** is a non-departmental government body with responsibility for environmental protection (including protecting and improving the quality of rivers) and water management (including water resources, flood defence, fisheries, conservation, recreation and navigation) in both England and Wales. It operates via 8 regional offices and 26 area offices. In the field of flood defence, the Environment Agency has the duty to exercise general supervision over flood defence and implement government policy. These obligations include advising planning authorities and developers on flood risk, conducting surveys of flood-risk areas, monitoring the condition of flood defences, assessing flood risk and reporting on progress towards achieving and maintaining government high-level targets for flood defence. In addition, the Environment Agency has permissive powers to disseminate flood warnings and to maintain, operate and build flood defences on the principal watercourses (so-called "main rivers"). Permission from the Agency is required by anyone wishing to carry out works which affect a watercourse or flood defences, ensuring there is no increased risk of flooding. As of April 2006 the Environment Agency has been given responsibility for flood policy delivery. In connection with this move, the Government plans to transfer to the Environment Agency the operation of the national capital investment system for flood risk management.[3]

The Environment Agency shares the responsibility for flood defence in England and Wales with 29 flood defence committees, 228 internal drainage boards and over 400 local authorities. Where and how flood defence funding is spent is decided not by the Agency but by 9 regional **flood defence committees** supported by local flood defence committees. Flood defence committees were set up under statute to raise funds, mainly from local authorities, and to determine priorities for regional/local flood defence programmes following advice from the Environment Agency. Regional flood defence committees, in particular, are influential by virtue of their power to make decisions on investments. They comprise a chairman and members appointed by Defra, two members appointed by the Environment Agency and a statutory majority of councillors from local authorities.

Operational tasks of flood defence on all smaller rivers (so-called "ordinary watercourses") are the responsibility of internal drainage boards and local authorities. **Internal drainage boards** (IDBs) are independent bodies created under land drainage statutes with a history of land reclamation and drainage dating back to the 13th Century (on the following Hodge and McNally 2000;

---

[3] Defra (2006): http://www.defra.gov.uk/environ/fcd/policy/frgrchng.htm. Accessed on 21.03.2007.

Interview 2). There exist 228 drainage boards concentrated in lowland areas with special drainage needs, covering some 1.2m hectares, including many of the ecologically most valuable wetland habitats in the country, such as the Somerset Levels and the Cambridgeshire Washes. Originally groups of local landowners pursuing their own drainage interests, they are today formal institutions with statutory responsibilities not only for land drainage and flood defence but also – following the Land Drainage Act of 1991 – for nature conservation and enhancement within the scope of their activities.[4] They undertake flood defence works on ordinary watercourses and improve and maintain drainage systems, as well as develop Water Level Management Plans (see below) together with the Environment Agency in consultation with nature conservation bodies. Internal drainage boards fund their activities primarily through drainage levies on farmers and other occupiers within the internal drainage districts and special levies on local authorities. Membership of internal drainage boards comprises those elected by the drainage ratepayers and members nominated by the local authorities in the drainage districts according to the proportion of drainage board income they contribute. Although efforts are being made to extend the membership of internal drainage boards to include representatives from trade and industry, they remain dominated by agricultural landowners. Their commitment to nature conservation is very varied (Hodge and McNally 2000: 114-5). Outside drainage board districts, flood defence works on ordinary watercourses are undertaken by over 400 **local authorities**, mainly district councils and unitary authority councils. They also fund most of the defence work on main rivers carried out by the Environment Agency via levies paid to regional and local flood defence committees. Local authorities have a particular influence over flood risk and floodplain restoration by virtue of their planning powers over land use and development (see Section 6.3.2).

On paper, the multi-functional responsibilities of Defra (including flood defence, agriculture and environmental protection) and the Environment Agency (including flood defence, water resource management and environmental protection) in conjunction with the operational tasks of the Agency through its regional and local offices provide a favourable institutional framework for delivering a more integrated, cross-sectoral approach to flood management. However, the fragmentation of responsibility for flood defence in England and Wales has created problems for effective flood defence measures, as highlighted by the response to recent flooding events. Criticism is levelled at the number of authorities with responsibilities for delivering flood mitigation (some 700 in England and Wales according to one survey, ICE 2001: 70), the unclear division

---

4 The IDBs are required to "further the conservation and enhancement of natural beauty and the conservation of flora, fauna and geological or physiographical features of special interest consistent with any enactments relating to their functions" (Land Drainage Act 1991, p.113).

of responsibilities between the various operating authorities, creating disputes between the parties, and the arbitrary and unnecessary categorisation between main rivers and ordinary watercourses. As the Environment Agency itself acknowledges: "We believe that the existing complex, 70-year old administrative arrangement surrounding flood management causes confusion to the public and hinders the effective delivery of flood defence services to them".[5] This criticism is hardly surprising given that the Agency is likely to benefit from ongoing Government reviews of the organisation of flood defence. Responsibility for so-called critical ordinary watercourses has already been transferred to the Environment Agency. Further possible reforms include the creation of single-tier flood defence committees and the amalgamation of the smaller internal drainage boards. Both the Association of Drainage Associations and the Local Government Association are campaigning hard, however, to retain operational flood defence at the local level as legitimised by local authorities (LGA 2001; Interview 2).

Policies and planning instruments

The policy framework for flood defence is set out in the Strategy for Flood and Coastal Defence in England and Wales, published in 1993 by MAFF and the Welsh Office. This document marks, in retrospect, a first step towards a more integrated approach to flood management, reflecting the influence of a forward-thinking flood management division within MAFF at the time (Interview 4). The guidance stresses the importance of taking a strategic approach to river management, going beyond isolated defence measures to consider the impacts of interventions up- and downstream and the interrelationship between watercourses, land use and development within a whole catchment (MAFF and WO 1993: 20-21). To this end, it advocated the creation of Catchment Management Plans and Water Level Management Plans. Catchment Management Plans were developed by the then National Rivers Authority during the course of the 1990s and amalgamated into some 120 **Local Environment Agency Plans** with additional information on waste management and air pollution following the formation of the Environment Agency in 1996 (Slater *et al.* 1994). Targeting primarily water and environmental quality issues, neither of these plans is of direct relevance to floodplain restoration.

Water Level Management Plans, by contrast, are regarded as one of the most significant instruments affecting the management of floodplains (Swash 1998: 5). Water Level Management Plans comprise a written agreement between interested parties on water level requirements for a range of activities, including agriculture, flood defence and nature conservation in a specified area

---

5 Environment Agency (no date): http://www.environment-agency.gov.uk/aboutus/512398/289428/656297/?version=1&lang=_e&lang=_e. Accessed on 21.03.2007.

(MAFF 1994). Launched by MAFF in 1994 they were intended in particular for designated sites of nature conservation, especially those national Sites of Special Scientific Interest (SSSIs) which today qualify as Special Areas of Conservation (SACs) pursuant to the EU Birds Directive, Special Protection Areas (SPAs) pursuant to the EU Habitats Directive and Ramsar Sites. The operating authorities prepared a list of some 600 sites where Water Level Management Plans were to be introduced by the end of 1998 (Swash 1998: 6). By 2001, however, it was estimated that only 20% of the envisaged plans for SSSIs had been completed. Reasons given for this delay include the inability of many drainage boards to raise even the modest 20% co-funding for the plans and the low priority accorded to water level management by the operating authorities. A third planning instrument – **Catchment Abstraction Management Plans** – emerged out of a review of the water abstraction licensing system in 1999. Catchment Abstraction Management Plans are a mechanism for time-limiting water abstraction licenses and balancing the needs of abstractors, other water users and the aquatic environment in consultation with the local community and interested parties. They set down strategies for the management of water resources at a local level, covering 129 areas in England and Wales.

In 1997 the Environment Agency set out its flood defence policies in relation to floodplains in a policy document entitled "Policy and Practice for the Protection of Floodplains" (EA 1997). The stated aim of the Agency was, and still is, to "secure and, where necessary, restore the effectiveness of floodplains for flood defence and environmental purposes" (p.3). The policy statements are concerned principally with protecting existing floodplains from development in order to maintain their flood defence and environmental functions. There are a few isolated references, though, to the need to restore natural floodplain areas at some locations where this is practicable (see Table 6.1).

Despite the call for an integrated approach to flood defence in the 1993 guidance and the introduction of new planning instruments to this end, it is generally conceded that, especially prior to the 1998 and 2000 flooding events, most flood defence schemes have largely ignored the knock-on effects of flooding downstream (NAO 2001: 32). Nature conservation groups are particularly critical of the continued dominance of hard engineering solutions and the neglect of environmental and biodiversity considerations in practice. The Government's commitment to applying the various new planning instruments – in particular the Water Level Management Plans – has, until recently, been limited (NAO 2001: 23). It has thus been relatively easy for farmers to resist attempts to permit their land to be flooded, whether for flood defence or nature conservation benefits (Adams *et al.* 2004: 1932-3).

In the wake of the 1998 and 2000 flooding events a large number of policy review papers have been produced by the statutory agencies concerned with flood defence criticising past practices of flood management. These include the

Agricultural Select Committee's report on flood and coastal defence of August 1998, the Environment Agency's "Easter Floods Action Plan" of November 1998, the Agency's report "An Environmental Vision – Reducing Flood Risk" of December 2000, Defra's "National Appraisal of Assets at Risk from Flooding and Coastal Erosion in England and Wales" (2001) and Defra's consultation document "Making space for water" (2004). Whilst the focus of many of these reports is on improving flood forecasting, flood warning, standards of flood defences and management structures, two common points of criticism relevant to floodplain restoration stand out: the lack of a strategic approach to flood defence issues in general and the poor delivery of policy objectives by operating authorities "on the ground".

To help overcome the perceived policy delivery gap MAFF published in November 1999 a series of high-level targets on flood defence (Defra 2001). These 14 targets, introduced in April 2000, require the operating authorities to ensure and demonstrate delivery of the government's policy aims and objectives as set out in the 1993 strategy, reporting on progress made in a number of key fields, including the preparation of Water Level Management Plans, biodiversity and development in areas at risk from flooding. Further government specifications relating to individual flood defence projects are given in Flood Defence Project Appraisal Guidance papers published by MAFF in 1999 and 2000. One of these, the Environmental Appraisal of March 2000, sets out how schemes should take account of environmental objectives and EU Directives and seek opportunities for environmental enhancement. Nature conservation organisations such as Natural England, the Royal Society for the Protection of Birds and Wildlife Trusts have been critical, however, of the lack of consideration of reversing past damage to biodiversity when revising flood defence schemes (NAO 2001:53). In April 2005 a revised set of High Level Targets was introduced, replacing those of 2000, which takes greater account of the Government's emergent flood risk management strategy.[6]

A major planning initiative of considerable potential value to floodplain restoration is the introduction of **Catchment Flood Management Plans**. Catchment Flood Management Plans are a high-level strategic planning framework for the integrated and long-term management of flood risk across a whole catchment. They are intended to give a general assessment of risks, opportunities and constraints, areas of uncertainty, pressures and issues affecting flood management. On this basis a set of flood-risk management policies for relatively large catchments of between 1,000 and 5,000 km^2 are to be developed, taking a 50-100 year perspective (Halcrow Water 2001; EA 2004). These policies are then to be specified in Strategy Plans for smaller sub-catchments

[6] Defra (2005): http://www.defra.gov.uk/environ/fcd/hltarget/default.htm. Accessed on 21.03.2007.

and subsequently translated into operational schemes and actions at a local level. The expectation is that Catchment Flood Management Plans, by taking a catchment view of the issues, will be able to identify optimal areas where flood storage is needed and feasible. The ambitious programme was tested with several pilot studies, initially for the following five catchments: Somerset Levels and Moors (Parrett), Yorkshire Derwent, Irwell, Upper Severn and Medway. Some 80 catchment flood management plans are planned, covering the whole of England and Wales. Many of these had, by early 2007, been issued for consultation.[7] However, even prior to their implementation criticism was voiced about the likely cost-effectiveness of the plans, given the scale of the planning exercise and uncertainty as to what mechanisms are available to ensure delivery of the strategies they prescribe (NAO 2001:25).

The most recent Government policy initiative is the consultative exercise "Making space for water", launched in 2004 (Defra 2004; Defra 2005). The new strategy is intended to pursue a more holistic approach to managing flood (and coastal erosion) risks in England over the next 20 years, replacing the 1993 policy framework. Future flood risk management will consider all sources of flooding and embed flood risk management across a range of government policies (Defra 2005). It will also take account of the social, environmental and economic pillars of sustainable development and long-term drivers such as climate change. Decisions are to reflect the uncertainty surrounding flooding events and take a precautionary approach "where appropriate".

6.3.2 Land-use planning

Policy on land-use planning (including development in areas at risk from flooding) in England is determined by the Department for Communities and Local Government (DCLG) – formerly the Department for Transport, Local Government and the Regions – working closely with Defra on matters relating to flood risk. In Wales both functions are covered by the National Assembly for Wales. Planning authorities are responsible for protecting the flood defence interests of people whose property is affected by development proposals. They are required to follow guidance provided by the Government in drawing up development plans, as specified in regional planning guidance notes, and in implementing land-use policy and operating the planning system, as given in planning policy guidance notes. The Environment Agency is a statutory consultee on development plans and advises planning authorities on the implications of development proposals for flood risk. Where a development proposal is inconsistent with the Environment Agency's floodplain policies the

---

7 Environment Agency (2007): http://www.environment-agency.gov.uk/subjects/flood/1217883/1217968/907676/999493/?version=1&lang=_e. Accessed on 21.03.2007.

Agency will object, giving reasons and advising the planning authority on how to overcome the objection. The Environment Agency has produced maps of floodplains – so-called "flood maps" – indicating areas at risk from flooding with the intention of raising awareness of the flood risks associated with land use.[8]

The Environment Agency, however, possesses no powers to veto development on grounds of flood risk. It may create byelaws prohibiting development on riparian strips with a width of usually 8 metres and the filling in of a floodplain, but in practice these restrictions can be circumvented by granting planning permission (Gardiner 1998: 18). In general, local planning authorities are very reluctant to include policies in development plans aimed at restricting development on a floodplain and frequently disregard Government policy and Environment Agency recommendations to this end (cf. Gardiner 1998: 19; Defra 2005). It is very rare for a local planning authority or a developer to be held accountable for increases in flooding downstream resulting from the loss of a functional floodplain.[9] At the same time, some local authorities are critical of the Environment Agency being too lenient towards development on floodplains in the past (Interview 10).

Criticism of the contribution of past land-use and development planning practices to the flooding events of 1998 and 2000 prompted a review of planning guidance to strengthen consideration of flood risk at all stages of the planning process. Planning Policy Guidance Note 25 (PPG 25) was introduced in 2001 with the aim of improving coordination between land-use/development planning and the operational delivery of the Government's flood and coastal defence strategy (DTLR 2001). PPG 25 was mainly of relevance for protecting existing floodplains from development, rather than creating openings for floodplain restoration, and provides strong justification for maintaining high standards of flood defence for existing settlements (NAO 2001: 43). Even so, the original draft of PPG 25 was revised substantially – taking account of severe criticism during consultation and lessons learned from the 2000 floods – so as to strengthen restrictions on development in functional floodplains and to recognise the value of floodplains and washlands for flood defence, habitat creation and landscape amenities (see Table 6.1; Interview 7). However, concerns were raised about how PPG 25 was applied in practice, in particular how planning authorities interpreted the guidance to suit their own

---

[8] The government definition of a floodplain is: "All land adjacent to a watercourse over which water flows in times of flood or would flow but for the presence of flood defences where they exist" (DoE Circular 30/92). The flood maps are accessible under http://www.environment-agency.gov.uk/subjects/flood/826674/830330/927042/927049/927084/927104/927150/. Accessed on 21.03.2007.

[9] Functional floodplains are defined by the government as "unobstructed or active areas where water regularly flows in time of flood" (DTLR 2001: Pt.23), i.e. that part of the floodplain which is not defended.

interests with little consideration for Environment Agency recommendations (Defra 2005: 21). In response, the Government replaced PPG 25 in December 2006 with Planning Policy Statement 25: Development and Flood Risk (DCLG 2006). This establishes a sequential procedure for assessing flood risk in planning developments, comprising flood risk appraisals at regional and local levels. However, it makes little explicit reference to the creation or restoration of functioning floodplains as a means of minimising flood risk.

6.3.3 Nature conservation

The key player of nature conservation policy is **Natural England** (formerly **English Nature**), the statutory body responsible for the conservation and enhancement of wildlife in England.[10] Natural England's tasks include principally the management of national nature reserves in its ownership, the designation of protected sites and rules for their management and the implementation of tasks specified for the agency in the UK Biodiversity Action Plan (see below). Natural England is also active as a consultee on programmes and projects affecting nature conservation and funds projects with grant aid. By exploiting its involvement in various policy and programme consultation processes Natural England has proved effective recently in strengthening the nature conservation component in other policy fields, notably flood defence, agriculture and forestry. Natural England seeks out and promotes the nature conservation dividend of ongoing policy reforms and in return helps make these politically more acceptable and – by unlocking new funding sources – economically more viable (Interview 3). Natural England's policy towards floodplains and their restoration is set out in the position statement "Action for Floodplain Biodiversity" (EN 2001a). Here, English Nature, as it was then called, argued that floodplains fulfil a wide variety of functions and need to be managed "in ways that work with nature wherever possible" in broad consultation with stakeholders and within a strategic vision for the whole catchment (EN 2001: 4). Natural England has funded several studies and reports on the benefits of floodplains and washlands for flood defence and biodiversity, notably "The effectiveness of floodplain ESA schemes in the Maintenance and Enhancement of Biodiversity" (2000a), "Habitat Restoration Project" (2000b), "Vision for Wetlands: towards Biodiversity-rich Floodplains" (2000c) and "Sustainable Flood Defence: the case for washlands" (2001b). A recent joint study with Defra, entitled "Integrated Washland Management for Flood Management and Biodiversity", is summarised in Morris *et al.* (2005).

---

[10] In October 2006 Natural England was created as an amalgamation of English Nature, the environment activities of the Rural Development Service and the Countryside Agency's Landscape, Access and Recreation division.

A further public agency relevant to nature conservation in England is the **Countryside Association**, which advises on policy for managing the countryside. Until recently the Countryside Association straddled the policy fields of nature conservation, landscape development, recreation, rural development and agriculture. In October 2006 its Landscape, Access and Recreation division was amalgamated under the new organisation Natural England (see above). The Countryside Association is involved in the Land Management Initiatives scheme to develop particular landscape and farming types by offering attractive targeted packages to farmers for a change in land use, including floodplain restoration (see Section 6.6). In Wales the **Countryside Council for Wales** generally speaking covers the responsibilities of Natural England and the Countryside Association: natural beauty, wildlife and recreation.

The principal instrument of nature conservation in the UK, as in most countries, is the designation and controlled management of **protected sites**. Many protected areas cover wetland sites which are important for the conservation and enhancement of wildlife and natural features of the landscape. These comprise Ramsar sites, Special Areas of Conservation (SACs), Special Protection Areas (SPAs) and national Sites of Special Scientific Interest (SSSIs). Whereas these designated sites are good for protecting and enhancing existing natural habitats they are far less effective as an instrument for creating a new habitat as in the case of floodplain restoration. As Ledoux *et al.* argue, the EU Habitats Directive is for this reason not just a driver, but also a prime constraint to the managed realignment of rivers (2005: 139).[11] Since the restoration of a floodplain on a designated site as a rule entails a change in the habitat(s) for which the site offers protection, such initiatives regularly encounter opposition from nature conservationists keen to maintain and enhance the existing habitat. Resistance by ornithologists to the restoration of floodplain woodland in SACs designated to protect the habitats of wading birds is a case in point. Where designated areas do prove useful for floodplain restoration is in attracting grants for restoration initiatives, such as the EU-LIFE fund (on funding schemes, see Section 6.4).

An important instrumental driver of floodplain restoration in England and Wales is the **UK Biodiversity Action Plan**, which features several wetland species and wetland habitats that may be compatible with functioning floodplains (Morris *et al.* 2005: 107). Delivering targets for key habitats has become a central objective of government policy on nature conservation since the UK Biodiversity Action Plan was initiated in 1994. For each habitat identified there is an action plan specifying targets for enhancing biodiversity by

[11] Ledoux *et al.* argue for a more flexible interpretation of the Habitats Directive to encompass habitat creation (Ledoux *et al.* 2005: 142).

a fixed date. Local action plans have been devised to help implement the national targets. Floodplains support six key habitats identified in the UK Biodiversity Action Plan, including wet woodland. The action plan for wet woodland envisages the restoration of 3,375 ha to native wet woodland and the colonisation and/or planting of 6,750 ha of wet woodland by 2015.[12] In addition, a number of demonstration sites to show good practice have been identified and around four large-scale sites (>50 ha) are to be identified for the re-creation of floodplain forests in the UK, taking hydrological, wildlife conservation, economic and amenity issues into consideration.

6.3.4 Agriculture and forestry

Recent crises affecting farming in Britain – notably BSE, swine fever and foot-and-mouth – in combination with pressure to reform the current practice of subsidising agricultural production have prompted vigorous debate on the future priorities of farming and rural land management. Even within the farming community recognition is growing that the EU Common Agricultural Policy is undergoing a long-term process of substantial reform. To quote Morris *et al.*: "Funds previously committed to support farm output are increasingly diverted to encourage land managers to deliver environmental benefits" (2005: 105). Some commentators anticipate changes to agricultural policy resulting in large areas of land coming out of agricultural production or serving additional functions, creating *inter alia* opportunities for floodplain restoration (EA 2002: 25). Others warn against exploiting the current restructuring processes to promote a myth that low-lying agricultural land could be usefully used for flood mitigation on a significant scale (ICE 2001: 27).

An English Nature study identified a number of general obstacles to private landowners carrying out habitat enhancement measures (EN 2000b: 38). These comprise: loss of income, uncertainty of obtaining funds, funds not covering the full cost, schemes not being tied directly to biodiversity action plan targets, inappropriate skills, the unwillingness of farmers to conduct work which prevents future reversal to productive agricultural land and their preference for schemes which impinge least on the farm business. There is some debate as to the importance of inadequate financial incentives (cf. Morris *et al.* 2000; Morris *et al.* 2005; Burgess *et al.* 2000; see Section 6.4). Land-use change is indeed costly: the transition from intensive to extensive pasture was calculated in 1998 to incur losses of £401 per hectare, and from extensive arable to extensive pasture of £378 per hectare (EN 2001b: 17). Morris *et al.* also calculate that washland development is unlikely on intensive grassland or arable land given the extent of income loss (2005: 109-110). It is also clear, though, that farmers

[12] http://www.ukbap.org.uk/UKPlans.aspx?ID=4#4. Accessed on 04.12.2006.

are often confused or outfaced by the bureaucracy involved and are in need of better information and communication on the various schemes available (Hodge and McNally 2000: 110). A study of income losses in the Somerset levels and Moors in south-west England shows these to be relatively small – at ca. €20 per hectare – in cases where grassland is flooded for about one week during the winter (Morris *et al.* 2002). As Burgess *et al.* point out, there is reluctance on the part of nature conservationists in general, and Natural England in particular, to appreciate fully the multiple identities and viewpoints of farmers (2000: 131). Considerable anecdotal evidence suggests that where direct contact and close relationships with farmers are sought by scheme managers or intermediary organisations, such as Fisheries and Wildlife Advisory Groups, the willingness of farmers to enter a scheme and maintain new standards of land stewardship is much greater (Burgess *et al.* 2000, Hodge and McNally 2000).

In terms of forestry, interest is growing amongst nature conservationists and foresters in restoring a "lost component" of Britain's flora: floodplain woodlands (Kerr and Nisbet 1996: 1). There exist today only fragments of ancient floodplain forest, estimated at 25-30,000 ha, primarily in Scotland and the New Forest, with a further 25-40,000 ha of recent wet woodland. A scoping study commissioned by the Environment Agency put the case for restoring natural floodplain woodland in 1996 (Kerr and Nisbet 1996). This report argued the benefits of floodplain woodland restoration in terms of the value for pollution control (acting as a buffer for diffuse pollutants), flood mitigation (offering added roughness and infiltration), nature conservation (restoring a lost ecosystem), timber production, improved fisheries and landscape enhancement. It also pointed out the main risks emanating from restoring floodplain woodland, including losing some control of flooding in previously defended areas, the possibility of greater evapotranspiration exacerbating regional water shortages, threats to existing wetland areas with an already high conservation value and problems for navigation resulting from creating multiple braided channels and obstructions from woody debris. These arguments are subjected to scientific scrutiny in the flagship publication of the EU-funded project FLOBAR2 (Hughes 2003).

The call for the restoration of floodplain woodland resonates with recent developments in forestry policy and a number of instruments directed at habitat enhancement. The Forestry Authority – the policy wing of the Forestry Commission – today has a broad remit going beyond timber production and including conservation and recreation interests. Regarding floodplain woodland restoration the Forestry Authority has issued a guide to the management of wet woods and offers grants for the creation of wet woodland via the Woodland Grant Scheme. Funding for wet woodland is also available from the Defra Farm Woodland Premium scheme and the Environmentally Sensitive Areas scheme (see Section 6.4). There remain, however, a number of obstacles to the restoration of floodplain woodland in practice. One of these is the limited economic incentive for landowners to convert agricultural land to woodland given the low return for timber in recent years. Another is resistance within the Environment Agency to the presence of trees on functional floodplains. Environment Agency policy currently restricts the planting of trees near watercourses in order to prevent the creation of obstructions to flood flows. There are signs, however, of attitudes within the Agency – at least in policy-making circles – becoming more accommodating to the notion of restoring floodplain woodland (Interviews 4 & 9).

6.4 FUNDING PROGRAMMES AND ECONOMIC INSTRUMENTS

Restoring floodplains can be a costly activity. Persuading stakeholders to alter entrenched practices of land and river use for biodiversity or flood defence benefits is very difficult without providing attractive financial incentives. Over the past decade, however, strategic re-thinking of policy towards flood protection, nature conservation and agriculture has created openings for improved incentives for both the creation and maintenance of functional floodplains.

Flood protection

In the field of flood protection a number of flood defence reviews since the 1998 and 2000 flooding events have pointed to the need for a substantial increase in spending on flood defence but also, significantly, to a reappraisal of funding priorities and methods of assessing funding needs. Within this ongoing debate the case is being made for using washlands or functional floodplains as a cost-effective form of flood mitigation. The total value of assets at risk from flooding in England is, according to a series of recent studies, £626 million annually (HR Wallingford 2002, cited in Ledoux *et al.* 2005: 129). A survey in 2000 estimated that around 40% of the Environment Agency's flood defence assets were in only a fair, poor or very poor condition, requiring major capital investment in the future (cited in NAO 2001: 2). Other studies have argued that, with annual investment in the maintenance of all flood defence assets at ca. £80 million, the condition of existing flood defence infrastructure will almost inevitably decline (ICE 2001: 44). The Institution for Civil Engineers has called for an increase in expenditure levels on asset maintenance "perhaps as much as double the present" (ICE 2001: 44). The Government has increased funding on flood and coastal erosion risk management in England substantially, from £310 million in 1996/97 to an estimated £564 million in 2005/06 (Defra 2005: 17). The bulk of funding for flood and coastal erosion management comes in the form of grants from Defra to the operating authorities. In the financial year 2004/05 Defra contributed some 80% of total public funding (£502 million).[13] Most of this – 88% – was allocated to the Environment Agency, the rest to local authorities and internal drainage boards, who raise additional funds locally. The grant-aid rate for drainage board capital works is generally 45%, rising to 80% in internationally important conservation areas (SPAs, SACs and Ramsar sites) and 50% in other nationally important SSSIs. The Government recently announced an increase to 100% for priority SSSIs (Defra 2005: 26).

The procedures and priorities of funding flood defence have come under scrutiny in recent years. A Government Comprehensive Spending Review revealed in the late 1990s that the cost-benefit analysis of expenditure on capital works was not providing a clear picture of the optimal forms of flood defence and maintenance. In practice, schemes were approved on the basis of past experience and traditional procedures. In 2002 Defra/HMT (the Treasury) conducted a Flood and Coastal Defence Funding Review exploring a number of options to improve existing practices. In terms of improving conditions for floodplain restoration, the most significant proposals were to levy a charge on developers building on floodplains and to raise a floodplain levy on existing property. Development on floodplains is already becoming more costly following the major flooding events, as insurance companies seek ways of avoiding costly claims for flood damage in the future. It is estimated that the 1998 floods prompted claims of ca. £400 million (DTLR 2001) and the 2000 floods claims exceeding £700 million (ICE 2001: 69). Insurance companies are responding by tailoring insurance policies to reflect the individual risk of a customer's property, even to the extent of withdrawing insurance cover in particularly vulnerable areas. This is having a negative effect on property values (DTLR 2001: Pt.32).

Efforts to exploit the current reappraisal of funding priorities to create opportunities for floodplain restoration, however, face several problems relating to the mechanisms of flood defence funding (O'Riordan 2002; Ledoux *et al.* 2005). Firstly, cost-benefit analyses as currently practised are heavily weighted in favour of benefits to which an economic value can readily be ascribed, such as property values. Environmental benefits, being much harder to monetarise, tend to be underestimated. The Government promises, in its new strategy for flood risk management, to redress this imbalance (Defra 2005: 25). Secondly, schemes for flood retention, especially those involving the relocation of defences, can be considerably more expensive than conventional flood defences. The additional benefits for biodiversity, recreation etc. are difficult to justify economically when they are funded out of a budget earmarked solely for flood protection purposes. Thirdly, using washlands or functional floodplains for flood

---

[13] Defra (2006): www.defra.gov.uk/environ/fcd/policy/funding.htm. Accessed on 04.12.2006.

defence purposes would in general require a greater proportion of funding for maintenance than for capital works (EN 2001: 3). Since flood defence expenditure on capital works and maintenance is administered separately this would necessitate a major structural reform of existing funding mechanisms. The search is for effective "packages" which target multiple objectives, rather than single sources of funding (Ledoux *et al.* 2005: 140). One solution proposed is to use flood defence funds for the capital cost of washland construction combined with agri-environment schemes to fund subsequent management costs (EN 2001: 41; Morris *et al.* 2005: 112).

Nature conservation and agriculture

Payments to farmers from agri-environment schemes are often used to protect or enhance natural habitats by means of altered land management practices (Burgess *et al.* 2000). One of the most effective means of habitat restoration in England and Wales has been the Environmentally Sensitive Areas (ESA) scheme (EN 2000b: 35; Burgess *et al.* 2000). Payments from ESA and countryside stewardship schemes are among the most common sources of funding for habitat restoration projects such as floodplain restoration. The ESA scheme was set up in 1987 to encourage farmers to safeguard areas of high landscape, wildlife or historic value in accordance with EU regulations. There are 22 designated ESAs in England, covering 10% of the agricultural land area (Swash 1998: 6; EN 2000a: 1). Eight of these ESAs have management provisions relating to riverine floodplains or wetlands (EN 2000a). Farmers in designated ESAs enter 10-year voluntary management agreements specific to each ESA in return for an annual payment per hectare of land in the scheme. The ESA scheme is the only one which allows for large-scale restoration projects on land which is privately owned or farmed. However, its potential contribution to enhancing biodiversity on floodplains is not being fulfilled, owing to a low take-up of the higher tiers of management prescriptions, inadequate monitoring and a lack of coordination with Water Level Management Plans and flood defence agencies (EN 2000a: iii-iv, 65). For example, the uptake of the higher tiers (2 and 3) of the ESA scheme in the Somerset Levels – i.e. those involving raised water levels – was considerably lower than for the lower tiers, amounting to only 24% of the area of uptake (Hodge and McNally 2000: 111).

The Countryside Stewardship scheme has proved a more flexible and better targeted instrument than the ESA scheme (EN 2000a: iii). Its objectives are the conservation and enhancement of English landscapes and habitats, including access to them. Co-funded by the EU as part of the England Rural Development Programme and administered by the rural development services within Defra, the Countryside Stewardship scheme has proved effective in creating linear habitats in intensively farmed areas (EN 2000b: 35).[14] The weaknesses of this scheme are that it offers only relatively low levels of payment, requires considerable administration and, from a nature conservation perspective, is not weighted in favour of biodiversity targets given its multi-functionality (Ledoux *et al.* 2005: 133). In 2005 the Environmentally Sensitive Areas scheme and Countryside Stewardship scheme were combined under the new scheme Environmental Stewardship.[15] Interestingly, flood management is one of two secondary objectives of the new scheme. The environmental priorities supported under the funding category Higher Level Stewardship include the restoration of wet grassland and traditional water meadows (Defra 2005a: 8). A further incentive is the new single farm payment regime in England, in which annual income support is "decoupled" from production but requires compliance with good farming practice (Morris *et al.* 2005: 111).

6.5 FORMS OF COOPERATION, CONSULTATION AND EDUCATION

Given the multiple functions of floodplains and the cross-sectoral nature of their management, any schemes for floodplain restoration rely heavily on effective cooperation and consultation between and within the responsible organisations. Environmental policy and management in the UK is generally accredited with exhibiting a fairly high level of consultative procedures and practices relative to most other European countries (Lowe and Ward 1998; Héritier *et al.* 1996: 77-102; Smith 2001: 157-166). This can be attributed in part to the statutory and other requirements for consultation between the responsible government agencies, interest groups and the general public. These have been extended in recent years to cover not only planning procedures but also, significantly, funding programmes. In order to gain funding from many nature conservation, flood defence or land management schemes, applicants need to demonstrate they have liaised with, or gained the support of, the principal affected agencies.

Beyond the legal obligations, a shift in the policy and culture of government agencies in recent years has made them much more pro-active in consulting with other parties and seeking partners for policy development and implementation. The Environment Agency – a central actor in floodplain restoration for its multiple responsibilities – is an illuminating case in point. Constantly criticised in the past for failing to consult adequately with those affected by its actions, it

[14] Between 2000 and 2006 a total of £500 million was allocated under the Countryside Stewardship scheme to nearly 17,000 agreement holders. http://www.defra.gov.uk/erdp/schemes/css/default.htm. Accessed on 06.12.2006

[15] Defra (2005): http://www.defra.gov.uk/erdp/schemes/es/default.htm. Accessed on 21.03.2007.

is increasingly playing the role of facilitator on projects, bringing together partners such as Natural England, the Countryside Association or the Royal Society for the Protection of Birds from the outset. Internally, the main government agencies are also seeking to improve coordination and cooperation between their own departments. Defra, the Environment Agency and Natural England are all undergoing internal restructuring to facilitate greater integrated thinking on policy (Interviews 3, 4, 7 & 8). This applies to the flood defence and water resources divisions of the Environment Agency where, for instance, conservation officers and landscape architects are being used to advise flood defence engineers. Within Natural England economists are being employed to identify windows of opportunity for nature conservation emerging from the debate on the future of agriculture (Interview 3). However, there are signs that the emergent cooperative culture of central offices is not filtering down effectively to operators at the regional and local level. Whilst at the national level government agencies are becoming more process-oriented, representatives in local area offices often continue to operate according to established practices (Interview 4) or are over-burdened by new policy requirements (see the UK case studies in Chapters 9 and 12 of this volume).

6.6 PROGRAMMES AND PROJECTS TARGETED AT FLOODPLAIN RESTORATION

Floodplain restoration is still in its infancy in England and Wales (Adams and Perrow 1999: 90; Adams *et al.* 2004: 1934-5). Only very recently has interest in floodplain restoration grown beyond a small circle of experts, in response to the flooding events, giving rise to new projects involving floodplain restoration as part of an integrated strategy for flood mitigation, biodiversity enhancement and rural development. The roots of floodplain restoration in England and Wales lie in the River Restoration Project, set up in 1990 with funding from the EU LIFE programme (Boon *et al.* 1992). In the context of this project two demonstration schemes – on the River Cole and the River Skerne – were implemented to gain experience on river restoration and publicise the possibilities, primarily from a nature conservation perspective (Eden *et al.* 2000). Following completion of these projects in 1997/98 the River Restoration Centre was established as a clearing house for knowledge acquired from river restoration projects nationally and internationally.[16] A small number of these projects in England and Wales do involve the removal of flood defences to create washlands but they are still largely site-based and small-scaled (Adams *et al.* 2004: 1938; Ledoux *et al.* 2005: 130). Examples of larger-scale schemes include the Parrett Catchment

---

[16] River Restoration Centre at www.therrc.co.uk. Accessed on 21.03.2007.

Project (see below), the floodplain forest project on the Ouse at Milton Keynes[17] and the Tir Coed woodland project (Tir Coed 2001).

Opportunities for creating naturally functioning floodplains are emerging today in response to new programmes – such as the Catchment Flood Management Plans and Land Management Initiatives – and policy drivers to meet biodiversity targets and more strategic approaches to flood defence. A scoping report for the River Severn Strategy, developed within the framework of a Catchment Flood Management Plan for the Severn, provides one example. It calls for a strategic, catchment-oriented approach to flood management including as possible options the use of washlands and natural floodplains for flood storage and biodiversity enhancement (EA 2002: 17). It even considers the development of floodplain woodland, arguing the merits for achieving national biodiversity targets for wet woodland and alleviating the effects of flooding (EA 2002: 22-23). Also on the Severn is the Countryside Commission's Severn-Vyrnwy Wetland Land Management Initiative. The initiative aims to manage water more sustainably in the catchment by taking an integrative approach to the needs of agriculture, conservation and flood defence. It has developed from a project to create wet grassland as a breeding ground for wading birds into a more broad-based, cooperative exercise in new ways of managing the rural landscape (Interview 9). A catchment-based approach was also one of the distinctive novel features of the Parrett Catchment Project, launched in 2000 following a successful EU-funded "Wise Use of Floodplains" project along the River Parrett (Parrett Catchment Project 2001). The Parrett Catchment Project, replaced in April 2007 by the Somerset Water Management Partnership, was in many ways indicative of a new approach to floodplain restoration, reflecting many of the recent policy drivers in England and Wales. Apart from taking a catchment perspective on flood management, the project encouraged the active engagement of a wide range of stakeholders, addressed water level management in the context of land stewardship tasks and aspired to contributing to UK Biodiversity Action Plan targets. It is investigated in detail later in this book (see Beveridge and Moss in Chapter 12).

Schemes of this kind are being supported by a variety of research projects funded by programmes of key national agencies. One of these is the Rural Economy and Land-Use Programme supported by the UK Research Councils, currently funding a research project on the integrated management of floodplains.[18] Natural England has its own research programme in support of UK Biodiversity Action Plan targets and the research agency of the Forestry

---

[17] The Parks Trust Milton Keynes (2006): http://www.theparkstrust.com/parks-trust/DisplayArticle.asp?ID=33092. Accessed on 21.03.2007.

[18] Rural Economy and Land-Use Programme: http://www.relu.ac.uk/research/projects/SecondCall/Morris.htm. Accessed on 21.03.2007.

Commission – Forest Research – is funding research on the role of floodplain forests.[19] In January 2007, Defra announced the commencement of a series of pilot projects on flooding and coastal erosion, including two relating to floodplain restoration: Farming Floodplains for the Future, and Restoring Floodplain Woodland for Flood Alleviation.[20]

6.7 CONCLUSION: DRIVERS AND CONSTRAINTS OF FLOODPLAIN RESTORATION

This policy analysis has identified growing interest in floodplain restoration in diverse policy contexts, ranging from flood risk and water resource management to nature conservation and spatial planning. Restoring floodplains, or – to use the current jargon – giving more space to rivers, has become a priority strategic objective for many central government agencies engaged in water and/or land management. This can be attributed not only to the multiple environmental and economic benefits this can provide but also to growing recognition of the inadequacies and disadvantages of 'hard engineering' attempts to control rivers. Floodplain restoration, less than 10 years ago largely the prerogative of nature conservationists, has today become an integral part of mainstream policy for river and land management, at least for the national Government.

To conclude, the institutional drivers and constraints of floodplain restoration in England and Wales are summarised here, as identified through analysis of key policy documents, research reports and 11 expert interviews.[21] These interviewees generally agreed that the most powerful drivers and constraints affecting floodplain restoration lie in the field of flood management. This is explained by the influence of flood defence policy and practice on floodplain restoration-related issues and the scale of flood defence funding. Growing flood risk awareness following the recent flooding events in the UK has substantially accelerated a shift in approach from flood defence to flood risk management, which is creating openings for floodplain restoration, if as yet largely unexploited. Policy makers are taking a more integrated, holistic approach to flood management, looking at the connectivity between land and water across a

---

[19] Forest Research (2005): http://www.forestresearch.gov.uk/pdf/FRSBnews_05Feb.pdf/$FILE/FRSBnews_05Feb.pdf. Accessed on 21.03.2007.

[20] Defra (2007): http://www.defra.gov.uk/news/2007/070104a.htm. Accessed on 21.03.2007.

[21] See also the drivers and constraints of managed river realignment described by Ledoux *et al.* (2005: 135-138). They identify the following five "problem areas" (2005: 130): traditional institutional arrangements, the limited body of casework, the EU Habitats Directive, policy on financial compensation and current cost-benefit arrangements and patterns of funding.

whole catchment, taking a broader perspective of costs and benefits of different flood defence options, and seeking to understand how the various relevant systems work in conjunction. Growing interest in the flood retention capacity of floodplains is to be understood in this context. Nevertheless traditional attitudes and practices of flood defence – in particular the persistence of hard-engineering and site-based solutions and their funding mechanisms – are proving substantial constraints to floodplain restoration.

Biodiversity enhancement is generally seen as the second most important area of drivers of floodplain restoration. Foremost amongst these is the UK Biodiversity Action Plan with its targets for the restoration of individual habitats. This is an important factor for the recent – if, as yet, limited – rise in interest in wet grassland and woodland. Designated nature conservation areas are more significant for the protection of existing habitats rather than the creation of new ones, thereby constraining some schemes for floodplain restoration. The funding resources available for nature conservation and enhancement purposes, though, are the most frequently used instruments for a floodplain restoration project.

Whether recent re-thinking on the future of agricultural production and the reform of subsidy regimes will constitute a driver for floodplain restoration is, as yet, unclear. The farming lobby is no longer as influential over flood defence policy as it once was and is, to a relatively limited degree as yet, seeking additional sources of revenue through land stewardship and conservation enhancement schemes. Against these potential drivers for floodplain restoration must be set, however, the inadequate incentives for changes to land-use forms, the uncertainties farmers associate with government subsidies and the heavy dependence of large-scale floodplain restoration schemes on the cooperation of landowners. In the field of forestry, where government influence is strong given the large proportion of state-owned forests, a policy shift towards greater diversification is creating new interest in wet woodland to improve the biodiversity and amenity value of forests.

An important constraint to floodplain restoration remains the limited connectivity between land-use planning and flood protection. The planning guidance on development and flood risk (PPG 25 / PPS 25) marks a major step towards protecting existing functional floodplains from further encroachment from development but is unlikely to offer any significant opportunities for changing land-use forms to the benefit of floodplain creation. More general constraints on floodplain restoration include the difficulty in combining funding from a variety of different sources, the limited time scales of funding programmes and scepticism towards the long-term impacts of floodplain restoration – especially on a large scale – given the limited experience of such schemes in the UK.

The important question is whether these constraints are the main reason why the number of floodplain restoration projects successfully implemented in England and Wales still remains remarkably low. For, despite the emergence of powerful new drivers, the practice of restoring floodplains is not keeping pace with the policy rhetoric. Many responsible for promoting – or, at least, considering – floodplain restoration in the key (national) agencies claim that the shift in thinking is not "filtering down" effectively from the strategic level to the operational level. They criticise local and regional stakeholders – both within and beyond government agencies – for failing to adapt established practices and ways of thinking to meet the new policy signals. Given the permissive – rather than binding – powers of many relevant instruments for restoring floodplains in England and Wales, much does indeed depend on the discretion of key public agencies.

However, there is mounting anecdotal evidence to suggest that where local initiatives are taken to launch ambitious restoration schemes of the kind being encouraged by central government, they are running into major implementation problems. Over-complexity, it would appear, is emerging as a critical constraint on the successful implementation of new-generation floodplain restoration schemes (cf. the conclusions of Ledoux *et al.* 2005: 141). In these cases it is not that local stakeholders fail to appreciate recent shifts in government policy, but that they are not able to follow them as they – and central government – would like. The requirements for stronger consultation, larger partnerships, collaborative funding and more complex procedures of approval and appraisal are making the more ambitious and holistic restoration projects more time-consuming and more costly. Considerable effort is today required to win support in the community and amongst stakeholder groups, to accommodate the individual criteria of different funding programmes and to conduct the necessary feasibility studies prior to implementation. This is in part a product of the sheer complexity of the schemes themselves, but also of the stipulations being placed on them by central government. The concern for accountability, cost effectiveness and policy integration has generated planning and auditing procedures which are placing a huge additional administrative burden on project managers at the regional and local level. This is creating new tensions between central government policy and local or regional practice which is, ultimately, jeopardising policy implementation. This critical issue is investigated in greater depth in subsequent case studies of the Long Eau scheme and the Parrett Catchment Project (see Chapters 9 and 12).

6.8 REFERENCES

ADA (Association of Drainage Authorities) (2001) *Flood Defence and Land Drainage in England and Wales*. www.ada.org.uk. Accessed on 12.11.2001.

Adams, W.M and Perrow, M. (1999) Scientific and institutional constraints on the restoration of European floodplains. In *Floodplains. Interdisciplinary Approaches*, (eds. S. Marriott, J. Alexander and R. Hey), Geological Society, Special Publications, 163, London, pp.89-97.

Adams, W.M., Perrow, M.R. and Carpenter, A. (2004) Conservatives and champions: river managers and the river restoration discourse in the United Kingdom. *Environment and Planning A* **36**, 1929-1942.

Boon, P.J., Calow, P. and Petts, G.E. (eds.) (1992) *River Conservation and Management*, John Wiley, Chichester.

Burgess, J., Clark, J. and Harrison, C.M. (2000) Knowledges in action: an actor network analysis of a wetland agri-environment scheme. *Ecological Economics* **35**, 119-132.

DCLG (Department for Communities and Local Government) (2006): *Planning Policy Statement 25: Development and Flood Risk*, The Stationary Office, London. http://www.communities.gov.uk/pub/955/PlanningPolicyStatement25DevelopmentandFloodRisk_id1504955.pdf. Accessed on 21.03.2007.

Defra (Department for Environment, Food and Rural Affairs) (2001) *Flood and coastal defence. High level targets.* www.defra.gov.uk/environ/fcd/hltarget/hltarget.htm. Accessed on 23.04.2002.

Defra (2001) *Flooding and Coastal Erosion. Defra Grant Aid for Capital Projects.* www.defra.gov.uk/environ/fcd/policy/grantaid.htm. Accessed on 12.11.2001.

Defra (2004) *Making Space for Water: Developing a new Government strategy for flood and coastal erosion risk management in England. A consultation exercise.* Defra, London. http://www.defra.gov.uk/corporate/consult/waterspace/consultation.pdf. Accessed on 04.12.2006.

Defra (2005): *Making Space for Water: Taking forward a new Government strategy for flood & coastal erosion risk management.* http://www.defra.gov.uk/environ/fcd/policy/strategy/vision.htm. Accessed on 04.12.2006

Defra (2005a) *Environmental Stewardship. Look after your land and be rewarded.* Defra, London.

DTLR (Department for Transport, Local Government and the Regions) (2001) *Planning Policy Guidance Note 25: Development and Flood Risk.* www.planning.dtlr.gov.uk/ppg25.htm. Accessed on 15.11.2001.

EA (Environment Agency) (1998) *Environment Agency Response to the Independent Report on the Easter 1998 Floods. Action Plan.* Environment Agency, Bristol.

EA (2001) *Policy and Practice for the Protection of Floodplains.* Environment Agency, Bristol.

EA (2002) *River Severn Strategy. A Flood Management Strategy for the River Severn Corridor.* Scoping Report, January 2002. Environment Agency, Solihull.

EA (2003) *Effective Flood Management: Position Statement.* http://www.environment-agency.gov.uk/commondata/105385/effmanpos_908050.pdf. Accessed on 04.12.2006.

EA (2004) *Catchment Flood Management Plans. Volume 1 – Policy Guidance.* http://www.environment-agency.gov.uk/commondata/105385/cfmp_policy_ 891477.pdf. Accessed on 04.12.2006.

Eden, S., Tunstall, S.M. and Tapsell, S.M. (2000) Translating nature: river restoration as nature-culture. *Environment and Planning D* **18**, 257-273.

EN (English Nature) (2000a) *The Effectiveness of the Floodplain ESA Schemes in the Maintenance and Enhancement of Biodiversity.* English Nature, Peterborough.

EN (2000b) *English Nature Habitat Restoration Project.* Final report. English Nature Research Reports No.377.

EN (2001) *Action for floodplain biodiversity.* Pamplet. English Nature, Peterborough.

EN (2001) *Sustainable Flood Defence. The case for washlands*. English Nature Research Reports No.406.

Gardiner, J.L. (1998) Floodplain management in the United Kingdom. In *United Kingdom Floodplains* (eds. R.G. Bailey, P.V. José and B.R. Sherwood), Westbury, Otley, pp.17-26.

Halcrow Water (2001) *Catchment Flood Management Plans. Interim guidelines for consultation and pilot catchment studies*. Report for the EA and MAFF, March 2001. Halcrow Water, Swindon.

Halcrow, with CSERGE and CRU (2002) *Managed Realignment Review. Report to Defra and the Environment Agency*. Policy Research Project FD 2008.

Héritier, A., Knill, C. and Mingers, S. (1996) *Ringing the Changes in Europe. Regulatory Competition and the Transformation of the State. Britain, France, Germany*. Walter de Gruyter, Berlin/New York.

Hodge, I. and McNally, S. (2000) Wetland restoration, collective action and the role of water management institutions. *Ecological Economics* **35**, 107-118.

HR Wallingford (2002) *National Flood Risk Assessment 2002. Executive Summary. Report to the Environment Agency*. Report EX 4722.

Hughes, F.M.R. (2003) The flooded forest: guidance for policy makers and river managers in Europe on the restoration of floodplain forests. FLOBAR 2, Department of Geography, University of Cambridge, Cambridge.

ICE (Institution of Civil Engineers) (2001) *Learning to Live with Rivers. Final report of the Institution of Civil Engineers' presidential commission to review the technical aspects of flood risk management in England and Wales*. Institution of Civil Engineers, London.

Kerr, G. and Nisbet, T.R. (1996) *The Restoration of Floodplain Woodlands in Lowland Britain. A scoping study and recommendations for research*. Environment Agency, R&D Technical Report W15, Bristol.

Ledoux, L., Cornell, S., O'Riordan, T., Harvey, R. and Banyard, L. (2005) Towards sustainable flood and coastal management: identifying drivers of, and obstacles to, managed realignment. *Land Use Policy* **22**, 129-144.

LGA (Local Government Association) (2001) *LGA Position Paper on Flood Defence*. www.lga.gov.uk/lga/executives/lga-executive/13_09_01/item9.pdf. Accessed on 23.04.2002.

Lowe, P. and Ward, S. (1998) *British Environmental Policy and Europe*, Routledge, London.

MAFF (Ministry of Agriculture, Fisheries and Food) and WO (The Welsh Office) (1993) *Strategy for Flood and Coastal Defence in England and Wales*. MAFF, London.

Morris, J., Gowing, D.J.G., Mills, J. and Dunderdale, J.A.L. (2000) Reconciling agricultural economic and environmental objectives: the case of recreating wetlands in the Fenland area of eastern England. *Agriculture, Ecosystems and Environment* **79**, 245-257.

Morris, J., Vivash, R.M., Alsop, D., Lawson, C.S., Leeds-Harrison, P.B. and Bailey, A.P. (2002) *Economic Basis and Practicalities of Washland Creation on the Somerset Levels and Moors. Report to Wise Use of Floodplains Project*, Cranfield University, Silsoe, Bedford. www.floodplains.org.uk/pdf/somerset/Washlnads%20Report.pdf. Accessed on 04.12.2006.

Morris, J., Hess, T.M., Gowing, D.J.G., Leeds-Harrison, P.B., Bannister, N., Vivash, R.M.N. and Wade, M. (2005) A framework for integrating flood defence and biodiversity in washlands in England. *International Journal of River Basin Management* **3**(2), 105-115.

Newson, M.D. (1992) River Conservation and Catchment Management: A UK Perspective. In *River Conservation and Management* (eds.. P.J. Boon, P. Calow and G.E. Petts), John Wiley, Chichester, pp.385-396.

NAO (National Audit Office) (2001) *Inland Flood Defence*. The Stationary Office, London.

O'Riordan, T. (2002) On evaluating possible new management arrangements for coastal defence. In *Redesigning the Coast: A Report of a Workshop*, (ed. T. O'Riordan), CSERGE Working Paper PA 02-01, University of East Anglia, Norwich.

Parrett Catchment Project (2001) *The Parrett Catchment Project – An Action Strategy for Integrated Catchment Management*. Report of 17 January 2001.

Petts, G.E. (1998) Floodplain rivers and their restoration: A European perspective. In *United Kingdom Floodplains,* (eds. R.G. Bailey, P.V. José and B.R. Sherwood), Westbury, Otley, pp. 29-41.

RRP (River Restoration Project) (1994) *Institutional Aspects of River Restoration in the UK*, RRP, Huntingdon.

Slater, S., Marvin, S. and Newson, M.D. (1994): Land use planning and the water sector. *Town Planning Review* **64**(4), 375-397.

Smith, R. (2001) The Development of Environmental Policy in Britain. In *European Union Environment Policy and New Forms of Governance* (eds. H. Heinelt, T. Malek, R. Smith and A.E Töller), Ashgate, Aldershot, pp. 157-166.

Swash, A.R.H. (1998) The environmental aspects of riverine floodplain management: What is MAFF's involvement? In *United Kingdom Floodplains,* (eds. R.G. Bailey, P.V. José and B.R. Sherwood), Westbury, Otley, pp. 3-9.

Tir Coed (2001) *Tir Coed. Woodland, Water & Flooding*. Expert workshop held at the National Assembly for Wales, 5th July 2001.

UK Biodiversity Action Plan (1994) *Habitat Action Plan – Wet Woodland.* www.ukbap.org.uk/plans/habitats/NBNSYS0000004607.htm. Accessed on 14.02.2002.

Walker, S., Sargent, B. and Waterworth, J. (2000) Responsibilities and Strategies of UK Organisations. In *The Hydrology of the UK: a Study of Change* (ed. M. Ackermann), Routledge, London, pp.227-243.

LAWS

Water Resources Act of 1991, accessible under http://www.opsi.gov.uk/ACTS/acts1991/Ukpga_19910057_en_1.htm. Accessed on 24.07.2007.

Land Drainage Act of 1991, accessible under http://www.opsi.gov.uk/ACTS/acts1991/Ukpga_19910059_en_1.htm. Accessed on 24.07.2007.

Environment Act of 1995, accessible under http://www.opsi.gov.uk/acts/acts1995/Ukpga_19950025_en_1.htm. Accessed on 24.07.2007.

Town and Country Planning Act of 1990, accessible under http://www.opsi.gov.uk/ACTS/acts1990/Ukpga_19900008_en_1.htm. Accessed on 24.07.2007.

INTERVIEWS

Interview 1: Project manager at River Restoration Centre, 27 November 2001.

Interview 2: Chief Executive of Association of Drainage Authorities, 30 November 2001.

Interview 3: Senior officer at English Nature, 3 December 2001.

Interview 4: Coordinator at Department for Environment, Food and Rural Affairs, Flood Management Division, 4 December 2001.

Interview 5: Flood defence officer at Environment Agency, Anglian Region, 4 December 2001.

Interview 6: Woodland ecologist, forestry policy officer and rivers ecologist at Countryside Council for Wales, 5 December 2001.

Interview 7: Strategic Environmental Assessment coordinator at Environment Agency, Midlands Region, 6 December 2001.

Interview 8: Flood defence officer at Environment Agency, North West Region, 7 December 2001.

Interview 9: Severn-Vyrnwy Project Officer at Countryside Agency, 10 December 2001.

Interview 10: Lecturer at Oxford Brookes University, 11 December 2001.

Interview 11: Senior officer at World Conservation Monitoring Centre, 13 December 2001.

Part C

EARLY GENERATION SCHEMES OF FLOODPLAIN RESTORATION

7

The restoration of a floodplain on the Upper Rhine: managing the interface of large-scale policy and small-scale implementation

Sylvia Kruse

7.1 INTRODUCTION

Alluvial floodplains did not play a significant role in science or society until the loss of floodplains and retention areas along rivers was linked to a decline in the biodiversity of riverine ecosystems, and to rising flood risk and vulnerability from extreme flood events. The Rhine has lost more than 80% of its natural retention areas within the last centuries due to intensive and multiple usage of the river basin. Since the 1980s the restoration of floodplains along the Rhine has arrived on the political agenda.

This case study analyses the restoration process of the Rheinaue floodplain near Rastatt as an example of an early generation floodplain restoration (FPR) scheme. A principal characteristic of natural alluvial ecosystems is the connectivity of the water bodies between the main river and its backwaters. The decisive habitat conditions of the alluvial area can only be maintained if there is a functioning water discharge (cf. Tittzier and Krebs 1996). Before the introduction of floodplain restoration in the "Rheinvorland Süd" project area, the alluvial ecosystem was highly influenced by anthropogenic changes in the water regime of the River Rhine. The water discharge conditions in particular had been disrupted substantially by technical structures such as dykes, barrages, roads etc. The main objective of the "Rheinvorland Süd" project was to improve the ecological and hydrological conditions in the Rheinaue floodplain. The main instrument was a catalogue of interventions that were funded by ecological compensatory measures.

The project was first initiated in 1987 and the planning and implementation phase began in 1993. The analysis of almost twenty years of experience in floodplain restoration in the "Rheinvorland Süd" project highlights both the influence of general policy drivers and constraints as well as specific emblematic issues which suggest transferable recommendations for floodplain restoration.

The present case study is based on the analysis of project-specific documents (e.g. contractual agreements, management plans and evaluation schemes) as well as oral expert interviews with representatives of the actor groups involved.[1]

7.2 THE PHYSICAL CONTEXT AND USE OF THE RIVER

7.2.1 Physical characteristics

The Rhine is the largest river in Central Europe. From its source in the Swiss Alps to its mouth in the North Sea, the river basin of the Rhine comprises over 185,000 km^2 spread over a length of 1,320 km (BMU 1998). The majority of the catchment area lies in Germany (100,000 km^2) and the rest is spread between France, Switzerland and the Netherlands (20,000 to 30,000 km^2 each), Austria and Luxemburg (2,500 km^2 each), and Italy, Liechtenstein and Belgium with very small shares. The river basin can be divided into six parts: the Alpine Rhine, the Higher Rhine, the Upper Rhine, the Middle Rhine, the Lower Rhine and the Rhine Estuary.

---

[1] This contribution is based on the Case Study Report "Rheinvorland Süd" which was compiled by Manuela Siener in the EU-project "Floodplain Biodiversity and Restoration" (FLOBAR2). Further thanks go to Wolfram Grönitz (Karlsruhe) and Jochen Monstadt who contributed many helpful remarks to the chapter.

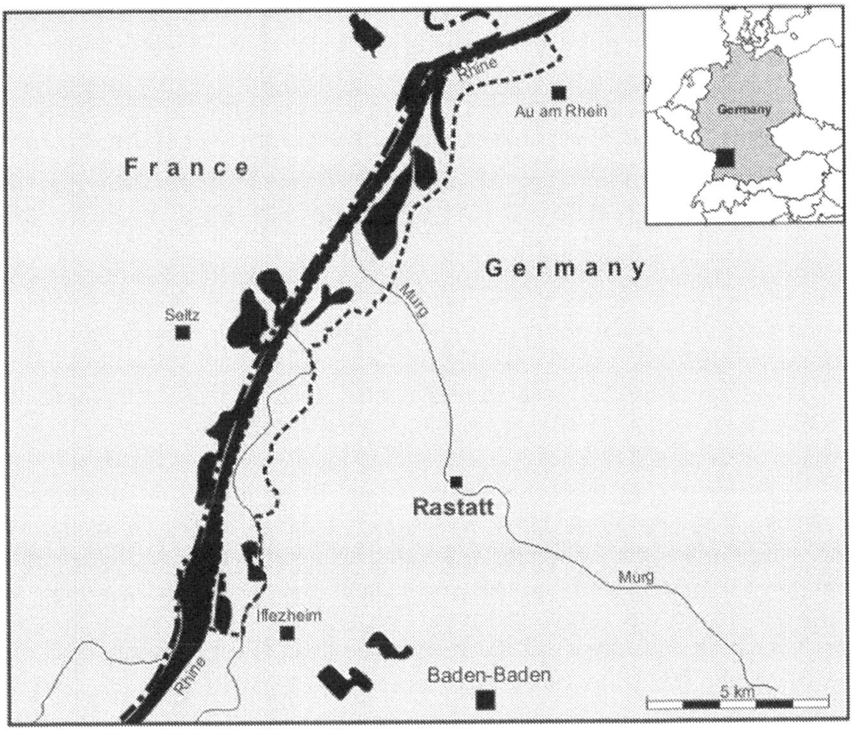

Figure 7.1: Map locating the Rheinvorland-Süd project area.

The "Rheinvorland Süd" project area is located in the Upper Rhine (see Figure 7.1); the latter covers 400 km from Basel to the city of Mainz. The climate in the Upper Rhine area is mild and characterized by an average temperature of 18°C in summer and 0°C in winter and a high amount of rainfall. Typically, the high water level is attained between May and August, although there can be variances in the late winter when widespread rainfalls within the catchment area concur with an early thaw period. The coinciding of a high flow of the River Rhine with extensive rainfalls in the catchments of its tributaries typically leads to extreme flood events along the Upper Rhine (cf. Tittzier and Krebs 1996).

7.2.2 Use and management of the river

The river basin of the Rhine has a long history of various forms of land use and has been subject to management and structural modification for several centuries. Over time, the river and its floodplains have been used for settlements, forestry, fishing, agricultural and industrial purposes, drinking water and waste water disposal and the shipping industry. Whereas deforestation had already taken place on a significant scale in the Middle Ages, the meandering and furcating riverbed has only been regulated systematically since the beginning of the 19th Century (cf. Tittzier and Krebs 1996). Under the large-scale Rhine correction measures headed by Colonel Tulla of Baden, the river banks of the Upper Rhine were reinforced with dykes, meanders were transfixed and furcations were technically regulated by barrages (cf. Löbert 1997). The main aim was to prevent existing settlements and land use from regular flooding and to reclaim new land for new settlements. At the beginning of the 20th Century the systematic development of the Upper Rhine was continued with the construction of several barrages along the Upper Rhine for power generation (Bernhardt 2003).

The consequences for flood protection were ambivalent as Tulla's measures improved the protection of the Upper Rhine area in some cases but had serious effects on the Middle Rhine area. When taken together with development in the 20th Century, the consequences for the Upper Rhine area were far-reaching: retention areas and alluvial plains were heavily reduced, the river flow speeded up, and erosion and sedimentation became a major problem as the riverbed deepened and sedimentation hindered shipping as well as water discharge in high water situations. The course of the river was shortened, so that in cases of high water levels the peaks of the flood of the River Rhine coincided with the peaks of the tributaries and caused extreme flood events along the Middle Rhine. Today only 6% of natural habitat conditions have been maintained in the Upper Rhine area (Galusser and Schenker 1992).

7.3 THE POLICY CONTEXT

7.3.1 National policy context

A change in the national policy context of river basin management (RBM) was precipitated by increasing pollution, chemical accidents and rising flood risks as well as extreme flood events in the late 1980s and during the 1990s. As a consequence, national legislation has been reformed and floodplains as well as flood protection are today more frequently referred to in the Federal Water Act, regional planning legislation and the Federal Nature Conservation Act (cf. Monstadt and Moss in Chapter 4 of this volume). Apart from national legislation, a catchment-oriented approach, and therefore transnational agreements, became more important for the general policy context in Germany. One example is the INTERREG Community Initiative launched in 1990 aimed at cross-border cooperation in the area of regional planning, in particular in the management of water resources. In the late 1990s initiatives under the Habitats

Directive and the EU Water Framework Directive (WFD) had direct and indirect influences on RBM and floodplain restoration activities in Germany (cf. Dworak and Görlach 2005; Maltby and Blackwell 2005). For the project "Rheinvorland Süd", though, changes in the national and transnational policy context only influenced a later phase of the project, as it had been initiated in 1987 and started in 1993 (cf. Section 7.4).

7.3.2 Initiatives in the catchment

Although many initiatives concerning the national policy context only affected the later project phase, various initiatives in the catchment of the River Rhine influenced the initiation, implementation and progress of the project's activities.

Growing flood risk awareness

In 1968 the "International Flood Study Commission of the River Rhine" was set up to examine the impact of the systematic development of the Upper Rhine area. The findings showed that, especially since the increasing canalization and the construction of the barrages between Basel and Iffezheim, the potential for extreme flood events had risen massively. Thus France and Germany agreed in a bi-national contract in 1982 to restore the level of protection against flooding that had existed prior to the systematic development begun in 1955. The proposals listed included several retention measures for flood control with an overall capacity of 212 million m^3. In the planning phase, a number of technical and non-technical problems emerged. For example, the environmental impact assessment showed that the construction and operation of the polders and weirs would cause major impacts on the alluvial ecosystems. The state of Baden-Württemberg therefore agreed to the Integrated Rhine Programme (IRP) in 1988. The IRP pursues two major objectives: the re-creation of technical flood protection and the restoration of the ecological state of the alluvial plains along the Upper Rhine (cf. Oberrheinagentur 1995 & 1996; Plate 2000). Whereas France fulfilled its obligations within the appointed timetable, the IRP in Germany is still behind schedule; so far, only three polders have been built.

Rising concern for water quality and flood risk management

Another transnational initiative in the catchment aims to improve the water quality of the River Rhine. As a result of increasing water pollution and the accident at the Sandoz chemical company in Basel in 1987, the International Commission for the Protection of the Rhine (ICPR) was founded (cf. Huisman *et al.* 2000). The transnational commission agreed upon the Rhine Action Programme, which aimed at improving the biological and hydrological conditions of the river (IKSR 1987). In 1989 alluvial plains were included in the ecological master plan (IKSR 1991). Only after the flood events of 1993 and

1995 was an Action Plan on Floods adopted by the ICPR in 1998. Moreover, hazard and risk maps were published in the so-called Rhine Atlas in 2001 showing the extent and the inundation depth for different flood risk levels (ICPR 2001). The Rhine Atlas, with information about the existing flood hazard, is accessible to the public through the Internet (Cf. http://www.iksr.org).

However, water quality *and* flood protection issues have only recently been summarized in the objectives of the ICPR in an integrated way and connected to the transnational activities of the WFD. The restoration of floodplains plays a central role in the programme "Rhine 2020" (IKSR 2001; Disse and Engel 2001).

Local initiatives

Parallel to the early phases of the IRP, the World Wide Fund for Nature and Floodplain Institute (WWF-Aueninstitut) carried out research activities on Floodplain restoration in 1987 and 1988. These two studies were commissioned by the Regional Council of Karlsruhe (Regierungspräsidium Karlsruhe) in order to analyse the possibilities of improving the ecological and hydrological conditions in the Rheinaue floodplain between Rastatt-Wintersdorf and Au/Rhein (WWF Aueninstitut 1992). In this context the WWF also carried out several pilot measures to demonstrate the positive effects of floodplain restoration activities on the specific hydrological and ecological conditions in the Rheinaue. Apart from the case study "Rheinvorland Süd", which was directly initiated by these research activities, in the following years local and state authorities, environmental associations and projects funded by the EU or the IRP implemented several floodplain restoration activities in the Upper Rhine area (cf. Jupe *et al.* 1999; Regierungspräsidium Karlsruhe 1996).

7.4 BRIEF DESCRIPTION OF THE PROJECT

The "Rheinvorland Süd" project is located near the city of Rastatt in the federal state of Baden-Württemberg in south-west Germany. The "Rastatter Rheinaue" nature reserve is part of the project area. The project activities cover an area of approximately 660 hectares between the barrage at Iffezheim and the municipality of Au (Rhine) (from km 334 to km 354) (see Figure 7.1). Around 17% of the project area consists of aquatic biotopes; only a few hectares are used for agricultural production (WWF Aueninstitut 1992: 5 ff.). The rest of the area is used for forestry and is mainly the property of the municipalities.

The "Rheinvorland Süd" project area is the southern part of a broader initiative of the Land of Baden-Württemberg, the Regional Council of Karlsruhe and the World Wide Fund for Nature to improve the hydrological and ecological conditions within the Rheinaue between the barrage at Iffezheim and the city of Mannheim (cf. Fischer 2005).

7.4.1 Purpose and objectives

The project's major objectives concentrate on hydrological and ecological issues within the project area. The main aim is to improve the ecological and hydrological conditions within the alluvial area of the Rhine. The water movements within the alluvial floodplains have been dammed up by constructional flood protection measures like dams, barrages etc. and disturbed by the sinking groundwater levels as a result of the increasing deepening of the riverbed. At the same time sedimentation developed within the alluvial plains, which led to increasing siltation. As a consequence, flora and fauna were significantly modified.

Therefore the aim was to develop and implement a catalogue of measures to improve the discharge rate within the floodplains. These included reconnecting dead arms with the main river, improving the permeability of roads and footpaths and lowering existing dams. Improving the ecological state of vegetation by (re-)planting endemic and adapted tree species was not an explicit goal of the project but was implemented in parallel by the forestry office responsible.

7.4.2 Origins and development

The first initiatives with regard to the "Rheinvorland Süd" project began in the late 1980s, whereas the main phase took place between 1993 and 2003. The initiation phase of the project was influenced by three parallel activities.

The first was headed by the WWF Floodplain Institute that had carried out research activities in 1987/88 on possibilities for improving the hydrological and ecological conditions in the alluvial meadows and forests within the Rheinaue floodplain. These pilot studies were commissioned by the Regional Council of Karlsruhe and the federal state of Baden-Württemberg, and later on became the basis of the "Rheinvorland Süd" project.

The second impulse for the project was a financial one based on the "Rastatt Declaration" of 1987. This contract pooled finances for compensating measures linked with the construction of a new plant of the then Daimler Benz AG in a fund for restoration measures that are implemented within the alluvial area of the Rheinaue floodplain. The government of Baden-Württemberg committed itself to provide DM 30 to 40 million (€15 to 20 million) for the implementation of ecological compensatory measures between Lahr and Karlsruhe, which includes the project area.[2]

The third initiative relating to the project was the Integrated Rhine Programme, which was adopted by the Land of Baden-Württemberg in 1988 (cf. Section 7.3.2).

In June 1992 a working group at the Regional Council of Karlsruhe in charge of the improvement of the discharge conditions within the floodplain along the Rhine decided to establish the project group "Rheinvorland Süd – Improvement of the Hydrological and Ecological Conditions within the Rheinaue". This new sub-group consisted of representatives of different actor groups: the Regional Council of Karlsruhe (Regierungspräsidium Karlsruhe), the Karlsruhe Agency of Water Management (Amt für Wasserwirtschaft und Bodenschutz Karlsruhe), the Baden-Württemberg State Office for the Environment[3] (Landesamt für Umweltschutz), the Forestry Office of Karlsruhe (Forstdirektion Karlsruhe), the District Office of Rastatt (Landratsamt Rastatt) and the WWF Floodplain Institute. This group, together with two engineering consultancies, conducted the detailed examinations necessary to implement restoration measures. The first meeting was held in October 1993 and from then on the project group met regularly two to four times a year.

Based on the preparatory studies by the WWF the project group developed a specific catalogue of 69 restoration measures, which were prioritised for implementation (cf. Section 7.4.3). In the early years the project group was very successful in planning and implementing measures to improve the alluvial area. The first licensing procedures for 12 measures started as early as 1994; 21 measures were completed in 1995. In the following years it became more difficult and time-consuming to obtain the funds from the Land of Baden-Württemberg and for some measures approval was not granted. So far, 49 measures from the original catalogue have been implemented, 3 are at the planning stage and 17 measures have been cancelled. While the work of the project group became less intense for the project area "Rheinvorland Süd", activities in other areas in the region intensified when financial sources were found (e.g. funding from the EU for the northern part of the Rheinaue floodplain).

---

[2] The fact that the state government paid these funds is extremely unusual. The "Rastatt Declaration" was the minimum consensus upon which the environmental associations and the state government could agree, and the state government decided to pay the compensatory measures in order to ensure that Daimler-Benz would set up the new plant in Rastatt instead of looking for another site.

[3] In 2006 it was transformed into the State Institute for Environment Measurements and Nature Conservation Baden-Wuerttemberg.

Monitoring the ecological and hydrological effect of the measures implemented did not play any role in the early phase of the project. Only in recent years has the Baden-Württemberg State Office for the Environment begun to analyse the ecological effects within the project area (2001-2003). The evaluation has been restricted by the limited database on the habitats before the restoration measures. Nevertheless, monitoring efforts for patches with adequate data have, for example, shown that the permeability of some of the dead arms rose from 45 to 365 days per year after the measures were introduced. As a consequence different sizes of substrate are to be found once again, indicating successful restoration of the water body. Also the number of fish species and the size of their populations have increased significantly (cf. Fischer 2005: 317 ff.).

The budget for the estimated measures was set at DM 13 million (about €7.5 million) for the period 1993-2001. The funds originated from the Daimler-Benz Fund as well as from other sources, such as fishery taxes or compensation measures from a gravel pit within the Rastatt floodplain. According to the interviewees it became more difficult to get funding for ecological floodplain restoration measures for the project area, as the Daimler-Benz Fund was also dedicated to other projects and after the extreme flood events of 1993, 1995 and 1998 there was a tendency to implement more technical flood protection measures.

7.4.3 Physical interventions

In order to improve the discharge conditions within the "Rheinvorland Süd" project area, the measures that were taken attempted to reconnect the alluvial water bodies to the main regime of the river. Four types of intervention were carried out in order to reactivate those parts of the area that were only connected to the main regime in times of very high water levels (cf. Fischer 2005: 313 ff.; Regierungspräsidium Karlsruhe 1996):[4]

- First, footpaths and roads were lowered and fords were constructed to enable discharge into cut-off meanders and bayous even at low water levels. Such measures had already been carried out by the WWF Floodplain Institute as demonstration projects in the late 1980s and early 1990s. This type of intervention has been one of the most frequently implemented;
- Secondly, several new passages and ducts have been built, or existing ones have been widened, to allow the water to enter the alluvial areas more easily. This kind of intervention has been the second most frequent form of implementation;

---

[4] More details of the implemented measures can be found in publications by the authorities responsible, which relate to similar projects (Jupé *et al.* 1999; Grönitz 2005).

- A third type of intervention has been the building of new bridges, when footpaths had to be kept passable even at higher water levels. Due to high construction and maintenance costs this type of intervention was rarely chosen;
- The removal of artificial structures and barriers in the project area is the fourth type of intervention used in this project catalogue. This kind of intervention is possible when the respective road or barrier is no longer needed. In one case the plan to remove a public road crossing the project area could not be implemented until additional flood retention areas within the IRP had been established.

With the given time and financial resources it was impossible to implement the relocation of dykes although the project group sympathized with this measure. The implementation would have been too complex and time-consuming for the working group. Especially after the extreme flood events of 1993, 1995 and 1998 there was a political preference for implementing controllable retention areas (e.g. polders), rather than creating additional flooding areas by relocating dykes.

7.4.4 Organisation and actor involvement

While the Regional Council of Karlsruhe officially headed the project, the "Rheinvorland Süd" project group coordinated and conducted the decision-making processes for the project. As already mentioned above (cf. Section 7.4.2), it consisted of representatives of all the government authorities relevant to the planning and implementation process of the specific measures. Furthermore, the Water Authority (Gewässerdirektion)[5] became an important actor since it was in charge of implementing the measures agreed on by the project group.

In the early phase of the project the main task of the working group was to prioritize the specific measures proposed by the WWF. On the basis of this catalogue of measures they elaborated detailed implementation plans for each one. In this phase, the group tried to gain planning permission for up to ten measures in advance so that construction work could be carried out as soon as money was available. In most cases the office authorizing permission was the District Office of Rastatt. Due to the involvement of representatives of all the relevant administrations as well as the nature conservation and environmental protection associations throughout the planning process, the authorizing procedure was normally conducted very quickly. The concerns and objections of the various actors and interest groups were generally discussed and clarified

[5] The Water Authority for the southern part of the Upper Rhine, formerly Upper Rhine Agency, was integrated into the Regional Council of Karlsruhe in 2003.

during the project group meetings and – perhaps even more importantly – in bilateral and multilateral informal meetings. These informal meetings were usually arranged when one or more members of the project group had concerns about the design or potential consequences of a certain measure. When these concerns could not be removed, the group either tried to adapt the measure's design or delayed or abandoned the measure as a whole. This intense preparatory phase ensured that by the time a measure had reached approval stage any conflicts had usually already been resolved.

7.4.5 Instruments used

In the different phases of the "Rheinvorland Süd" project various instruments were used on different operational levels. On the level of the planning process one important instrument for the funding of the project catalogue was the connection of compensating measures[6] to the project objectives. In the main, funds for the compensating measures for the Daimler-Benz factory facilitated the financing of the project catalogue. However, compensating measures for other building activities in the region were also implemented within the project area.

One legal instrument used during the project work was planning permission from the Rastatt Environmental Agency which, as the local water authority, had to grant permission for every single measure. Planning permission was generally given quickly because of the preparatory work carried out by the working group. Due to the limited scope of the single measures, broad approval procedures for the declaration of the plan have not been necessary so far. The members of the project group regarded this as a positive effect of the project design, especially when bearing in mind that planning approval procedures are much more time-consuming and cost-intensive than the less complex permission procedure for small interventions in the watercourses.

In terms of nature conservation, several protection areas within the larger region around Rastatt were established or extended in order to support the aim of ameliorating the run-off conditions in the Rheinaue floodplain.

Another group of supporting methods important for the development of the project are tied to the informal level of communication and social interaction. The planning and implementation process was facilitated by the interdisciplinary and intersectoral constitution of the project group (cf. Section 7.6.2). Regular meetings and bilateral communication helped to create a constructive atmosphere between the members of the project group, so that even with conflicting views the group managed to arrive at decisions everyone was able to support. Municipalities, which are the owners of the land on which the measures were implemented, were

[6] Section 8 of the German Nature Protection Act covers impact regulation. It requires that interventions causing environmental impacts on nature or landscape that cannot be avoided need to be compensated, offset or redressed through substitution.

informed during the planning process but mostly played no active role in the decision-making processes. The participation and information of a broader public was only pursued in a later phase of the project.

7.5 PRINCIPAL DRIVERS AND CONSTRAINTS OF FLOODPLAIN RESTORATION ON THE PROJECT

The following section will focus on the principal drivers and constraints of the "Rheinvorland Süd" project. The leading question is which drivers and constraints became most relevant over the course of the project and how they relate to the characteristic drivers and constraints of floodplain restoration (FPR) in Germany (cf. Monstadt and Moss in Chapter 4 of this volume).

7.5.1 Drivers

Flood protection

Flood protection can be seen as a principal driver although the main aim of the "Rheinvorland Süd" project is not flood protection but the improvement of the hydrological and ecological conditions via the implementation of compensating measures – one of the strongest instruments of nature conservation. Nevertheless, growing flood awareness and a growing interest in the flood retention capacity of floodplains formed the background of almost twenty years of planning and implementation of floodplain restoration measures in the alluvial plains of Rastatt. The implementation of the Integrated Rhine Programme (IRP) in Baden-Württemberg in 1988 was an expression of the tendency to change from purely technical flood protection to the regional coordination of technical protection and both ecological conservation and the restoration of the alluvial plains as areas of natural water storage (cf. Section 7.3). Nevertheless, the "Rheinvorland Süd" project operated largely independently (especially financially) of the implementation of the IRP.

The extreme flood events in 1993, 1995 and 1998 along the Rhine and its tributaries raised public awareness of flood risks. However, it was only in 1998 that the International Commission for the Protection of the Rhine (ICPR) passed the Action Plan on Floods and thus adopted a catchment-based approach not only to the ecological development of the Rhine but also to flood protection issues. Since then the motto "give more space to the Rhine" and the aim of restoring floodplains has become part of the catchment-oriented international activities of the ICPR. Also, the EU initiated supporting programmes for integrated flood protection concepts (e.g. INTERREG and LIFE programmes).

For the "Rheinvorland Süd" project this international policy shift along the catchment of the Rhine played only a secondary role, as the project's design was already established before the major international change in flood risk awareness and interest in catchment-based approaches to water management. Its development as a pilot project had tended to be site-oriented – initiated more out of local activities to restore the alluvial ecosystems than to contribute to flood protection strategies. More important than the policy shift in flood protection were local issues, industrial development and access to financial resources for flood restoration (cf. Section 7.6).

Water management

Comparable to the policy change on flood protection, recent developments in water management, e.g. the EU Water Framework Directive (WFD), as well as the link between flood protection and water management in the Federal Water Act and the subsequent reforms in the State Water Acts (cf. Monstadt and Moss in Chapter 4 of this volume) were introduced only when the project had already reached an advanced phase of implementation. Therefore the recent developments for the more extensive protection and restoration of floodplains in water management legislation only show an impact on the follow-up projects in the Upper Rhine region.

In substance, the project has high relevance for the water flow regime at the local level and aims at improving the condition of the waterbodies' structure. The high interest shown by the different local and regional water management, nature conservation and forestry actors involved in the restoration of the water quality function of floodplains at the local level was an important driver for the project. Therefore the evaluation of the "Rheinvorland Süd" project can provide valuable suggestions for appropriate measures at the local level for restoring waterbodies and water quality in floodplain ecosystems.[7]

Nature conservation

In the policy field of nature conservation the designation of protected areas proved to be very relevant for the "Rheinvorland Süd" project. With the declaration in 1984 of the "Rastatter Rheinaue" nature reserve, which is part of the project area, the city of Rastatt had already shown its great willingness to protect the alluvial areas along the Rhine. Further designations of smaller nature reserves in the Upper Rhine region reinforced this tendency. The pursuit of biodiversity targets outlined in the Rhine Action Programme of the ICPR in 1987 was another main driver for the initiation and design of the project, as was the prospect of endangered species, e.g. salmon, as a lead species, becoming

[7] On the relevance of the WFD for floodplain restoration cf. Bohn 2004, Korn *et al.* 2005.

endemic again (cf. IKSR 1987). There was also high interest in natural floodplain forests, but the main focus of the project was on the discharge conditions and technical measures.

The important role of nature conservation as a driving policy priority became relevant with regard not only to the project's objectives but also to the project's funding. The greater part of the funds originate from compensatory measures as an instrument of nature conservation (cf. Sections 7.4.5 and 7.6.1).

Forestry

Forestry played a more important role than agriculture since the forestry authorities are the main landowners and land users in the project area. One aim of the forestry authority involved in the project group was to enhance the diversification of the existing alluvial woodlands and to increase the natural riparian habitat. The primarily mono-cultural forest was dominated by hybrid poplars. From the beginning of the "Rheinvorland Süd" project, which aimed at the improvement of discharge conditions and therefore impacted on the habitat conditions rather than on the habitat itself, the forest was supplemented with other tree species (e.g. ash trees, oaks, silver willows). The strategy was to establish a mixed population within the floodplains over the next ten to fifteen years. Parts of the hybrid poplar population were thus successively removed as soon as the hardwood stock had grown, which will lead (with the new planting) to higher biodiversity on the floodplains and to vegetation appropriate to the location. Hybrid poplar, which is much more resistant to flooding than the hardwood species, was not, however, taken out completely. This helped to ensure that in the case of an extreme flood event the municipalities, as the owners of the forests, will not suffer a complete write-off. Thanks to this argument, the forest wardens were able to persuade the municipalities to support this reinsurance strategy (Staatliche Forstämter Karlsruhe und Rastatt 1996). With this strategy, a balance between economic and ecological interests was created and the common conflict of interest between forestry and floodplain restoration was reduced to a minimum.[8] The main reason for this supportive situation was not so much the policy context as the personal commitment of the head of the forestry directorate of Rastatt, who was at the same time the commissioner for nature conservation (cf. Section 7.6.2).

---

[8] The EU project "FOWARA – Problems in the realisation of forested water retention areas" (2003-2005) analysed specific conflictual interdependencies between forestry and floodplain restoration along the Upper Rhine (www.fowara.org).

Spatial planning

Regional planning legislation aiming at the spatial coordination of multiple uses (e.g. nature conservation, industrial use, agriculture) within floodplains did not play a major role in the case of "Rheinvorland Süd" as the project area lies within a nature reserve. Nevertheless the assessment of the environmental impact of land use and the compensating measures as planning instruments played an important role in the initiation and implementation process of the project, since the funds derived from the environmental compensation for the construction of the Daimler-Benz factory (cf. Section 7.6.1).

7.5.2 Constraints

Floodplains generally provide multiple functions for human use as well as for natural habitats. Affecting a wide range of stakeholders, floodplain restoration is potentially highly controversial (cf. Moss and Monstadt in Chapter 1 of this volume). In the case of "Rheinvorland Süd" this common constraint did not play a crucial role as part of the project area was already designated as a nature reserve where land use was strictly regulated.

Therefore, the usual objections to floodplain restoration from the farming and forestry community were minimised, as the relevant actors practised close linkage between land-use management and nature conservation. Objections came from the local population, which used the floodplains for recreation and fishing, as free access to fishing areas and walks was temporarily restricted by construction activities and increasing water levels. Since the project group tried to find ways of negotiating or mediating between the opposing parties, they rarely faced resistance on the part of the local municipalities or population.

Another constraint linked to controversial usage patterns in floodplain areas was that for shipping reasons the restoration of the discharge conditions for the backwater was not allowed in some places in order to prevent the main river from being drained in times of low water levels. Therefore permission for the lowering of dykes and barrages connecting the backwater with the main river in order to attain discharge conditions throughout the year and achieve good ecological and hydrological conditions was not granted by the shipping authority (cf. Fischer 2005: 317).

Another constraint commonly linked to floodplain restoration is the persistence of technical flood defence. Although flood defence was only an indirect aspect of the objectives of the "Rheinvorland Süd" project, and was instead pursued in the broader context of the Integrated Rhine Programme (IRP), in the Upper Rhine area a "technical turn" is named by the interviewees as a distinct obstacle during the implementation process. This "technical turn" took place parallel to the extreme flood events along the Rhine in 1993, 1995 and 1998. Through these events flood protection and catchment-oriented coordination became more important on the one hand. But on the other hand the relationship between technical protection measures and ecological improvement of the natural retention areas embedded in the IRP shifted to an imbalance. Politically, priority was placed increasingly on the implementation of technical protection measures and the construction of polders as controllable retention areas rather than on the ecological recovery of natural alluvial habitat conditions. For the project activities this meant that financial support became more difficult to obtain in the advanced implementation phase. Nevertheless, some interviewees hinted that even if technical measures were implemented more often, there would always be some money left for introducing ecological measures as well:

> "The technical measures certainly help us towards restoring the alluvial meadows. We just have to make sure that the engineering does not become an end in itself but is embedded in the ecological idea" (Interview 7).

Finally, the high conflict potential associated with the creation of large polders coupled with a rather low political priority in Baden-Württemberg led to the situation that, so far, only three of the thirteen contractual flood retention areas agreed on in the IRP have been implemented.

Territorial and sectoral fragmentation played only a partially restrictive role in the case study. The measures implemented were conceptualized only for the project area and followed a single-issue scheme; potential benefits for other policy areas were not taken into account. The aim was to quickly implement small-scale measures that did not include complex licensing procedures. In the main, a conflict-free path was followed which meant that large-scale measures with the biggest anticipated influence on restoration were not implemented, though they were closest to some of the actors' hearts. Nevertheless, the project was set up in the context of other restoration measures in the Upper Rhine region (e.g. "Rheinvorland Nord") and the planned retention areas of the IRP, even if there was no complete concept for the catchment as a whole.

7.6 SELECTED EMBLEMATIC ISSUES

The implementation of floodplain restoration measures not only depends on the national policy context and institutional drivers and constraints. Each project is also formed by the specific existing constellation of financial and personnel resources, of windows of opportunity, of chances that were taken or missed, of obstacles that hindered or were surmounted and of discourses associated with the project. The following section explores project-specific emblematic issues in order to increase our understanding of the characteristic aspects of the process behind the "Rheinvorland Süd" project. We suggest that these issues play a key role in understanding what the problems of floodplain restoration can be, which solutions can be sought and how projects develop. As the “Rheinvorland Süd” project forms a small part of the restoration measures taken in the context of the Integrated Rhine Programme (IRP), these characteristic issues can, on the one hand, suggest some possible starting-points and individual conditions for the successful implementation of floodplain restoration and, on the other hand, point out existing problems that can be avoided in future projects.

7.6.1 Windows of opportunity for the initiation phase

Two issues created the windows of opportunity for the project: the atmosphere of change and substantial funding.

Atmosphere of change

The initiation phase of the “Rheinvorland Süd” project coincided with different factors which together formed a constructive atmosphere of change. This was first the designation of the “Rastatter Rheinauer” nature reserve in 1984 and therefore the growing willingness to protect the existing floodplains from further damage. In 1986 the Sandoz accident then created high awareness of the poor ecological state of the Rhine and led to the establishment of the International Commission for the Protection of the Rhine (ICPR). A year later the international contract on flood protection of 1982 was implemented in the form of the Integrated Rhine Programme (IRP) at the regional level in Baden-Württemberg, including floodplain restoration as one main pillar of the contract. Furthermore, the actual start of the project in 1993 was influenced by the extreme flood event in the same year and the subsequent floods of 1995 and 1998 and the resulting increase in public flood risk awareness.

Substantial funding

In this general atmosphere of change another window of opportunity opened: the substantial compensatory funding from the Daimler-Benz Fund and the “Rastatt Declaration” in which the authority agreed with leading environmental groups on the funding of specific floodplain restoration activities in the area of Rastatt. This money functioned as

> “the motor that got the project going […]; otherwise these measures would never have been implemented” (Interview 9).

As a compensating measure for the construction of the Daimler-Benz factory, the money was dedicated to the improvement of the ecological and hydrological conditions in the Rastatter Rheinaue. This was a crucial advantage for the implementation of the “Rheinvorland Süd” project as the planned retention areas within the IRP are the subject of controversial discussion. Though the conservation and re-creation of floodplain areas is an explicit goal of the IRP, it has often become the victim of budget shortages and heavily delayed timetables. For the "Rheinvorland Süd" project these aspects became less of a hindrance when the Daimler-Benz Fund was established. The substantial funding enabled the actors to organise and carry out the project in a relatively short time-span and therefore avoid or bypass political disputes and sectoral disruptions. Nevertheless, shortages of money became a more important issue in the later phase of the project, when the money from the Daimler-Benz Fund had run out and the project group had to revert to other financial sources. Consequently, progress on implementation slowed down in the later phase of the project.

7.6.2 Exploiting opportunities

Whilst the availability of sufficient amounts of money within a supportive atmosphere created the opportunity to implement floodplain restoration measures, the next step was to transform this opportunity into action. Two factors played a crucial role: readily available concepts and the interdisciplinary project group.

Concepts in reserve

Due to the early involvement of the WWF Floodplain Institute and its mapping efforts, the status of the "Rheinvorland" area had been well-known and well-documented since the mid-1980s. The resulting catalogue of measures under the authority of the Regional Council of Karlsruhe explicitly attempted to demonstrate the ecological and hydrological specifics of the area as well as the potential for improvement. Pilot studies had tested the effect of single measures such as the lowering of footpaths or the construction of fords and prepared a site-based catalogue of possible measures. The availability of the proposal of measures both from WWF and the scientific studies carried out enabled the representatives of the different administrative bodies to refer to these documents using them as a basis for further development. According to a member of the project group, the availability of a conceptual framework was of crucial importance:

> "What I have learnt from the whole story is the following: you have to have some concepts ready in the drawer. If they are there, the money will come from somewhere" (Interview 6).

Overcoming fragmentation: the role of the project group

More important for the transformation of the opportunities into action and the further development of the "Rheinvorland Süd" project was the establishment of the project group. According to the interviewed group members two of the most distinctive features and factors for the progress of implementation were the composition of the working group and the personal enthusiasm of some of its members.

As already mentioned, the common constraint of sectoral fragmentation did not apply to the "Rheinvorland Süd" project (cf. Section 7.5.2). The key actors in the relevant administrative bodies already knew each other in the initiation phase of the project, so that as soon as the money was available the project group could be set up. Therefore from the beginning of the project in 1993 all relevant administrations, agencies and nature protection associations were equally represented in the project group (cf. Section 7.4.4). The members of the group knew each other from their daily work or former cooperation. Therefore they did not have to put much effort into trust-building measures. A constructive atmosphere from the beginning of the project enabled the group to concentrate on the substantive aspects and challenges and to avoid or minimize conflicts between the various interest groups:

> "[It is a] great experience to see that groups which normally fight each other are capable of sitting together and [...] working towards the same aims" (Interview 2).

The Daimler-Benz Fund had provided the working group with additional financial resources for the sectoral budgets of the institutions involved. Therefore money did not create competition amongst the different authorities; none of them was forced to cancel, delay or limit another project in order to implement measures in the "Rheinvorland Süd".

Personal commitment and enthusiasm was another aspect that helped to overcome sectoral fragmentation. The majority of the members of the working group came from the Rastatt region and were highly interested in the improvement of floodplains, linking personal and professional aims with the implementation of the project. This led to high personal continuity within the project group. The personal commitment was matched by a strong willingness to work on an interdisciplinary basis, which led to learning processes for adequate solutions as shown exemplarily in the following quote:

> "Many constructive proposals came from the engineers, and it was interesting for the nature conservationists to realize: 'Look, it is possible to cooperate with them, and they make constructive proposals which are even in accordance with our wishes as to how to solve the problems on-site'" (Interview 6).

The constructive and consensus-oriented atmosphere was another factor playing a crucial role for both the continuity of the project and the way the overall project aim was pursued. Certain members of the group played a significant role in mediating different interests and integrating municipalities as the most important landowners in the discussion of the single measures. Due to the professional experience and the good reputation of these key actors, the approval of the municipalities for the implementation of certain measures on their land was often obtained comparatively easily. This consensus-oriented atmosphere helped to identify various problems and obstacles at a very early stage and to acknowledge this in implementation or to abandon individual measures that were too controversial.

The work of the project group during the planning and implementation phase of the project was preparatory. Meetings and decisions were held on an informal basis. Final permission was given by the licensing authorities. The intensive preparatory work of the working group facilitated a quick and smooth licensing procedure and was therefore responsible for the comparatively fast implementation of the individual measures. While the whole working group met several times a year in the early phase but only once a year in the later phase, the communication flow also took place in bilateral or small informal meetings. Sometimes meetings were held on-site. These forms of communication and cooperation turned out to be less time-consuming and more effective than "normal" administrative practice.

On the whole the interdisciplinary, consensus-oriented and informal work of the project group can be viewed as a form of cooperative planning. Classic floodplain restoration problems with institutional interplay and interdependencies between different institutions (cf. Moss 2003) only played a secondary role in the "Rheinvorland Süd" project. The crucial importance of cooperative planning for the floodplain restoration process, together with the financial support from the Daimler-Benz Fund, shows that personnel and financial resources were a lot more important for the process than policy changes. These had only an indirect influence through the general supportive atmosphere. However, the work of the project group also shows that the most pressing problems of a high level of complexity and controversy were avoided to the benefit of the quick and smooth implementation of single-issue measures.

7.6.3 Professional consensus instead of public support

Despite the routine and knowledge which the working group developed during the following years, almost no attempts were made to promote the project in public. Especially in the early stages of the project, effort and attention were focussed on the internal processes – from the negotiation and passage of the catalogue of measures to interdisciplinary cooperation and consensus at a professional level. The reasoning behind this was that the limited funds should be used for the actual technical measures rather than public relations or monitoring processes. Group members were also rather glad that they did not attract much resistance and media attention so that the assigned measures could be implemented smoothly.

In the later stages of the project most members of the group considered the lack of public information a significant deficit of the project:

> "The public relations work is deficient because we are doing good things, and it all happens quietly. [...] Perhaps the project has wasted the opportunity to achieve effective publicity by presenting details in a form comprehensible to the layman" (Interview 8).

Besides this lack of knowledge transfer of the ecological dimensions into the public realm, scientific and professional information about the proceedings, measures and effects of floodplain restoration in the Rastatter Rheinaue was also deficient. A single brochure was published in the mid-1990s, but it neither presented the current status of the work nor explained the cumulative impact of the project. For a few years now the project group has tried to address this deficit (cf. Grönitz 2005; Jupé *et al.* 1999; Regierungspräsidium Karlsruhe 1996).

The lack of public relations and specialised information, but also the complex context of floodplain restoration, led to a lack of interest within the administrative and political realm in general. Members of the "Rheinvorland Süd" project group saw a deficiency but also an opportunity in the project as a pilot study:

> "It is a pity that in the end absolutely no-one in the administration was interested in the project... Assessing this complex system is difficult even for the affected population on the spot, and politicians in the state parliament who come from the dry areas of the Swabian Alb are then expected to evaluate the effects of the multifaceted – highly technical, very expensive – IRP. In this respect, the small sand-table exercise "Rheinvorland Süd" could be a model for establishing public relations, and creating an understanding of these criteria on the part of politicians" (Interview 8).

7.6.4 Many cents make a dollar: small-scale measures and their evaluation

In contrast to the IRP the "Rheinvorland Süd" project concentrated on many separate small-scale actions to improve the discharge and natural habitat conditions of existing and former alluvial meadows. As previously mentioned, the project catalogue included sixty-nine separate measures. The main pillar of the IRP includes the creation of thirteen large-scaled retention areas and therefore needs to deal with complex licensing procedures and controversial usage patterns as well as associated public and political resistance. Whereas only three of thirteen retention areas have been created within the IRP since 1988 in Baden-Württemberg, the "Rheinvorland Süd" project was much more successful in implementing the listed measures, last but not least because of minor resistance and inexpensive licensing procedures. Due to the limited scope and the small scale of these single interventions the project as a whole has never been questioned.

Especially in the latter project phase, when the money from the Daimler-Benz Fund had run out and funds from the state's budget had to be earmarked to continue the project, the demonstration and scientific proof of the positive ecological and hydrological effects became very important. Consequently, the Regional Authority of the Environment began to systematically monitor the effects of certain measures in 2001. An evaluation of the ecological results of the project and a substantial database were contracted out to an engineering office (cf. Fischer 2005). The results of these monitoring efforts have confirmed the substantial ecological improvements achieved by the measures (cf. Section 7.4.2).

7.7 CONCLUSIONS AND RECOMMENDATIONS

In conclusion, the "Rheinvorland – improvement of the hydrological and ecological conditions in the Rheinaue" project shows four distinctive characteristics:

- The floodplain restoration measures implemented were based on conceptual preparatory work that had been established by the pilot studies of local environmental associations;
- The project could only be implemented when substantial funding from compensatory measures became available;
- An interdisciplinary project group was established to prepare and accompany the measures' implementation. The group was highly motivated to achieve an improvement in the ecological and hydrological conditions of the alluvial meadows of the "Rheinvorland Süd". This led to active and constructive cooperation between the representatives of the different authorities and associations;
- The national and catchment-based policy context – the Integrated Rhine Programme (IRP) – indirectly created a supportive atmosphere. Whilst the IRP comprises two equally important parts – ensuring protection against flood events on the one hand, and improving the ecological status of the floodplains and alluvial meadows on the other – the project clearly contributed to the latter of these two aspects. According to the members of the project group this "silent" support can be understood as approval of the way the project moved forward and of the results that were achieved step by step.

What recommendations can be drawn from the analysis for policy development and project management?

7.7.1 Policy development

First, with reference to the Daimler-Benz Fund that triggered the implementation of the floodplain restoration measures in the case of the "Rheinvorland Süd" project, it can be stated that substantial budget funding (or the creation of a foundation) is as necessary to ensure the medium and long-term financing of restoration projects as support by policy instruments. The effects and the sustainability of floodplain restoration projects are highly dependent on both mid-term timeframes and ongoing maintenance, which highlights the need for a stable financial framework. At the same time, single sources of funding should be embedded in long-term regional or national policy initiatives on floodplain restoration in the catchment.

Secondly, the implementation of substantial funding and the need for a floodplain restoration policy framework go hand in hand with the requirement of continuous monitoring and evaluation activities. Only if the effects of already implemented floodplain restoration measures are constantly monitored and assessed, can existing activities in floodplain restoration compete with other riverbed-related activities in the political arena. This must be considered in the context of flood protection measures as well as nature conservation.

Furthermore, the relationship between the Integrated Rhine Programme (IRP) and the organisation of the "Rheinvorland Süd" project itself suggests another aspect that is of general importance for policy development: managing the interface of large-scale policy and small-scale implementation. This means that policy-making efforts can be limited to the formulation of overall goals and to controlling the achievement of these goals, whilst the application and implementation of the specific projects can be dedicated to those actors at the lowest possible administrative level. Despite the conflictual course of implementation of the IRP, for the "Rheinvorland Süd" project the programme served as the political framework in which general goals were fixed, leaving the design and the implementation of the project to the working group on site.

7.7.2 Project management

The existence of a cross-sectoral, interdisciplinary working group seems to be one of the main factors behind the success of the "Rheinvorland Süd" project. Generally speaking, the installation of an informal planning body outside the regular administrative routines that is responsible for preparing, coordinating and accompanying the implementation of the various measures creates a supportive background for floodplain restoration. From the perspective of an early generation restoration project, cooperative planning of the conceptual framework and the site-based pursuit of single measures also seem to be essential for the smooth implementation of floodplain restoration projects. This addresses especially the fact that the constellation of interests and stakeholders of floodplain restoration is usually highly diverse, that institutional arrangements often do not consider the physical boundaries of the river basin (issue of fit) and that existing interdependencies between different institutions affected by, and involved in, floodplain restoration are complex.

Furthermore, the availability of a conceptual framework outlining the objectives and the applied method helps to channel available funding for the improvement of the hydrological and ecological conditions in floodplains. It is therefore necessary to initiate and support preparatory studies and pilot implementation, and to develop concepts to establish a vision of medium and long-term development goals that can be substantiated and specified as soon as funding is available.

Whilst the lack of constant monitoring within the "Rheinvorland Süd" project reflects a "scientific" deficit, the lack of public relations work suggests the need to secure public support and democratic legitimisation. The attempt to inform the public about the aims of the project and the means – both technical and financial – in order to achieve these aims must be an integral part of floodplain restoration projects from the very beginning. Therefore project management schemes should require monitoring and participatory processes as obligatory elements.

7.8 REFERENCES

Bernhardt, Ch. (2003) *PCCP processes in History: The Model of the Upper Rhine Region.* UNESCO-IHP technical document, PCCP series, Paris, (CD-ROM) http://unesdoc.unesco.org/images/0013/001332/133286e.pdf.

BMU (Bundesministerium für Umwelt, Naturschutz und Reaktorsicherheit) (ed.) (1998) *Wasserwirtschaft in Deutschland.* Reihe Umweltpolitik, Bonn.

Bohn, C. (2004) Die Bedeutung der Auen für die Umsetzung der Ziele der EU-Wasserrahmenrichtlinie. In *Tagungsband "Interdisziplinäre Methoden des Flussgebietsmanagements" March 2004* (ed. J. Möltgen, D. Petry), IfGIprints 21, Institut für Geoinformatik, University of Münster.

Disse, M., Engele, H. (2001) Flood Events in the Rhine Basin: Genesis, Influences and Mitigation. *Natural Hazards*, Vol. 23, Nos. 2-3 (2001), pp. 271-290.

Dworak, T., Görlach, B. (2005) Flood risk management in Europe – the development of a common EU policy. *Intl. J. River Basin Management* **3**(2), 97-103.

Fischer, H.-J. (2005) Dokumentation von Renaturierungsmaßnahmen und ihren ökologischen Wirkungen im Rheinvorland. In *Naturschutz und Landschaftspflege in Baden-Württemberg* (ed. Landesanstalt für Umweltschutz Baden-Württemberg), Karlsruhe.

Gallusser, W. A., Schenker, A. (1992) *Die Auen am Oberrhein: Ausmaß und Perspektiven des Landschaftswandels am südlichen und mittleren Oberrhein seit 1800*, Birkhäuser, Basel.

Grönitz, W. (2005) *Impulse für den Naturschutz am Oberrhein – 15 Jahre Integriertes Rheinprogramm.* Landesanstalt für Umweltschutz Baden-Württemberg, Info 2/2005.

ICPR (International Commission for the Protection of the Rhine) (2001) *Rhine Atlas 2001*. Koblenz.

IKSR (Internationale Kommission zum Schutz des Rheins) (1987) *Aktionsprogramm „Rhein"*. APR-Bericht No. 1, 18 pages + appendices, Strassburg.

IKSR (Internationale Kommission zum Schutz des Rheins) (1991) *Ökologisches Gesamtkonzept für den Rhein*. Bericht No. 24, Koblenz.

IKSR (Internationale Kommission zum Schutz des Rheins) (2001) *Rhein 2020. Programm zur nachhaltigen Entwicklung des Rheins*, Koblenz.

Huisman, P., de Jong, J., Wieriks, K. (2000) Transboundary cooperation in shared river basins: experiences from the Rhine, Meuse and North Sea. *Water Policy* **2**, 83-97.

Jupé, A., Manke, J., Wendel, G. (1999) Maßnahmen zur Verbesserung der ökologischen Verhältnisse im Rheinvorland. *Wasserwirtschaft*, **89**, 176-181.

Löbert, T. (1997) Die Oberrheinkorrektion in Baden – Zur Umweltgeschichte des 19. Jahrhunderts. *Mitteilungen des Instituts für Wasserbau und Kulturtechnik der Universität Karlsruhe (TH)*, Issue 193, Karlsruhe.

Korn, N., Jessel, B., Hasch, B., Mühlinghaus, R. (2005) *Handlungsempfehlungen und Maßnahmenvorschläge für die Fließgewässer- und Auenentwicklung auf Grundlage der Wasserrahmenrichtlinie.* NNA-Berichte, Vol. 18, No. 1, Schneverdingen, pp. 115-122.

Maltby, E., Balckwell, M. (2005) Managing riverine environments in the context of new water policy in Europe. *Intl. J. River Basin Management* **3**(2), 133-141.

Moss, T. (ed.) (2003) *Das Flussgebiet als Handlungsraum. Institutionenwandel durch die EU-Wasserrahmenrichtlinie aus raumwissenschaftlicher Perspektive,* LIT-Verlag, Münster.

Oberrheinagentur (ed.) (1996) *Rahmenkonzept des Landes Baden-Württemberg zur Umsetzung des Integrierten Rheinprogramms.* Materialien zum integrierten Rheinprogramm, Vol. 7, Lahr.

Oberrheinagentur (ed.) (1995) *Grundsatzpapier Auenschutz und Auenrenaturierung.* Materialien zum integrierten Rheinprogramm, Vol. 4, Lahr.

Plate, E. (2000) Management strategies for the development of the Upper Rhine. *Water International* **26**(1), 51-61.

Regierungspräsidium Karlsruhe (ed.) (1996) *Verbesserung der Abflussverhältnisse im Rheinvorland,* Karlsruhe.

Staatliche Forstämter Karlsruhe und Rastatt (eds.) (1996) *Naturerlebnis Rheinauen zwischen Rastatt und Karlsruhe,* Karlsruhe/Rastatt.

Tittzier, T., Krebs, F. (1996) *Ökosystemforschung: der Rhein und seine Auen*, Springer, Berlin.

WWF Aueninstitut (ed.) (1992) *Untersuchungen zur Verbesserung der wasserwirtschaftlichen und ökologischen Verhältnisse in der Rheinaue zwischen Rastatt-Wintersdorf und Au a. Rhein,* Rastatt.

INTERVIEWS

Interview 1: Water Authority North (Gewässerdirektion Nord), 1 July 2002.

Interview 2: Engineer's office, 1 July 2002.

Interview 3: Floodplain Institute of the World Wide Fund for Nature (WWF-Auen-Institut), 2 July 2002.

Interview 4: Environmental Agency Rastatt (Umweltamt Rastatt), 3 July 2002.

Interview 5: Environmental Agency Karlsruhe (Bezirksstelle für Naturschutz und Landespflege Karlsruhe), 15 July 2002.
Interview 6: Baden-Württemberg State Office for the Environment (Landesamt für Umweltschutz) 15 July 2002.
Interview 7: District Office of Rastatt (Landratsamt Rastatt).
Interview 8: Forestry Office of Karlsruhe (Forstdirektion Karlsruhe), 15 July 2002.
Interview 9: Association of Nature Conservation Baden-Württemberg (Landesnaturschutzverband Baden-Württemberg), 15 July 2002.
Interview 10: City of Rastatt (Stadtverwaltung Rastatt), 16 July 2002.
Interview 11: Regional Council of Karlsruhe (Regierungspräsidium Karlsruhe), 16 July 2002.

8

Restoring an alluvial forest and ancient meander on the Garonne: linking environmental protection to fishing and recreation

Maria Hagemeier and Axel Klaphake

8.1 INTRODUCTION

The River Garonne is characterised by serious degradation of its floodplain ecosystems due to the fragmentation of wooded areas, the replacement of natural habitats by cultivated ones and the loss of river dynamics in combination with the aggravation of low water conditions. This is the principal reason why floodplain restoration on the Garonne is seen as a major task for water and nature protection policy.

This case study of the floodplain next to the village of Bourret on the River Garonne is an example of a site-based, small-scale restoration project, the main objectives of which are biodiversity conservation in combination with other targets (e.g. recreation, angling). The project therefore serves to illustrate typical institutional and instrumental features which characterise the planning and implementation of this type of project in France.

Moreover, only a few restoration projects have been implemented on the Garonne in recent decades, so that it appears useful to study the principal drivers and constraints as well as the transferability of the approach used in this particular case. In contrast to other regions, the physical conditions for floodplain restoration are comparatively favourable as a complete isolation of the river from its floodplains has not taken place due to the fact that no coherent dyke programmes were implemented in the past. In addition, this example addresses an important property rights aspect of floodplain restoration in France: in the Garonne valley most of the ecologically interesting riparian areas that are regularly flooded are state property and managed by an administrative body at the departmental level that has neither the legal obligation nor the resources to support the restoration of riparian ecosystems. A special characteristic of the Bourret case is the general agreement of the main actors on the fact that something has to be done for the floodplains of the Garonne. There are, however, serious differences of opinion as to what precisely should be done.

8.2 THE PHYSICAL CONTEXT AND USE OF THE RIVER

8.2.1 Physical characteristics

The Garonne extends over a length of 580 km and drains a basin of 57,000 km^2. The river originates in the Pyrenees mountains and flows northward through the town of Toulouse where it meets the Canal du Midi, linking the Garonne to the Mediterranean Sea. After passing Bordeaux the river flows into the Atlantic Ocean.

The river section relevant to this case study extends over 100 km in the middle section of the Garonne, located between 20 km upstream and 80 km downstream of Toulouse in both the Haute-Garonne and Tarn-et-Garonne departments (see Figure 8.1). The fluvial landscape in this river section constitutes a transition zone between the rather rapid water flow upstream and the slow meandering river with a wide floodplain downstream of the confluence with the River Tarn (see ECOSYS 2001). In this part of the river the mean annual flow of water is about 200 m/s and the floodplain enlarges to a maximum width of around 3-4 km and a maximum spread of 500-2000 km^2, which is flooded on average every 10 years. The area that experiences annual flooding reaches a mean width of only approx. 350 m. Occurring floods are sudden and violent (maximum daily flow of 2,500 m/s for the 50-year flood, 4,500 m/s for the 100-year flood). In contrast, the mean daily flow in the summer period is very low (approx. 50 m/s) making low flow management an important issue (see ECOSYS 2001; Decamps *et al.* 1988).

Even though various anthropogenic pressures affect the ecosystem downstream of Toulouse (see Section 8.2.2), it remains a fairly diverse riparian ecosystem with impressive biodiversity, mainly because of the absence of systematic embankment programmes. The valley is characterised by many ancient meanders that are currently completely or partly disconnected from the main river (see Décamps *et al.* 1988). These zones play a key role for the development of juvenile fish, natural forest dynamics and bird habitats (see Belaud and Lim 1994). In addition, several studies have pointed out their ecological value in terms of water purification and flood prevention capacity.

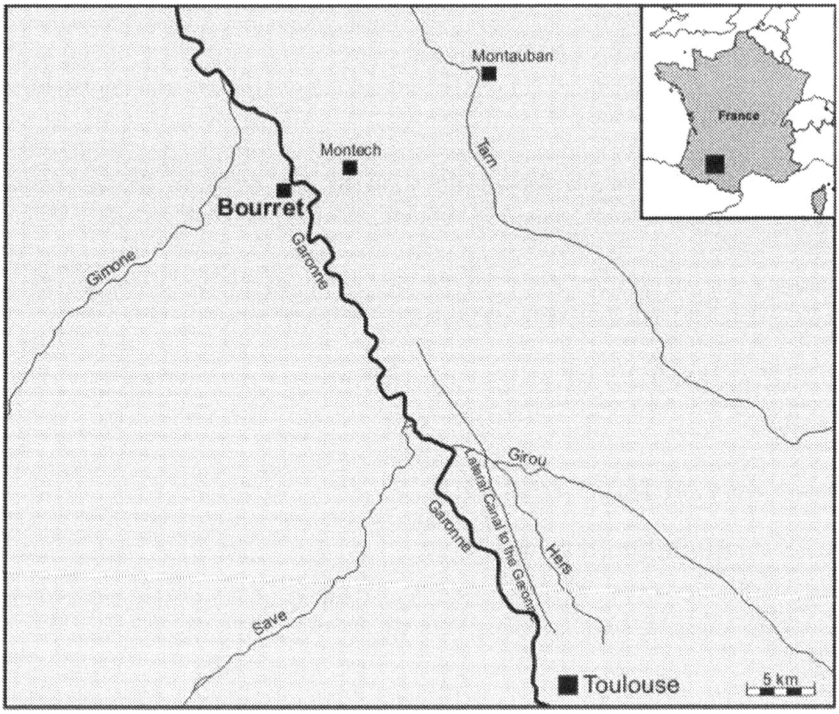

Figure 8.1: Map locating the Bourret project area.

8.2.2 Use and management of the river

Nevertheless, the riparian ecosystems are far from being pristine and floodplains are affected by various man-made pressures. One important factor is the general loss of the natural fluvial dynamic and the heightening of flow velocity resulting from dam construction, river regulation and the stabilisation of eroding banks. Moreover, urban sprawl in the conurbation of Toulouse has profoundly modified the fluvial landscape. Still more important for the river's morphology and the rural floodplains are massive gravel and sand extractions, which have contributed to a significant drop in the water level and the drying-up of areas which used to be regularly flooded (see Décamps *et al.* 1988). Water quality in this river section decreases from acceptable (upstream of Toulouse) to poor (20 km downstream of the city) because of urban waste water. It increases again beyond the 20 km mark downstream of Toulouse.

One of the most adverse man-made effects is the aggravation of low water conditions in dry summer periods when the flow is regulated by dams to produce electricity and when irrigation increases. During the summer periods, water releases are jointly managed by the local communities, the Water Agency Adour-Garonne and the national power company EDF (Electricité de France), which is the principal dam operator in the basin. In addition, changing land use has major adverse effects on floodplain ecosystems, in particular on the preservation and development of the natural alluvial forest (see Amigues and Desaigues 2002). Areas periodically flooded are by now predominantly converted to agriculture: today, 52% of the surface flooded on average every ten years downstream of Toulouse is used for agriculture, whereby intensive irrigation agriculture (maize, orchards etc.) dominates (see ECOSYS 2001). Nowadays, the natural alluvial forest is highly fragmented and only covers 4% of the decennially flooded area in this part of the Garonne (ECOSYS 2001). Moreover, the site has, in part, the characteristics of a 'no man's land' with unauthorised human interventions (motocross, illegal waste dumps etc.)

8.3 THE POLICY CONTEXT

8.3.1 National policy context

From 1990 to 1995, the European Community financed the programme 'Garonne Vivante' that was carried out by the non-governmental organisation Nature Midi-Pyrénées (NMP). This programme aimed to support the implementation of the European Union's Birds Directive, to raise awareness about the loss of riparian habitats and to identify floodplains worthy of protection and restoration. The NMP – in cooperation with the Water Agency Adour-Garonne and other administrative bodies – conducted a study on a strategic approach to the management of the floodplains in the middle section of the Garonne (NMP 1995). This study can be regarded as a first rough blueprint for an integrated preservation and restoration programme (see ECOSYS 2001). NMP also defined some conditions for restoration measures, including the management of riparian forests and the restoration of ancient meanders. Another determining factor was the implementation process of the Habitats Directive that started in the mid-1990s. The low-flow channel and parts of the main river bed (e.g. alluvial forests) between Toulouse and the Tarn confluence were proposed as a NATURA 2000 site to the European Commission by the regional environmental administration DIREN (Direction Régionale de l'Environnement Midi-Pyrénées).

National wetland policy was also an important contextual factor. Referring to the 1992 water law, which introduced the notion of wetlands into French water law, the national government adopted a national wetlands action plan (Plan d'action gouvernemental de sauvegarde et de reconquete des zones humides) in 1995 (see Monstadt in Chapter 5 of this volume). As a result of this plan the floodplains downstream of Toulouse were chosen by the Ministry for the Environment in 1996 as one of the 87 major national wetlands. Furthermore, the Garonne valley was selected as one of ten national wetlands in which the national environmental ministry finances large scientific studies in order to enhance the knowledge base for an integrated management approach (the so-called Programme National de Recherche sur les Zones Humides – PNRZH).

Downstream of Toulouse, the corresponding research programme 'Zones humides riveraines de la Garonne' was carried out from 1997-2000 by an interdisciplinary research group named GIS (Groupement d'intérêt scientifique) with the principal objective of developing a decision-support system. This programme was regionally co-financed by the Adour-Garonne Water Agency, the Midi-Pyrénées Region and some European funding.

8.3.2 Initiatives in the catchment

The most important planning instrument for floodplain restoration is the SDAGE (Schéma Directeur d'Aménagement et de Gestion des Eaux), the general function of which is to integrate all water-related plans and programmes (Comité de bassin Adour-Garonne 1996; see Monstadt in Chapter 5 of this volume). The SDAGE for the Garonne basin was adopted in 1996 and includes the objectives of preventing floods and protecting freshwater ecosystems. It recommended the maintenance of the flood prevention capacity of the floodplains downstream of Toulouse and declared large parts of the floodplains between Toulouse and the Tarn confluence a 'zone verte' (green zone). This status restricts economic and infrastructural activities and endorses restoration and protection measures. Moreover, it aims to coordinate water planning with the implementation of the EU Habitats Directive. The Water Agency Adour-Garonne, which is in charge of the implementation of the SDAGE, subsidises restoration measures as well as relevant scientific studies.

There were some earlier, parallel floodplain restoration projects which were concerned with the rehabilitation of ancient meanders and the restoration of natural alluvial forests. These initial restoration projects were not very effective and mostly had a narrow perspective in ecological terms as their main objective was to restore habitats for fish reproduction without taking into account the general impacts on riparian ecosystems and fluvial dynamics (see ECOSYS 2001). The restoration projects were mostly implemented with the support of the Fédération de Peche (FDP), an organisation that represents the interests of fishermen and anglers in the region, in collaboration with the local communities and different administrative bodies at the departmental level. Additionally, there were some previous experiences relating to ecological forest conversion and the restoration of alluvial forests, e.g. the restoration of the alluvial forest at Ramier de Bigorre (52 ha) on a site which was formerly used for intensive forestry and gravel extraction. The principal objectives of this pilot project were the initiation of a natural forest dynamic, bird protection and ecological education.

8.4 BRIEF DESCRIPTION OF THE PROJECT

The Bourret site is a floodplain of the Garonne characterised by an ancient meander. The site extends to approx. 100 ha and is located in the middle section of the Garonne some 60 km downstream of Toulouse (see Figure 8.1). Due to the general loss of fluvial dynamic in the river basin and the lack of natural erosion at the river banks, the frequency of flooding has significantly fallen – today complete flooding of the area occurs only once every five years – and has contributed to the separation of the main river from its secondary branch. Statistically, however, the bed of the ancient meander (8 ha) is still flooded for several days every year although the amount of water and the length of the flooding is insufficient to maintain its ecological function as a fish habitat and to ensure the ecologically important interaction between land and water in the area (see CAREX 2000).

8.4.1 Purpose and objectives

In the mid-1990s the mayor of the village of Bourret – a small local community of around 400 inhabitants – and the DIREN Midi-Pyrénées launched a programme for the restoration of the floodplain. The objectives of this programme focus on ecological and recreational issues (Interviews 3, 6, 8). The most important aims are the improvement of the ecological quality of the forest, the reconnection of the ancient meander to the river and the upgrading of the area for recreational purposes. Additionally, ecological studies and ecological monitoring were to be carried out in order to enhance the knowledge base and, in particular, to evaluate the possible transferability of the management approach.

8.4.2 Origins and development

The rehabilitation of the ancient meanders in the middle section of the Garonne had already attracted attention in the 1980s. Since there is a disconnected arm of the river every 1.5 km between Toulouse and the Tarn confluence, their restoration was perceived as a major concern if the ecological quality of the river in terms of fish habitats was to be enhanced. However, the early projects to restore old arms of the river, such as the first attempt to restore the old arm at Bourret in 1986, failed. Furthermore, they were largely dominated by fish reproduction interests, giving little consideration to the development of the ecologically important riverine forests.

The origins of the present Bourret project are more diversified, however. Once again, fishing interests played an important initial role because the Conseil Supérieur de la Pêche (CSP) was involved in the project (Interviews 3, 8). The CSP is an administrative body which is in charge of the surveillance and promotion of fishing activities and even implements measures aimed at protecting fish species. Interestingly, an assistant of the mayor of the commune of Bourret was a former representative of the CSP and the FDP and strongly supported the idea of re-attempting to reopen the old arm. In order to get financial, organisational and legal support, the local community contacted the relevant administrative bodies at the departmental and regional levels.

The DIREN, which is able to support such restoration projects financially via national or European funding, generally supported the project idea but aimed at a broader ecological perspective and the involvement of all the relevant stakeholders. In particular, the DIREN was of the opinion that a more coherent restoration programme for the entire floodplain should be carried out because of the legal obligations resulting from its declaration as a NATURA 2000 site. Since, currently, monocultural poplar plantations largely predominate, the DIREN was particularly seeking to demonstrate the ecological and economic feasibility of the conversion of monocultural forests and considered the Bourret site potentially suitable as a pilot project (Interview 3). The Water Agency Adour-Garonne was also involved in the project because of its water expertise and its authority to subsidise floodplain restoration in the 'green zone' in the Garonne valley. Equally, the nature protection organisation NMP pushed the project because it had already identified the site as ecologically important. Further, the particular role of the DDE (Direction départementale de l'Equipement) is worth mentioning, as it is the formal proprietor of the site. Most of the Bourret site has the legal status of a Domaine Public Fluvial (DPF) with the result that the DDE is in charge of the management of the area.

To promote the project and to gain knowledge about the possible ecological and economic (cost) impacts, three important preparatory studies and plans were made on behalf of the local authority and the DIREN. First, a rough forest management plan with basic recommendations for reforestation was drawn up

(ONF – Castre 1998) and refined by the large private forest cooperative COFAGAR (Coopérative Forestière Garonnaise). Second, the NMP was engaged to carry out an ecological study and to map relevant ecological units (NMP 1999). Third, a private consulting company prepared a feasibility study on the reopening of the abandoned meander (CAREX 2000).

In order to enhance the involvement of, and the cooperation among, the stakeholders, the DIREN promoted the establishment of a project coordination committee in which all the relevant administrative bodies and stakeholders were to be represented. Principally, the various actors agreed on a rough management plan with the general objective of developing the site into a mosaic of various ecological units and biotopes (open wetlands, re-timbered 'natural' alluvial forest, spontaneous vegetation, poplar plantation etc.) and to enhance accessibility for the public. Further, the re-connection of the old arm, which is the key interest of the local actors, was envisaged but no actions have taken place on this aspect so far (see below).

8.4.3 Current status

The first step towards the implementation of the project was the DDE's decision to lease the area to the local authority free of charge for a period of 15 years and to shift the management of the area to the local authority of Bourret. However, neighbouring alluvial forests are private property and do not belong to the project site, which is limited to the area of the DPF.

Although the whole project was initially heavily dominated by the wish to restore habitats for fish species and to facilitate flooding of the area, currently there are major obstacles that emerged during the debates which took place during the period of the study in 2002/2003 (Interviews 3, 4). One of the key obstacles is still the financing of the project. Apart from the technical and related financial problems, political support for the project decreased significantly for a while but then rose again. Environmental and water policy actors emphasise the pilot character of the project and stress the positive experiences and the commenced restoration of the natural alluvial forest as proof of the feasibility of floodplain restoration projects in the Garonne valley (Interviews 1, 3, 8). Moreover, the upgrading and restoration of the ancient meander is on the political agenda of the departmental executive body (Conseil Général de Tarn-et-Garonne 2006). The Bourret site is also integrated into the regional water management plan SAGE for the Garonne Valley, which was newly developed in 2006. The SAGE 'Vallée de la Garonne' pays special attention to the following issues: first, rising flood risks due to global climate change have to be faced at the Garonne; second, nature conservation in the alluvial corridor of the Garonne should be strengthened; third, the water quality should be enhanced. The SAGE proposes a variety of actions for the different targets.

8.4.4 Physical interventions

The COFAGAR and NMP made a joint application for financial support for reforestation and ecological monitoring in 2000. In 2001 the two first stages of forest restoration were implemented but plantings only took place in an area of around 7 hectares (Interviews 3, 8), making the project a mini-floodplain restoration effort rather than an area-wide rehabilitation.

The ecological monitoring of the area is also ongoing. In addition, early activities to upgrade the area for recreational activities (clean-up etc.) were started but are not yet finished, so that the recreation value of the area is still low, while illegal waste dumps spoil the landscape. In cooperation with the Ecole Forestière Française Nancy (French Forestry School in Nancy) the COFAGAR developed signs and posters to inform future visitors about the natural characteristics of the area. However, because the local community failed to provide the necessary infrastructure (parking spaces, waste bins etc.) and to carry out a full clean-up of the area, the upgrading of the site for recreational purposes is still incomplete. Here, the low financial and staff resources of the local community as well as the relatively low involvement of the citizens are the limiting factors (Interview 8).

Regarding the reopening of the old arm, no physical interventions have taken place so far. The restoration of the ancient meander is, however, on the current political agenda of the Tarn-et-Garonne departmental executive (Conseil Géneral de Tarn-et-Garonne 2006). According to this agenda, different measures to upgrade the ecological status of the old arm at Bourret are proposed, but no final decision about their implementation has yet been taken.

8.4.5 Organisation and actor involvement

The planning and implementation of the Bourret project reveals broad actor involvement, ranging from state actors and administrations from different decision-making levels to non-governmental organisations. With the establishment of the project committee a participatory and coordinating instrument was used to bring the actors together and to handle potential conflicts. However, a complicated and time-consuming decision-making process – the first restoration measures took five years to implement – and the commissioning of various experts to carry out preparatory studies is seen as disproportionate by some participants, in particular in view of the rather small area in which forest restoration was actually achieved (Interview 8).

Local level

At the local level, the local authority of the commune of Bourret was one of the key drivers of the project. In particular, the enthusiasm of individual actors (the mayor and his assistant) owing to their strong interest in the re-creation of habitats for fish reproduction was crucial for launching the project (Interviews 3, 8). Further, the local actors succeeded in convincing local citizens to support the project by offering them the prospect of a new recreation area.

Regional level

The comparably favourable local conditions were combined with a broader perspective on the rehabilitation of the Bourret site because of the active role of the regional environmental administration (DIREN). The support of the forestry actors has also been important. The ONF actively promoted the project via technical expertise in the initiation phase although its degree of involvement in the implementation stage can be regarded as low. Additionally, the active involvement of the COFAGAR was also an important – if surprising – contribution. In general, the COFAGAR acts in the economic interests of private forest owners and has little experience with the recreation of natural alluvial forests. In addition, with the involvement of fishery policy actors (CSP, Fédération de la Peche) administrative and political support for the implementation of the reconnection of the ancient meander could be gained.

Departmental level

The role of the administrative bodies at the departmental level is also worth mentioning because the support of the DDE was crucial. Any restoration activity in the watercourse has to be authorised by the state executive delegate in each district (the Prefet). The approval of the Bourret project by the DDE was an important formal condition. Further, the DDAF contributed to the planning of the restoration measures, and its technical support has been of value. However, the main focus of the DDAF in water-related environmental policy is on water quality objectives rather than on the restoration of riparian ecosystems for biodiversity protection purposes.

The Water Agency Adour-Garonne participated in the project but its support was limited to co-financing a preparatory study and attendance at the meetings of the project committee (Interview 1). The NMP also had an important role and can generally be judged a key actor in floodplain restoration in the Garonne valley downstream of Toulouse. With its ecological and forestry expertise and good reputation the NMP is widely seen as a cooperative and competent non-state environmental actor (Interviews 1, 3, 4).

8.4.6 Instruments used

Apart from the leasing contract between the DDE and the local community for a 15-year period, which was a key pre-requisite for its implementation, the project benefits from the legal and planning status of the area and the corresponding obligations involved. In addition, economic instruments play a certain role, in particular funding from different sources (on national policy instruments affecting floodplain restoration in France, see Monstadt in Chapter 5 of this volume).

With regard to its planning status, the Bourret site was designated a ZNIEFF (Zone Naturelle d'Intéret Ecologique Forstique et Faunistique) type II because of its biological potential. Because the ZNIEFF status is an informational rather than an implementing instrument, its contribution is indirect via consideration in local land-use plans and environmental impact assessments. In addition, the site is part of the 'corridor alluvial' of the SDAGE, of the proposed NATURA 2000 site in the Garonne valley and of the wetlands of national importance programme (Interview 4). Water and environmental policy actors therefore have strong legal authority for the restoration measures and can apply diverse funding instruments. However, neither the national nor the international legal status involves an obligation to carry out specific restoration measures: for example, a management plan for the NATURA 2000 site has not yet been drawn up (Interview 4). Further, the site benefits from its protection status under the PRI (Plan de Prévention des Risques) with the effect that no buildings are allowed that reduce its flood prevention capacity.

For the forestry sector, the most important planning instrument is the ORF document (Orientations Régionales Forestières) in which the general national forestry objectives (e.g. multi-functionality) are specified for the regional level. The ORF is also intended to serve, at least in theory, as a cross-sectoral instrument which integrates the water planning provisions which are relevant to forestry into the forestry plans. Responsible for drawing up the ORF is the Commission Régionale de la Foret et des Produits Forestiers (CRFP). The document for the planning period 2000-2006 states that natural riparian forests should be maintained and restored in order to use and enhance water purification and flood prevention capacity as well as to achieve positive impacts on biological diversity (see ECOSYS 2001). For the Bourret project the ORF might be a driver for the implementation of alluvial forest restoration.

As far as the management of the DPF by the DDE is concerned, no specific legal obligations or binding planning documents for the restoration of floodplains exist (Interview 9). In the department Tarn-et-Garonne the DDE is currently developing a scheme for valuing floodplains (Schéma de Valorisation des Champs d'Expansions des Crues) with the objective of defining actions to upgrade the ecological and social value of the DPF.

The orientation of fishery policy towards the reconnection of old arms is based on the departmental planning instrument relating to pisciculture management[1] which envisages measures such as the reopening of the ancient

[1] The relevant planning instrument is the SDVPF (Schéma Départemental de Vocation Pisicole et Halieutique), the general objectives of which are specified in the PDPG, which are the departmental

meanders and, interestingly, the removal of rock fills to recreate river dynamics in favour of the natural emergence of floodplains.

In the Bourret case, different funding sources have been used to finance the preparatory studies, the forest restoration measures and the ecological monitoring. For the latter, 100% financial support was promised by the DIREN. For the forestry measures, the DIREN and the regional authority (Conseil Regional Midi-Pyrénés) agreed to provide 80% financial support, while co-financing by the European Regional Development Fund (ERDF) was also used. Consequently, 20% of the reforestation budget (total costs for the two first stages of the project were estimated at approximately 500,000 FF (ca. € 76,220)) had to be paid by the local authority, an amount too high for the financial resources of a village of 400 inhabitants. This financial gap could be closed, however, because the local community was allowed to pay its share in kind (Interview 3).

8.5 PRINCIPAL DRIVERS AND CONSTRAINTS OF FLOODPLAIN RESTORATION ON THE PROJECT

The Bourret project is an example of site-based, small-scale floodplain restoration. As pointed out in the chapter on the general policy context of floodplain restoration in France (cf. Monstadt in Chapter 5 of this volume), this type of restoration project is frequently motivated by nature protection objectives and supported by the designation of protected areas according to international, European or national law. However, the analysis of the general policy context also shows that, at the project level, the biodiversity targets are regularly only one target among others, whereby recreation or even economic interests can also play an important role. Against this background, the Bourret project fits into the general picture of small-scale floodplain restoration projects in France.

8.5.1 Drivers

Fish reproduction

As described above, at the local level the main drivers of floodplain restoration were the interest in fish reproduction and the wish to enhance angling opportunities, whereas biodiversity and flood prevention objectives were not significant issues at the initial stage. In particular, the active involvement of individual local actors can be judged a key success factor. These local interests were supported by the fishery policy actors, whose general objective is to improve fish habitats, in particular via the restoration of ancient meanders. In any case, the willingness of the local actors to accept a broader restoration

plans for the protection of the aquatic environment and the management of pisciculture resources (Plans departementaux pour la protection du milieu Aquatique et la Gestions des Ressources Piscicole).

approach and a combination of the re-connection of the old arm of the river with forest conversion can also be traced to the fact that such a restoration package usually has a better chance of being accepted by environmental authorities which manage the necessary funds (Interviews 3, 8).

Ecological quality

The designation of sections of the floodplains between Toulouse and the Tarn confluence as a NATURA 2000 site was a driving force because this political decision on the part of the environmental administration offered the opportunity not only to mobilise funds but also to integrate biodiversity objectives into the project. The further implementation of the Habitats Directive in the Garonne Valley raises the question of necessary compensation for the costs of restoration or protection measures on private land (see CETEF 2001).

Equally, the integration of the floodplain into the alluvial corridor supports floodplain restoration because this entails potential financial and technical assistance by the Water Agency Adour-Garonne. However, the role of the Water Agency is not an active one; its involvement is limited to financial support for projects that are proposed and carried out by local or regional actors (Interview 1). With regard to the reopening of the ancient meanders, the Water Agency generally subsidises appropriate measures and has already financed a study on the feasibility of such projects in the department Tarn-et-Garonne.

Forestry

The active and trust-building involvement of the DDE Tarn-et-Garonne and of the forestry policy actors, in particular the COFAGAR, was an important driver for the Bourret project. Although the coordination of forestry and environmental policy in the Garonne valley is generally regarded as inadequate, the restoration of an alluvial forest on the DPF could be achieved because no important private or local interests were affected (Interviews 3, 8). The involvement of the COFAGAR demonstrates that the potential long-term economic interests of this organisation could be mobilised for the support of the project. But there are still unchanged poplar plantations on the Bourret site and in the surrounding area, which show the need for a further forest conversion strategy if alluvial forest restoration is to be achieved.

8.5.2 Constraints

Lack of financial and organisational resources

The lack of local financial and organisational resources is a principal constraint on the project. The principal difficulties relate to acquiring local co-financing of the necessary investments and maintaining political support. The involvement of

EDF, currently being considered, could possibly help solve this problem because of EDF's financial power.

Missing integration of flood protection

Flood prevention policy did not play any important role in the Bourret project, as the project site is rather small and unimportant with regard to flood prevention or flood mitigation. This fact can also be seen as a constraint. Nevertheless, some environmental actors justified the Bourret project and the financial support for it with the argument that the measures to be taken in the proposed project could have flood prevention effects. However, this assessment was not shared by flood control policy actors such as the relevant department in the DIREN or the responsible departmental administrations (Interview 2). Similarly, the flood prevention effects of the restoration of floodplains downstream of Toulouse were never precisely studied, so that the flood control argument tends to provide justification on a rhetorical level and might increase political support, but it can hardly be regarded as a product of factual reasoning. The Bourret project therefore fits well into the overall assessment that the relevance of flood control policy for the implementation of small-scale floodplain restoration projects, which are usually dominated by biodiversity or recreation targets, is comparatively low. Generally, flood control policy in the Garonne valley uses traditional technical measures and is aimed at the preservation of already existing floodplains. Ecologically oriented restoration is, as yet, not an integral part of flood prevention strategies.

Lack of support by the Water Agency

Although the Water Agency usually subsidises floodplain restoration measures (see above), it has not yet decided to give financial backing to the Bourret case. Therefore, the lack of support by the Water Agency can be regarded as a constraint on the project's implementation. The Water Agency has only a rather small budget for river restoration and does not see itself as a *maitre d'oeuvre* of the corresponding measures.

Plurality of actors

Finally, the necessary coordination of the numerous relevant actors and the fragmentation of competences between the various sectoral administrations (nature conservation, water, forestry) and between the different decision-making levels (local, state administration, departmental administration) caused the decision-making process to be long and complicated.

8.6 SELECTED EMBLEMATIC ISSUES

In order to tap the potential of the case study to illustrate important institutional features of floodplain restoration in France, three emblematic issues have been chosen for an extended discussion. First, the project presents an interesting property rights aspect which is at the heart of the implementation problem of floodplain restoration on the Garonne. Second, the case study can be used to outline some diverging perceptions regarding the objectives of floodplain restoration and, thus, it is possible to demonstrate that even between environmental actors different problem perceptions are apparent. Third, the case study offers the opportunity to outline the institutional fragility of small-scale floodplain restoration stemming, *inter alia*, from the weak local capacities to support such projects.

8.6.1 The property rights aspects of floodplain restoration

'Domaines Publics Fluvials' and the role of the DDEs

In current practice, the Garonne is a 'domanial' river with the consequence that responsibility for its management and the exercise of property rights are placed upon the state. In addition to the river bed and the river's body of water, land regularly flooded is part of the public domain (see Amigues and Desaigues 2002). Therefore, the majority of floodplains and wetlands close to the river have the legal status of 'domaine public fluvial' (DPF) with the consequence that the departmental administration DDE is in charge of its management. This legal tradition goes back to the ancient legal category of navigable watercourses and the extent of state property on the floodplain areas as defined in article 8 of the French Water Property Code.

Currently, in the middle section of the Garonne, approx. 1600 ha of floodplains are managed by the DDEs in the two relevant Départements (Haut-Garonne, Tarn-et-Garonne). The DPF areas represent around 60% of the total wetland area between Toulouse and the Tarn confluence (see ECOSYS 2001). In principle, this property rights structure could facilitate floodplain restoration because the relevant projects do not directly affect private property rights and the acquisition of land from private landholders or financial compensation for the abandonment of current land use would not be necessary. In addition, land use conflicts in the public domain could largely be avoided because mineral extraction, an important form of land use in the past, was forbidden in 1994. As a result industrial interests are no longer relevant in the DPF areas (see CETEF 2001). Despite these comparably advantageous conditions, floodplain restoration faces serious constraints.

Consequences of land leasing

In practice, the DDEs lease out the DPF areas if there are interested private parties or communities that intend a form of land use which does not harm the legal mandate of the administration. Today, half of the DPF area in the Département Tarn-et-Garonne is leased, mostly for agricultural and forestry purposes but hunting and fishing rights also play an important role (see

ECOSYS 2001; CETEF 2001). The maximum duration of the leasing contracts is 18 years, with the possibility of extension. The consequence of these contractual arrangements is that even though the Garonne's banks and floodplains are mostly in the public domain, political efforts to upgrade floodplains are regularly confronted with private or local interests.

This situation is further complicated by the fact that the legal definition of the DPF as 'regularly flooded areas' is a rather fuzzy concept for the assignment of property rights. Sometimes the DPF is not precisely recorded in the land register. Occasionally landowners do not even know whether, or to what extent, their land belongs to the public domain. Although most land along the river banks is state property, there is a broad agreement not to use legal entitlement to impose restoration measures on implicit private 'landowners' on DPF areas or on the holders of leasing contracts. Therefore, certain environmental provisions are only imposed on the occasion of the renewal of a contract (Interviews 3, 4). But even when environmental provisions are in place, the DDEs, due to their lack of financial resources, encounter many difficulties in controlling the behaviour of the riparian landowners efficiently. As a result, prohibited constructions and non-authorised agricultural practices are frequently observed along the river banks (Interview 8).

The special case of Bourret

Nevertheless, the Bourret project underlines that even within the framework of the existing allocation of competences, and in the face of uncertainty as to possible changes in the legislative framework, the implementation of floodplain restoration is possible. One of the success factors was the lack of economic interests in the area, as leasing contracts with private parties only concerned a small part of the project area (approx. 7 ha). Since the forests currently leased are not affected by the restoration measures, the private leaseholders agreed to the project. Moreover, without the readiness of the local authority to lease the area and to take on the co-financing of the restoration costs as well as general responsibility, the DDEs would not have supported the project.

8.6.2 Different perceptions of the 'right' restoration measures

The Bourret case demonstrates that floodplain restoration faces serious problems resulting from diverging perceptions with regard to the ecological objectives to be pursued. In the Bourret project all the relevant actors principally showed strong agreement on the fact that something had to be done for the floodplain, but there were clear-cut differences between the local authority, the forest administration (ONF) and the environmental actors (DIREN, NMP) about what precisely should be done and which targets should dominate. The most important controversial issues were the type of forest restoration measures and the manner of the reconnection of the old arm of the river.

Restoration of floodplain forest

There were major conflicts over forest restoration in the first stage of the project. The original proposal for the forest management plan by the ONF was dominated by a production-oriented point of view. Not surprisingly, this was criticised by environmental scientists (Interviews 7, 8). However, the measures that were finally proposed and implemented succeeded in reconciling environmental and economic interests to a certain degree. The solution which was finally chosen was, first, to restrict the area in which new plantations could be established, secondly, largely not to touch the existing monocultural poplar forests and, thirdly, not to establish a 'pure' natural alluvial forest but a semi-natural mixed forest. This solution satisfies the interests of the environmental policy actors because the project could serve as evidence that the ecological enrichment of riparian forests in the Garonne valley is possible without abandoning economic interests. With the ongoing monitoring programme on the ecological effects of the measures, which is carried out by the NMP, a key environmental actor has the chance to regularly evaluate the outcome of the project. Still, environmental scientists complain about the small extent of the area in which restoration measures have actually been implemented and are unconvinced of the general ecological effectiveness of the approach (Interviews 3, 7). For the local decision-makers, the floodplain should be restored primarily to restore the beauty and harmony of the landscape. In any case, neither forestry production interests nor ecological imperatives were the main arguments for the local authority to agree on the forest restoration.

The simple fact that a consensus on forest restoration in the Bourret case could be achieved can itself be regarded as a success. Traditionally, there is strong disagreement between the forestry actors and the environmental policy actors in the Garonne valley regarding the necessity and the economic consequences of environmental provisions in reforestation and forest management (Interview 3). The background is the general economic importance of poplar and the state and regional subsidies which, since the mid-1990s, have provided strong incentives to landholders to establish monocultural poplar plantations (see CETEF 2001). In the past, even in ecologically important areas, poplar cultures have been established with serious negative effects on bird protection and other nature protection objectives. These ecological impacts of poplar reforestation are heavily debated. While actors from the nature protection side complain about the impoverishment and homogenisation of the landscape, forestry actors stress the positive impacts on water purification and erosion protection in addition to economic arguments (see CETEF 2001). Against this background, it is worth mentioning that, currently, the existing regulatory framework (e.g. Code Rural, Code Forestier) only provides some general, weak environmental obligations for forest management, and economic incentives in the form of 'silvi-environmental measures' are not yet in place (Interview 3).

In recent times, this general conflict could be slightly reduced because of the existence of a consensus-oriented and trust-building working group consisting of the relevant water, forestry and environmental policy actors (e.g. DIREN, Water

Agency, NGOs, departmental administrations, forestry organisations, scientists). The function of the forum is to develop guidelines on the consideration of environmental aspects in the management of riparian forests in the Garonne valley. One intention of this working group is to reduce possible discrepancies in specific forest conversion projects and to objectify the relevant debates against the background of the declaration of parts of the riparian forests as a NATURA 2000 site. A common decision-support document of the forum was published in 2001 in order to give technical advice on the environmentally oriented restoration of riparian forests (CETEF 2001). The decision-making process in the Bourret case was able to benefit from this parallel process not only because of the final outcome but also due to the fact that important actors were both members of this working group and participants in the Bourret project committee. Thus, the Bourret project underlines both the continuing diverging perceptions regarding the role of biodiversity targets in forest conversion and the importance of a consensus-oriented forum and personal contacts to enhance the willingness of the actors to reach a compromise. However, the project also illustrates that the consistency of a biodiversity protection programme depends upon a sufficiently wide acceptance of the programme's aims by forestry and local actors (Interviews 3, 8).

Reconnection of the old meander

The reconnection of the ancient meander was heavily disputed due to different problem perceptions and cultural backgrounds. Apart from the financial and technical aspects mentioned above, the restoration could not yet be implemented because the reactivation of the meander is fundamentally contested on the part of the environmental actors. For example, environmentalists and the DIREN claim that the proposed technical method of reopening the abandoned channel would tend to create an artificial and static environment with the exclusive objective of recreating a specific fish habitat but not sufficiently taking into account a dynamic ecological perspective. In addition, the proposed technical solution could also necessitate 'unnatural' vegetation and further additional measures to protect neighbouring banks via rock fills, an implication that is clearly in contradiction to the general objective of stimulating the river dynamic as a whole and increasing the sediment flow. Some ecologists had even demanded the removal of the existing rock fills at the site, which were constructed to protect a road close by, and could, therefore, not easily accept additional rock fills. The key to this dispute is the different attitudes towards the development of a 'natural' environment and varying perceptions regarding the potential implications for its usability. While the local community represents the recreation interests of the citizens and tends to be in favour of a 'controlled' environment (Interview 8), the ecologists call for natural dynamics to be allowed to work and regard the artificial reconnection of old arms as a second-best solution. For them, a site-based reconnection of ancient meanders is insufficient and ineffective in ecological terms, preferring instead a more holistic approach of giving the river 'free space' to develop its own meanders naturally. Although the reconnection of the old arm is still on the agenda of the Bourret project, the problems resulting from diverging perceptions and additional financial and technical constraints make a short-term restoration unlikely.

8.6.3 Institutional weakness and adaptive management

Dependence on local conditions

The willingness to support floodplain restoration at the local level can be substantially enhanced if local benefits are associated with the project measures. The main reasons for the initiation of the Bourret project were the local objectives of creating habitats for fish reproduction and regaining recreation opportunities (Interviews 1, 3). In addition, local actors aimed to upgrade the site in order to obtain "proper" riparian land.

Successful restoration projects depend on the existence of favourable local conditions as there is no coherent strategy concerning the role of the environmental and water policy actors in the implementation of a basin-wide strategy (Interviews 1, 7). Of course, there are general provisions for the protection and restoration of floodplains in various planning documents (SDAGE, ORF etc.) and the designation of parts of the floodplains as a NATURA 2000 site, but these factors are not sufficient for the integrated management of floodplain restoration.

Lack of adequate instruments

One of the principal shortcomings is that it has proved impossible to develop an adequate instrumental approach to tackling the most important threats to natural alluvial forests (extension of irrigation agriculture and poplar plantations, general loss of river dynamic). Also, forestry policy has not effectively integrated floodplain restoration into its instruments and measures. So far, no compensation scheme for the initiation of the restoration of alluvial forests is in place and there is no consensus relating to spatial priorities in floodplain restoration. Additionally, no robust financial instruments for the support of local communities exist.

Environmental and water policy have also failed to develop an adequate approach to tackling the general loss of fluvial dynamic in the valley. For example, the implementation of low-flow management is largely considered to be ineffective in ecological terms (see Guilloy-Froget 2002) although with the low-flow management plan (plan de géstion des étiages) there is a collaborative management approach in place. Furthermore, as mentioned above, there is no convincing strategy for an ecologically oriented development of the DPF areas in the Garonne valley. To summarise, a coherent approach and an effective integration of the different relevant sectoral policies is not in place and the various forms of state intervention lack coordination and cooperation.

Lack of adequate institutional support

In the Bourret project, one key actor is missing who could promote and actively support floodplain restoration on a regional or inter-municipal scale (Interview 5). As mentioned above, the active involvement of the DIREN was a key driver for the initiation of the project but the DIREN is neither willing nor able to regularly push floodplain restoration at the project level.

There are two possible scenarios for enhancing institutional involvement in river restoration. First, more powers and financial resources could be allocated to the Water Agency, which was barely involved in the Bourret project for the reasons mentioned above. This scenario is not very realistic so long as the main powers for managing the ecologically interesting floodplains are allocated to the departmental administrations due to the legal status of the DPF. Since the Water Agency has no official power over the floodplains it does not see itself as a promoter of restoration projects but, rather, as a facilitator when the projects are already ongoing. Furthermore, the Water Agency's budget for floodplain restoration would have to be considerably increased – already the agency cannot financially support all the proposed measures due to its small budget for river management. The Bourret case illustrates that the limited involvement of the Water Agency clearly puts a constraint on project implementation. A further aspect is that the SDAGE, which represents the programmatic back-up for the Water Agency's activities, does not provide detailed restoration goals or site-specific or local land-use norms, because the SDAGE is a basin-wide programme and sets only rather general objectives. In principle, the basis for the Water Agency's activity could be substantiated and strengthened via the adoption of a SAGE (*schéma directeur d'amenagement et de gestion des eaux*). The SAGE is a water management plan which allows the introduction of sub-basin or even local provisions for floodplain development (see Monstadt in Chapter 5 of this volume). However, the SAGE procedure, which implies negotiations between the parties involved (state administrations, local communities, users and landowners), seems far too complicated and administratively costly for such a programme (Interviews 1, 4).

The second possibility would be to upgrade the role of an inter-municipal or inter-departmental organisation which might serve as an institutional link between the national and regional political objectives of protecting and restoring the riparian landscape, on the one hand, and the local implementation level (which is too small to effectively support floodplain restoration on its own) on the other hand. In the Bourret case, all the relevant actors agreed on the fact that the lack of local resources is one of the main constraints on project implementation and follow-up activities. In principle, the SMEAG (Syndicat Mixte d'Etudes et d'Aménagement de la Garonne) could fill this institutional gap because of its specific role in water management in the Garonne river basin. As an inter-departmental and inter-regional organisation with financial autonomy, the

SMEAG represents a functional organisation aiming to produce regional public goods in water management (Lahaye 1999). The SMEAG deals mainly with the management of low flows with the objective of ensuring water availability for the principal users.

In recent years, the SMEAG has extended its activities somewhat. For example, it attempts to coordinate floodplain and wetlands conservation and restoration with special emphasis on the floodplains downstream of Toulouse. Interestingly, the SMEAG aims to involve the relevant stakeholders in the development of a coherent approach to floodplain restoration downstream of Toulouse where experiences already gained at the project level are to be integrated (Interviews 5, 6). The SMEAG's activities have not yet had much effect, however. The SMEAG is particularly interested in the Bourret project because it believes that the project could not only demonstrate the potential creation of win-win solutions in floodplain restoration (recreation interests, forest conversion, fish reproduction etc.) but also the fruitfulness of co-operation by the actors involved (local authorities, environmental administration, forest cooperative etc.) (Interviews 5, 6). So far, the enlargement of SMEAG's floodplain restoration activities has been hampered by budget constraints and the small size of its staff. Moreover, since the SMEAG's activities were traditionally concentrated on the management of large dams and low flows, its 'new' role as a coordinator of the restoration of riparian ecosystems is not yet well established as far as its expertise and its acceptance by other relevant actors are concerned.

As the results of other approaches to floodplain restoration in France indicate, strengthening inter-local or inter-departmental task-specific organisations which support and cooperate with the other relevant actors (DDE, Water Agency, local communities etc.) might be a fruitful solution. Here, the option of allocating more tasks to the SMEAG, combined with adequate resources, might be part of a reasonable policy mix (Interviews 1, 5). Currently, the SMEAG appears to be prepared to take on a more important role in floodplain restoration but the 'scaling' of the optimal institutional approach is still to be resolved. For example, for some purposes the scale of the SMEAG's activities as a basin-wide organisation might be too large and too far away from the local actors to effectively promote river restoration and to guarantee the participation of local communities. There is, therefore, clearly a need for the development of a flexible multi-level perspective.

8.7 CONCLUSIONS AND RECOMMENDATIONS

8.7.1 Policy development

Many ecologically important floodplains in the Garonne valley belong to the public domain and therefore, in principle, the state has a great deal of influence over the use of the river and the development of the floodplain. However, the implementation of floodplain restoration measures is hampered by the fact that there are no specific legal obligations and the competent authority does not have sufficient financial and organisational resources. Furthermore, an action plan for the restoration of the state-owned floodplains has not been developed, which means that the implementation of the relevant projects depends on a number of favourable and rather coincidental conditions.

There is no consensus regarding spatial priorities in floodplain restoration in the Garonne valley and robust financial instruments for the support of the small local communities are not in place. Currently, the selection of sites tends to depend on the nature of property rights and the interests of the landowners, while ecological priorities play only a minor role. Floodplain restoration is therefore characterised by conflict avoidance and is largely concentrated on individual locations – such as the Bourret site – where resistance by private parties is low.

Inter-sectoral coordination must be substantially improved. Some progress has been made in recent years, but forestry, water, nature protection and flood control policy appear far from being sufficiently coordinated due to the lack of comprehensive plans and programmes as well as actors with a coordinating perspective.

The financial and organisational weakness of the small communities involved is a restrictive factor for the implementation of projects, so that not only additional financial support but also inter-communal cooperation on river restoration have appeared on the political agenda. Here, various approaches to institutional reforms are currently under consideration, but the 'scaling' of an adequate institutional approach is not a trivial task. The allocation of more tasks and authority to the SMEAG, combined with sufficient resources, might be an adequate step, which should be combined with the establishment of new forms of local governance between Toulouse and the Tarn confluence in the long term (e.g. regional nature park, *Pays*).

Apart from the necessity of developing a restoration strategy for the state-owned floodplains and implementing small-scale projects such as that at Bourret, adequate instruments have to be developed to tackle the most important threats to natural alluvial forests (extension of irrigation agriculture, poplar plantations and general loss of river dynamics). Possible measures might be the more specific use of agri-environmental instruments, the development of a compensation scheme for the initiation of biodiversity measures in forestry and the improvement of low-flow management.

8.7.2 Project development

The project underlines the importance of individual actors at the local level for the success of restoration projects and the necessity of offering benefits at the local level. In this specific case, the local interests in fish reproduction and angling opportunities were crucial factors. In general, the project illustrates that floodplain restoration depends heavily on local support and the mobilisation of local actors is therefore a crucial prerequisite. However, low local and organisational capacities were principal constraints on the project and the need for local co-financing of flood restoration projects will clearly be one crucial bottleneck for future schemes.

The establishment of the project committee was an important participatory and coordinating instrument used to bring the actors together and to handle conflicts. Interestingly, and in spite of the small area which was effectively restored, there was considerable actor involvement (departmental administrations, NGOs, forestry actors etc.) which resulted in a somewhat complicated and time-consuming decision-making process. Therefore, well-organised project management is clearly a matter of concern in order to avoid frustrations and the withdrawal of individual actors.

The project illustrates that even if all the relevant actors are strongly agreed on the fact that the restoration of a floodplain should take place, there are clear-cut differences about what precisely should be done and which targets (biodiversity, forestry, recreation etc.) should dominate. Here, external expertise and scientific assessments as well as the pursuit of a consensus-oriented strategy by the environmental actors appear possible approaches for preventing conflicts from jeopardising a restoration project of this kind.

8.8 REFERENCES

Amigues, J. P. and Desaigues, B. (2002) *A Cost-Benefit analysis of biodiversity conservation programmes in the Garonne valley*, OECD, Environment Directorate, Paris.

Belaud, A.. and Lim, P. (1994) Role et intérêt des 'bras-morts' de la Garonne. *Revue de l'Agence de l'Eau Adour Garonne* **60**, 15-20.

CAREX (2000) *Réhabilitation fonctionelle et écologique du bras mort de Bourret*, Aix-en-Provence.

CETEF – Centre d'Etudes Techniques et Economiques Forestier Garonnais (2001) *Préoccupations Environnementales et Gestion des Boisements Riverains de la Garonne*, Toulouse.

Comité de bassin Adour-Garonne (1996) *Schéma directeur d'aménagement et de gestion des eaux du Bassin Adour-Garonne*, Agence de l'eau Adour-Garonne, Toulouse.

ECOSYS (2001) *Pour une gestion globale et concertèe de la zone humide garonanaise de Blagnac à St. Nicolas de la Grave*. Final Report, Study commissioned by the EPTB Garonne, Saint Nazaire.

Décamps, H., Fortuné, M., Gazelle, F. and Patou, G. (1988) Historical influence of man on the riparian dynamics of a fluvial landscape. *Landscape Ecology* **1**(3), 163-177.

Guilloy, H. (2002) Dispersal, Germination, and Survival of Populus Nigra L. (Salicaceae) in Changing Hydrological Conditions. *Wetlands* **22**(3), 478-488.

Lahaye, N. (1999) *Gouvernance territoriale et espaces d'interet public: L'enjeu d'un developpement durable territorial*, LEREPS, Université des Sciences Sociales, Working Paper, Toulouse.

NMP – Nature Midi Pyrénées (1995) Gestion *globale et protection du fleuve Garonne*, Toulouse.

NMP – Nature Midi Pyrénées (1996) *Protection et restaurations des forets riveraines de la Garonne,* Plan de gestion forestier et itinéraires techniques du site du ramier de Bigorre (1997-2006), Toulouse.

NMP – Nature Midi Pyréneés (1999) *Diagnostic écologique et suivi scientifique du bras mort de Bourret*, Toulouse.

ONF - Office Nationale des Forets Castre (1998) *Réhabilitation de boisements alluviaux de Garonne : le Site de Bourret*, Castre.

SMEAG – Syndicat Mixte d'Etudes et d'Aménagement de la Garonne (2002) *Schema Directeur d'Entretien Coordonné du Lit et des Berges de la Garonne*, Toulouse.

Conseil Général de Tarn-et-Garonne (2006) *Cadre D'actions sur le Corridor Garonnais.*

SMEAG - Syndicat Mixte d'Etudes et d'Aménagement de la Garonne (2006) *Projet du SAGE. Vallée de la Garonne*, Toulouse.

INTERVIEWS

1. Water Agency Adour-Garonne, Resources and Rivers Division.
2. DIREN, Regional Environmental Administration Midi Pyrénées, Natural Risks and Flood Warning Division.
3. DIREN, Regional Environmental Administration Midi Pyrénées, 'Region Inventory and Environmental Policy, Forest and Environment'.
4. DIREN Regional Environmental Administration Midi Pyrénées, 'Region Inventory and Environmental Policy, Wetlands'.
5. SMEAG, Interdepartmental and Interregional Organisation for Water Management on the Garonne (Mrs. C.).
6. SMEAG Interdepartmental and Interregional Organisation for Water Management on the Garonne (Mr. L.).
7. Nature Midi Pyrénées, Toulouse.
8. Co.Fo.Gar Forest Cooperative Garonne.
9. DDE Departmental Administration for Public Works, Haute-Garonne.
10. DDE Departmental Administration for Public Works, Tarn-et-Garonne.
11. DDAF Departmental Administration for Agriculture and Forests, Haute-Garonne.

9

The Long Eau floodplain restoration project: "flying by the seat of your pants"

Timothy Moss

9.1 INTRODUCTION

This chapter analyses a scheme to restore a floodplain on two small sites on the Long Eau River in Lincolnshire, England.[1] The scheme, conducted between 1992 and 1996, entailed removing embankments and creating new ones further away from the river so as to allow periodic flooding of the formerly protected floodplain. Designed originally to enhance biodiversity in an area of low

---

[1] The author would like to thank Manuela Siener for her help in analysing data and drafting an early version of part of this chapter.

conservation value, the project produced, in addition, flood protection benefits by increasing floodwater storage capacity.

What makes the Long Eau scheme intriguing from an institutional perspective is how it could prove so successful – albeit on a small scale – within such a short period of time and at a very modest price. At a total capital cost of some £70,000, over 600,000 m^3 of flood storage were created, besides the additional benefits for wildlife, landscape quality and recreation (Interview 2). Given the problems facing floodplain restoration schemes today in generating support in multiple policy fields, seeking complementary funding sources and coping with growing bureaucracy, it is illuminating to explore what made an early generation scheme like the Long Eau project work. This chapter is interested in explaining the reasons for the relative success of the Long Eau scheme but also, secondly, what opportunities it missed to build on its strengths.

Researching a restoration scheme already completed by the mid-1990s has two important advantages. Firstly, it provides the opportunity to trace the scheme's development from the initial idea to full implementation and to reflect – with the benefit of hindsight – on its strengths and weaknesses. Secondly, it enables us to contrast the institutional and organisational contexts of floodplain restoration schemes in the early 1990s with those applicable today. Since all the key actors from the Long Eau project are engaged in similar activities today it has been possible to question them on the relative merits of implementing projects of floodplain restoration now and in the past. Interviews with these actors form the empirical basis of this chapter – by necessity, as well as by design – as there exists very little publicly available documentation on the Long Eau project (for a list of interviewees, see Section 9.8). Apart from a leaflet published by the Environment Agency (EA 1999) and brief references in presentations by those involved (Hall and Tartellin 1994; Smith 2001; Sisson 2001; Wardle 2001) the principal study has been by Morris *et al.* (2005), which investigated the Long Eau project briefly as one of five case studies of washland management in England.

The chapter begins with an overview of the physical and policy contexts of the Long Eau scheme. There follows a brief description of the scheme: its objectives, its development, the physical interventions, actor involvement and the instruments used. The results of this descriptive section are then summarised in terms of the principal drivers and constraints applicable to the project. On a deeper analytical level, the chapter reflects on four emblematic issues of the Long Eau scheme and concludes with lessons and recommendations on project management and policy development for floodplain restoration.

9.2 THE PHYSICAL CONTEXT AND USE OF THE RIVER

9.2.1 Physical characteristics of the river

The Long Eau flows from the Lincolnshire Wolds into the North Sea at Saltfleet Haven, to the South of the River Humber (see Figure 9.1). Despite its name, the Long Eau is only ca. 25 km long and drains a catchment of 22.3 km^2 of predominantly agricultural land in East Lincolnshire (on the following, EA 1996; EA 2007). Together with the Great Eau, which it joins some 6 km from the coast, it forms a catchment area of 112.3 km^2.

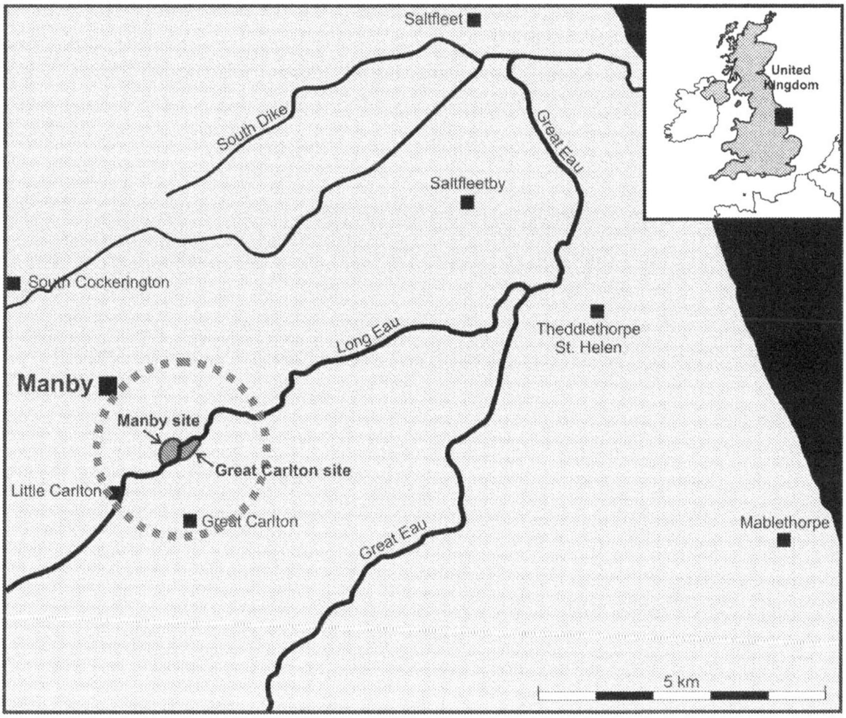

Figure 9.1: Map locating the Long Eau project area.

The Long Eau is part of the so-called highland river system which drains runoff from the upper catchment in the west to the coast across a low-lying coastal plain (EA 2007: 34). In this coastal plain, where the floodplain restoration scheme is located, the river has been embanked and channelled to protect adjacent land from flooding. Here, separate systems of field drainage exist which feed into main drains by gravity and pumping stations. This water is prevented from entering the highland river system by the river banks. When and where desirable, it is pumped into the river artificially.

The Long Eau is thus a highly regulated river. The floodbanks slope steeply into the water with no ledges at the water edge. The river bed is dredged to maximise capacity and the banks mown regularly to maintain their flood defence functions. Consequently, the river generally provides a very poor quality habitat for wildlife. By contrast, the biological quality of the water is generally good. In 2005 the Environment Agency's General Quality Assessment for the Steeping, Great Eau and Long Eau Catchment Abstraction Management Strategy showed that over 78% of the area had 'very good' to 'good' biological water quality (EA 2007: 38). Pressures on local water resources have lessened in recent years, largely as a result of a decline in demand by industry along the Humber. However, the Long Eau occasionally suffers from shortfalls in water quality caused by discharges from a sewage treatment plant and local septic tanks (EA 2007: 38).

9.2.2 Use and management of the river

The area around the Long Eau is essentially rural. Apart from the upper reaches, which include some wildlife corridor habitats – notably spring sites, areas of freshwater wetland and areas of marsh, woodland and hedgerow – the catchment is an intensively cultivated area. The agricultural land is primarily arable and fallow, with cereals such as wheat and barley as the predominant crops (EA 2007: 39).

This was not always the case. Until as late as the mid-20th Century the catchment of the Great and Long Eau had a long history of washlands, with the rivers flooding regularly onto the adjacent floodplains. Some of the older inhabitants – including one of the farmers/landowners involved in the Long Eau restoration project – can recall what the landscape was like before the river was heavily modified:

> "And I can remember when it all was flooded, when I was a boy, you see, because we used to go skating on it in the winter time when it all froze up, and it was lovely. Then it would take half the summer to dry out. [There were] a lot more birds: lapwings, snipe, plover, all sorts of redshank, golden-eye – a lot." (Interview 4)

From the 1940s onwards the river was embanked and canalised in order to enhance and protect adjacent agricultural land. The river was dredged during the 1960s, and between the 1960s and early 1970s the riverbed was deepened, widened and banked for purposes of water storage, land reclamation as well as flood defence. In this period, the banks alongside the rivers were heightened greatly. Major bank-raising schemes were being completed on the Great and Long Eau as late as the 1980s (EA 1996: 6). Up until the mid-1980s sizable grants were still being paid for draining land and ploughing up ancient grasslands (Interview 6). As along most of the rivers in Lincolnshire, the engineering work resulted in the dominance of a uniform, canalised and maintenance-dependent waterway system, and most of the landscape's diversity has been lost. Increased water abstraction, low river levels and a decline in water quality has led to the reduction and virtual extinction of many aquatic plants and to a general impoverishment of wildlife (cf. EA 1999; Lincolnshire Wildlife Trust 2000).

Despite the considerable efforts to provide flood-proof riverbanks the highland rivers continued to top their banks regularly, causing flood damage to crops on the floodplains. By the 1990s the National Rivers Authority responsible for flood defence on main rivers, such as the Long Eau, and the local drainage boards responsible for the lowland system were coming under growing criticism from the affected farmers for not maintaining river banks adequately (Interview 7). A particularly damaging flooding event in 1993/94 prompted a study into improved flood protection in lower reaches of four rivers (including the Great and Long Eau) which drain into the Saltfleet Haven (EA 1996).

9.3 THE POLICY CONTEXT

9.3.1 National policy context

Before turning to the floodplain restoration project itself it is important to assess the extent to which the origin and development of the scheme was framed and shaped by national and regional policy relevant to river and land management. The policy fields most pertinent to floodplain restoration initiatives primarily include flood defence, water protection, nature conservation and agriculture.

It is a distinctive feature of the Long Eau project that its origins – with one notable exception – were not informed by national policy. This was not because the key actors were unaware of, or deliberately sought to circumvent, policy initiatives. There existed at the time – 1992/93 – no policy targets in the fields of flood defence and river management which could have promoted floodplain restoration projects. The exception was the Countryside Stewardship Scheme, which had recently been launched in 1991 and was at the time managed by the Countryside Commission. This scheme, today termed Environmental Stewardship, offers grants to farmers for a period of 10 years in exchange for changes in the way their land is managed which enhance the environment and improve habitats. The Countryside Stewardship Scheme proved instrumental in gaining the support of the two farmers/landowners for the floodplain restoration measures on the Long Eau (see Section 9.4.5).

Since implementation of the restoration project a number of major shifts in government policy relating to nature conservation, flood defence and water management have created more favourable policy contexts for a project of this kind (see Moss in Chapter 6 of this volume; also Moss 2007). Although too late to influence the project implementation process, they have undoubtedly contributed to giving wider credence to the integrated way of approaching water and land use upon which the Long Eau project was founded.

9.3.2 Initiatives in the catchment

At the regional level of the river catchment or the County of Lincolnshire, it is also difficult to identify any policy or planning guidance which can be said to have had any direct influence on the initiation or development of the Long Eau project. Various engineering studies on the river's flood defence had been conducted previously (Interview 2). There had been a few isolated initiatives for restoring sections of the river, though not a floodplain. Most significant was a survey conducted immediately prior to the project which identified 18 potential sites for washland restoration along the Great and Long Eau, including the two sites under study here (cited in EA 1996: 9). The Lincolnshire Wolds – of which the Long Eau catchment is a small part – were designated as an Area of Outstanding Natural Beauty in 1973, but this status does not appear to have influenced the project directly.

As at the national level, several subsequent policy initiatives in Lincolnshire have highlighted the importance of the issues addressed by the project. This applies particularly to the Lincolnshire Biodiversity Action Plan (BAP), launched in May 2000, which includes rivers and wetlands as one of six identified habitat groups and comprises several plans for improving wetland biodiversity (Lincolnshire Wildlife Trust 2000). Chalk rivers, such as the Long Eau, are identified in the plan for Lincolnshire as a priority for conservation (EA 2007: 38). At the mouth of the Long Eau is a conservation site of international importance, the Saltfleetby-Theddlethorpe Dunes. It is designated as a Special Area of Conservation pertinent to the EU Habitats Directive and a Site of Special Scientific Interest (EA 2007: 37). Finally, a Catchment Abstraction Management Strategy covering the Steeping, Great Eau and Long Eau sub-catchments was drafted in 2007 and submitted for consultation. Designed to regulate water abstractions on the basis of water resource availability, it recommends continuing current licensing practice for the Great and Long Eau sub-catchments (EA 2007).

9.4 BRIEF DESCRIPTION OF THE PROJECT

9.4.1 Purpose and objectives

The objective of the Long Eau project was – in the words of an Environment Agency leaflet – the "restoration of floodplain storage to enhance flood protection and to improve wildlife habitat in the river channel, banks and floodplain" (EA 1999). By setting back or removing floodbanks and allowing periodic flooding of the formerly protected land the project was designed to contribute to the rehabilitation of the river and its floodplain. Of the two principal motives for the project – to improve flood protection and to benefit wildlife – the latter concern was the prime driver, especially in the early days. The principal actors were motivated above all by the desire to demonstrate how it was possible to achieve environmental enhancement in an area of very limited conservation value.

Rather than select a site of existing habitat potential which they could improve further, they deliberately chose one where any intervention was likely to show a marked improvement in biodiversity. As one of the key players put it:

> "...we started with clean sheets of paper if you like, in places where if you'd thrown in a shopping trolley it would have been an environmental enhancement. So you couldn't go wrong in these areas." (Interview 2)

The lowland section of the Long Eau catchment, with its uniform and canalised waterways and heavily drained arable farmland, fitted the criteria of low habitat value. In the words of a local conservation officer from the Environment Agency:

> "It's very much a man-made landscape of intensive agriculture and intensive drainage. So it's relatively easy to make a difference." (Interview 1)

Here it would be possible to achieve significant environmental enhancements merely by setting back the floodbanks and allowing the river to flood part of the adjacent land. This would allow a wetland habitat to develop on the new, shallower river margins and floodplain, which would be attractive to important flora and fauna. In addition to these aims directly related to the conditions of the watercourse, the project was also to provide both cover and feeding areas for ground-nesting birds and an increasingly diverse sward.

Since the restoration of connectivity between the river and its floodplain would inevitably have an impact on flooding, the rationale behind the project quickly acquired an additional flood protection dimension. It soon became clear that the key players were trying to demonstrate how it was possible not simply to enhance a low value habitat but to achieve cost-effective flood defence benefits by means of environmental improvements (Interview 2). At the time (the early 1990s) this objective was highly unconventional but it became a guiding motive for the duration of the project.

9.4.2 Origins, development and current status

It is widely held by the principal actors that the first step towards the future Long Eau project was made by a local officer of the Farming and Wildlife Advisory Group (FWAG) who drew up a report on prospects for schemes of river or floodplain rehabilitation on the Great and Long Eau in January 1993. On the basis of this report a selection of suitable sites was made. At around the same time a conservation officer at the regional branch of the then National Rivers Authority (now Environment Agency) became involved in the scheme. Together they organised a field trip to one of the potential sites in order to get local farmers interested in the idea of a floodplain restoration project. An officer from the Countryside Commission (CC) was present to inform the farmers of the payments available through the new Countryside Stewardship Scheme which could help offset some of the loss in revenue. Subsequently, the farmer/landowner at Great Carlton on the Long Eau showed interest in the idea. Already keen on nature conservation, he was attracted in addition to the prospect of receiving Stewardship payments for the less intensive use of land close to the river where the soil was heavy clay.

The four key actors at this stage – the FWAG officer, the NRA conservation officer, the farmer and the CC stewardship officer – began to liaise with each other regularly. From this point the project at Great Carlton moved forward rapidly. The FWAG officer surveyed the project area to assess its conservation potential. The NRA conservation officer gained the support of colleagues from the flood defence section of the NRA to draw up a full hydrological survey of the land, determining the location of the new floodbanks. A 10-year management plan was devised and the proposal for revenue under the Stewardship Scheme was approved. To cover the initial capital costs the NRA conservation officer had access to a budget for funding environmental enhancement projects over which he effectively had full control.

Under such favourable circumstances the restoration work was conducted within the space of a few months (see Figure 9.2). The existing floodbank was removed and a new one built further back from the river, creating an area of some 10 ha of open floodplain. At this site natural rises in the level of the adjacent land limited the height – and thereby also the cost – of the new embankments (on the physical interventions, see Section 9.4.3).

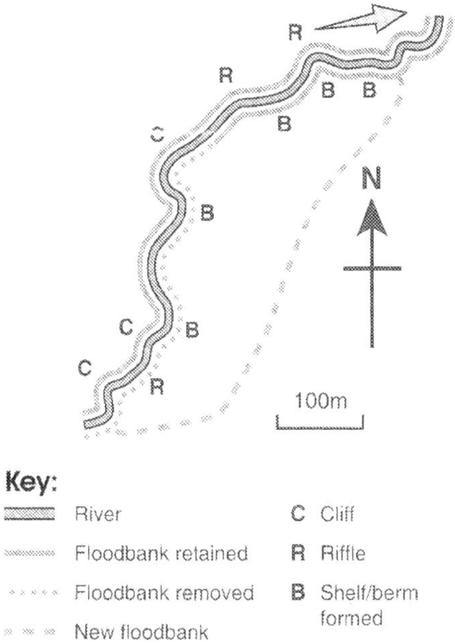

Figure 9.2: The site at Great Carlton. (Source: Environment Agency (1999), reprinted with permission).

The measures undertaken at the Great Carlton site persuaded the farmer who owned the land on the opposite side of the Long Eau at Manby to restore the floodplain on his land the following year (Interview 3; see Figure 9.3). Besides the attraction of Stewardship payments he was keen to find an alternative use for this land which was prone to frequent flooding and consequently not particularly profitable (Interviews 3, 4 and 7). Setting back the floodbanks at the Manby site proved a more complex task than at Great Carlton since a main drain – the Mablethorpe Upper Cut – ran through the middle of the floodplain and had to be re-routed further away from the river behind a substantial new floodbank (see Figure 9.3). This required the cooperation of the local Internal Drainage Board (IDB) – today the Lindsay Marsh Drainage Board – for consent to alter the water course (Interview 6). Even so, the small number of actors involved and their close cooperation ensured comparatively rapid implementation of the capital works as at the Great Carlton site.

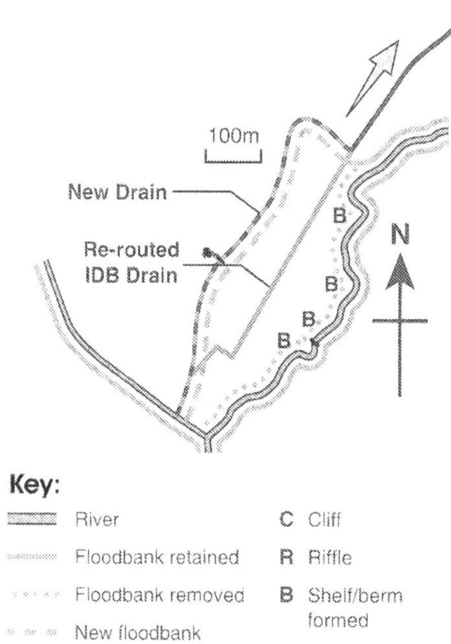

Figure 9.3: The site at Manby. (Source: Environment Agency (1999), reprinted with permission).

The Long Eau project has been completed for several years. All the physical interventions – primarily the setting back of floodbanks – were carried out at the Great Carlton and Manby sites between 1993 and 1996. All the representatives of the key agencies as they then were (FWAG, CC, NRA) agree that the principal achievement has been in improving flood protection. This is noteworthy given that the project was originally designed primarily for reasons of habitat enhancement and was started with conservation funding. Even the NRA conservation officer concedes today that "…the enhancements in flood defence were significant; environmental things came second" (Interview 2). The restoration measures on both sites created a total of more than 600,000 m^3 of additional storage space. The project area as a whole floods around 3-4 times a year for 3-4 days, but the lower lying ground on both sites can be a "semi-permanent lake" for most of the winter (Interviews 2 and 5). This additional flood storage capacity has increased the standard of flood protection provided in the area from a 1-in-10 year to a 1-in-30 year flood risk (Interviews 2 and 3).[2] This is corroborated by hydrological modelling over a 3 km stretch of the Long Eau at the two sites (EA 1999). Besides this reduction in flood risk, the flood level was lowered and – especially beneficial for the local drainage board – the water course became more manageable (Interview 6).

The conservation benefits are harder to quantify. General biodiversity has benefited greatly from the change in land use from arable to pasture and from the introduction of extensive forms of land management. Restored wetland habitats of wet grassland and open water have attracted a lot of wintering and migratory birds. Birdwatchers have recorded sightings of 47 species of birds at the Great Carlton site, including redshank, shellduck, snipe and avocet. It is disputed, however, whether the project has provided successful breeding sites for wading birds (Smith 2001; Interview 7).

Maintenance work on the sites has been limited and the costs low. The Environment Agency estimates the savings on the maintenance of the floodbanks to amount to approx. £400-500 a year (Interview 3). The maintenance that is required – such as the grazing of the new banks at the Great Carlton site – is largely covered by the Countryside Stewardship payments. These payments were based on a 10-year agreement, originally due to run out in 2003/04 (Interviews 2, 3, 4, 5 and 7).

9.4.3 Physical interventions

In general, the physical work carried out on both sites consisted of removing the floodbanks and thereby creating a flood storage area adjacent to the river. Especially at the Great Carlton site, natural rises in the level of the adjacent land could be used to maintain the level of flood protection. Only where this option did not exist, the removed flood banks had to be replaced with new, less steep embankments. Setting back the embankments now allows periodic flooding of the formerly protected land adjacent to the river (EA 1999). This has created a situation akin to a natural floodplain in which neither the inflow of water to the floodplain during a flooding event nor the subsequent outflow back to the river is regulated.

At the Manby site the work was more cost- and labour-intensive because the main drain had to be diverted behind the new, setback floodbank. In addition, a sewage treatment plant in the immediate vicinity needed protection from flooding from either water system by means of adequate embankments. The material excavated from the former embankment was used to form the new floodbank (Interview 3). Besides the setting back of embankments and the re-

---

[2] It should be noted that the existing 1-in-10 year standard already met the government's guidelines for agricultural area and therefore did not in itself justify the measures taken under the project (Interview 3).

routing of the main drain, a number of wet ledges and berms were created along the channel of the Long Eau in order to allow wetland marginal habitats to be established (EA 1999). This created ca. 2.2 km of wetland river margin on both banks of the river. In addition, riffles serving to increase the diversity of instream flow and bed material were introduced as a means of in-line habitat improvement; some cliffs for kingfishers were installed on the outside of the bends. The Stewardship scheme provided funding for fencing and gates to enable the farmer on the Great Carlton site to reintroduce grazing (Hall and Tartellin 1994). Public access to both sides of the river was made possible. On the Great Carlton site the farmer built at his own expense a bird hide with permission for access.

The Long Eau project also included the planting of some 600 trees, mostly along the ends of the meanders. These were intended primarily as a landscape feature (Interviews 1 and 2). At no time during the implementation of the project was there any consideration of using the opportunity to create a wet woodland habitat, either by planting trees or by allowing natural succession via periodic flooding (Interview 2). Woodland was widely regarded as not being a feature of the area historically (Interview 1). More importantly, it was felt by the key actors that trees would provide a habitat for predators of the threatened species of birds which the project was keen to bring back to the area. These species require wet grassland, which was – and still is – a habitat rapidly disappearing in Lincolnshire (Interview 7). The FWAG officer justifies the priority for wet grassland as follows:

> "If you look at Lincolnshire or the rivers we're talking about, there are very few species that I can think of that are still here that are threatened that live in wet woodland. Whereas there are a lot of species in Lincolnshire and other lowland lying parts of Britain – not just counting birds, incidentally – that need wet grassland, fennland, marsh" (Interview 7).

Interestingly, a few years after the completion of the project a self-generating copse of willows did grow on the Great Carlton site. It was removed, however, primarily for fear of providing protection for predators (Interview 5).

9.4.4 Organisation and actor involvement

A distinctive feature of the Long Eau project is its organisational simplicity. It never had a formalised structure or procedure, but operated essentially as an informal collaborative venture. The Long Eau scheme was conceived, planned and managed by less than 10 individuals. Leadership lay, *de facto*, with three officers from public agencies: the Farming and Wildlife Advisory Group (FWAG), the National Rivers Authority (NRA) and the Countryside Commission (CC) (Interviews 3 and 8). Each was highly motivated. The FWAG officer, a farmer himself who managed land under Stewardship schemes, was keen to develop a project with which wetland habitats could be created in combination with less intensive forms of agricultural production (Interview 7). The NRA conservation officer wanted to demonstrate how 'soft engineering' solutions of flood protection could not only enhance the environment but also provide more cost-effective flood defence (Interview 2). The Countryside Commission officer responsible for the Stewardship scheme, who was new to the area, was highly motivated to conclude "some cracking good agreements" with local farmers (Interview 8). Together, they complemented each other in expertise and influence, quickly developing a good working rapport.

These three actors were very successful in enrolling the support of other key players, notably a local NRA flood defence officer and a senior engineer of the local drainage board. This NRA officer recognised flood defence benefits from restoring the floodplain as well as the savings to be made on the maintenance of the floodbanks (Interview 3). The senior engineer of the local drainage board saw an opportunity to demonstrate the board's commitment to environmental enhancement as well as the operational benefits of reducing the flood risk at low cost (Interview 6). The two farmers/landowners on whose land the floodplain was restored were also instrumental in the successful implementation of the project. The farmer/landowner at the Great Carlton site, highly supportive from the beginning, saw benefits in the Stewardship payments, better wildlife and a new use for land that had not been easy to farm (Interview 5). His neighbour at the Manby site was appealed by the idea of restoring a landscape which he had known as a child and by the prospect of secure revenue for land which had been subject to frequent flooding (Interview 4).

An important motivating force for collaboration between these very diverse actors was that they carried out several of the physical interventions on the project themselves. Thus the local drainage board was contracted to re-route the main drain. The farmer of the Manby site was paid to remove the embankments on his land. The flood defence section of the NRA was also engaged to conduct some of the earth-moving works. Enrolling an organisation like the drainage board to conduct some of the physical interventions, rather than simply grant its statutory consent, is regarded as having been highly instrumental in gaining its support and avoiding serious difficulties during implementation (Interview 2).

The simplicity and individuality of the Long Eau project had, however, a disadvantageous effect on the involvement of the wider community. Internally, the project was highly inclusive. Externally, it remained relatively unknown. The parish council and the local community at large were not informed of the scheme, nor were they in any way encouraged to become involved in its implementation. Apart from the public access granted as part of the Stewardship agreements and the bird hide built on the Great Carlton site there was no concession made to the general public. It is estimated even by the key actors that local knowledge of the project and its achievements is likely to be very limited.

9.4.5 Instruments used

Implementation of the Long Eau project was greatly eased by the low cost of the work undertaken and the availability of several, complementary sources of capital and revenue funding which could be accessed rapidly. The initial source of funding for the project came from a funding budget under the auspices of the NRA conservation officer. He was largely free to spend money from this "nature conservation pot" on any project promoting habitat enhancement he chose (Interview 2). Since the funding only became accessible at the end of each calendar year and had to be spent by the end of the financial year there was huge pressure to get projects implemented and completed as fast as possible. In his own words:

> "I could use the money …at the end of the year. [This was] not necessarily the best time of the year to do them… but that's when the money appears, and so I would accept the money and we'd go with it… And so it was frenetic, quite manic… from Christmas till 4th March was just manic. We'd have an idea up to maybe February and then in four weeks we'd get out there, we'd have a scheme and create a washland." (Interview 2)

This conservation fund was not only a ready source of financing. Owing to the scope available to the NRA officer in allocating the money it could be used for funding improvements which went beyond direct nature conservation. In this way, it was possible to incorporate flood defence benefits in the remit. This was at the time a novel step, as the senior engineer of the local drainage board recalls:

> "On this occasion, it was money that had been set aside for environmental projects that actually had a flood defence benefit and that is very, very unusual" (Interview 6).

Physical construction work on both sites was also funded out of the NRA's flood defence budget. Interestingly, these funds did not come from the capital works programme which is targeted at a perceived flood defence need and requires feasibility studies to assess various options. Instead they came from the NRA's revenue budget earmarked for maintenance costs. This was possible because the setting back of the embankments was calculated to save maintenance costs in the long term. It is estimated that the cost for the flood defence budget of the new embankment on the Great Carlton site of ca. £2,000 was recouped in just 4-5 years owing to the savings on the maintenance of the old embankment of between £400 and £500 a year (Interview 3).

The Countryside Stewardship Scheme helped fund the changes in land management required to achieve the project's nature conservation objectives. Under the Stewardship scheme agreements are signed for a ten-year period during which time the farmer receives a regular payment in return for practising less intensive forms of agricultural production, such as managing riverside habitats, trees, bankside vegetation and other habitats and permitting public access (see Moss in Chapter 6 of this volume). The farmers/landowners at the Great Carlton and Manby sites entered into such agreements with the aim of providing cover of grassy areas for ground nesting birds and diverse sward. To this end stock is allowed onto the land only in the late spring. Light summer grazing by cattle is then permitted. No fertilizer is to be used on the land.

An important source of funding not usually documented in projects of this kind was the non- or low-costed labour of many of the key actors. The initial work in planning the project was not billed separately by the FWAG and NRA officers. The hydrological and engineering surveys were conducted in-house by the NRA and the local IDB. The two farmers dedicated part of their labour without payment (Interview 4). Heavy machinery belonging to either the drainage board or a farmer was put at the disposal of the operators free of charge. This helped reduce direct costs greatly.

Capital works for the Great Carlton site amounted to a mere £8,300. Of this, £3,350 was provided by the Countryside Commission and £4,994 by the NRA (Hall and Tartellin 1994). Revenue payments totalling £3,316 p.a. were paid to the farmer under his Stewardship scheme agreement. The cost of restoration work on the Manby site was considerably higher owing to the necessity of re-routing the main drain and protecting the sewage treatment plant from flooding, but even so amounted to only ca. £50-60,000 (Interview 2).

Interestingly, the Long Eau project made no use of planning instruments or instruments associated with the protection of designated sites as levers for accessing funds or securing environmentally beneficial practices of land management. As mentioned earlier, the key players deliberately sought sites of low habitat value. This was partly in order to achieve a greater net improvement in environmental quality. It was also, however, part of the strategy to avoid having a site in a designated area, as this would have required taking greater account of existing habitats and – more importantly – gaining consent for the physical interventions in a lengthy bureaucratic process. For this reason one of the interviewees regards the prospects of implementing a similar project to the Long Eau in a designated area as highly unlikely:

> "God help us, no. That's the last thing we want. ...The number of hoops you have to jump through if you are doing anything near an SSSI..., it puts you off." (Interview 3)

9.5 PRINCIPAL DRIVERS AND CONSTRAINTS OF FLOODPLAIN RESTORATION ON THE PROJECT

9.5.1 Drivers

Reflecting on the project as a whole, a number of drivers which helped the Long Eau project to emerge and develop can be summarised.

Several physical/environmental factors were instrumental in persuading the key actors to become involved in the project. The habitat quality of the area, and in particular of the river and its banks, was widely regarded as being very poor and in need of major enhancement – a powerful motive for the three *de facto* leaders of the project. A further physical factor was the incidence of flooding from the highland river system into the adjacent fields causing damage to crops. Despite heavy investment in the flood defence infrastructure the existing floodbanks were overtopping when the river was full, prompting local criticism of the both the NRA flood defence section and the local drainage board. The low productivity of some of the land designated for the project provided an additional motive for the farmers.

The organisational environment proved particularly favourable for a project of this kind. One of the most important drivers was the availability of adequate funding. Unusually, it was possible within the space of a few months to access funding from three different sources (the conservation fund of the local NRA, Countryside Stewardship and the flood defence revenue budget of the local NRA) which covered not only the (modest) cost of capital works but also the cost of maintenance and new land management practices – at least to the extent of providing an adequate incentive to the parties involved.

Secondly, the project offered multiple benefits to all the principal stakeholders at a relatively low cost. The farmers were receiving regular payments for the less intensive management of small, relatively unproductive parts of their farmland. The Countryside Commission was getting a wet grassland site without having to pay the capital costs. The flood defence section of the NRA was getting additional flood storage – i.e. better flood defence – for the modest price of the new embankments. The local drainage board was benefiting from the additional flood protection without incurring any direct costs, and received contracts for carrying out the construction work. Finally, the conservation officer of the NRA and the FWAG officer were able to demonstrate that it was possible to achieve habitat enhancement whilst at the same time improving flood protection and maintaining the viability of the affected farms.

9.5.2 Constraints

The Long Eau project encountered extraordinarily few constraints during its implementation. This can be attributed in part to its limited scope (in terms of site size, funding requirements and number of stakeholders), in part to the relative absence of procedural obligations at the time, but in part also to the largely conflict-free nature of the relations between the key players. During the early planning of the project, initial difficulties arose as a result of the unfamiliar nature of the task. The restoration of a floodplain was a novel experience for most of the organisations involved, as well as for the two farmers/landowners. The regional NRA had never restored a washland before, the CC was inexperienced in creating wet grassland habitats, the IDB was running a risk in diverting a watercourse and the farmers faced the uncertainty of capital depreciation of their land and a Stewardship agreement which would provisionally last only 10 years. Overcoming reservations was the principal obstacle to the project, as the NRA conservation officer recalls:

> "But the complexity at this early stage meant that the hardest thing was not the engineering or even finding the money, it was convincing everybody and making everybody happy that you could do these sorts of works." (Interview 2)

There were initial objections from the IDB that were resolved in negotiation (Interview 1). The farmer at the Manby site was at first sceptical, in particular at the prospect of losing good arable land on the northwest side of the existing main drain (Interview 8). Once the project had been decided upon, however, no further constraints were encountered. The farmer at the Great Carlton site cannot recall any problems at all during implementation (Interview 5).

9.6 SELECTED EMBLEMATIC ISSUES

In the following section four emblematic issues are presented which illustrate characteristics of the Long Eau project that are of particular value in raising our understanding of the contexts favourable to restoring floodplains on a site of this kind. Two of the emblematic issues relate to distinctive strengths of the Long Eau project, the third to a significant weakness and the fourth to important differences in the institutional environment of floodplain restoration projects today.

9.6.1 The right people in the right place at the right time

In the analysis of projects of this nature it is common practice to play down the role of individuals and to highlight instead the regulatory frameworks, financial incentives and the influence of organisations as prime determining forces. If given any specific consideration, the individual actors involved are generally viewed as agents of their respective organisation, pursing its interests to a greater or lesser degree. Where the distinctive role of an individual is self-evident, the case for exceptionality is often made. At the same time, however, it is frequently observed by those engaged in projects of this kind that the success or failure of a scheme can depend heavily on the skills of the leading participants – skills that cover not simply expertise in the given field but also social skills (on the "champions" of river restoration schemes, see Adams *et al.* 2004: 1935ff.).

The Long Eau project demonstrates in an unusually stark manner how important individuals can be in shaping a restoration project. The project was not the product of a shift in government policy or of a new strategy of regional organisations involved, although these factors did play a supportive role. It was essentially the creation of three people engaged professionally in habitat enhancement schemes who were each keen to develop good demonstration projects and two farmers/landowners who were willing to take the risk of changing the use of part of their land. Each of these actors brought with them expertise and knowledge – whether on suitable wetland habitats, available funding, flood defence or land management – without which the project could not have developed as it did. They also had at their disposal together the means with which to implement the project, in terms of statutory authority, funding sources and organisational resources. Apart from these complementary resources, they individually possessed the social skills to work together constructively and innovatively for the duration of the project. This applied particularly to the three officers from FWAG, the NRA and the CC who quickly developed a "strong, friendly working relationship" based on trust (Interview 8). The ability of being able to rely on one another to take the project forward was highly important in generating and maintaining the motivation of each to contribute their own part. It was a case, in the words of one of them, of "having three people at the right time in the right place" (Interview 2).

A further point worth making is that the motivation of the key individuals to become so closely involved in the project was only in part shaped by their professional or business interests. There were important personal motives involved as well that contributed to the project's advancement. To give one example, the officer from the Countryside Commission was in friendly rivalry with her colleague in a neighbouring county to conclude as many good Stewardship agreements as possible. Being new to the area, she was ambitious to overcome reservations amongst the farming community about the scheme. A second example is the farmer at Manby who had fond childhood memories of flooded landscapes along the river before it was regulated. These often incidental dimensions to personal motivation need to be recognised for providing important driving forces for projects of this kind, particularly one that involves considerable changes over a long period of time.

9.6.2 Organisational settings conducive to innovation and implementation

> "The strengths [of the project] included flexibility, getting on with each other, lack of paper work, flying by the seat of your pants, being supported by senior officers internally." (Interview 8)

The individuals involved closely in the Long Eau project could bring their collective skills and interests to bear because the organisational setting within which they operated offered adequate opportunities and incentives to innovate. Restoring a floodplain is even today an unusual activity; in the early 1990s it was regarded as highly experimental. None of the key actors had conducted a project of this kind before. Apart from the two farmers they all needed the backing of their respective organisations to pursue the scheme. In the case of all three leading actors their superiors were willing to support their initiative and to grant them the freedom to experiment. At that time it was more customary than today for officers in the involved agencies to exercise a substantial degree of discretion in deciding over what projects to become engaged in and which farmers to support. Considerable trust was placed in the ability of officers to make these decisions. These "innovative and experimental powers" (Interview 8) were particularly welcome for those who were keen to engage in unconventional projects such as that on the Long Eau. Such flexibility allowed the actors to respond quickly to changing demands in the course of the project.

This applied equally to the stipulations attached to the use of the relevant funding sources. The conservation fund at the disposal of the NRA conservation officer was – at £30,000 – by today's standards quite substantial and could be used for a variety of purposes generally conducive to environmental enhancement. It was, for instance, not a problem for part of the money to be used for flood defence benefits where habitat improvements could also be demonstrated. The approval of agreements under the Countryside Stewardship scheme was dependent largely on the discretion of the officer responsible, based on his/her own assessment of desirability and viability.

Above all, the Long Eau project was distinctive for its straightforward implementation. Given the minimal bureaucracy involved, the physical interventions of setting back the floodbanks could be completed within months. It was impressive for all the parties to witness the speed and effectiveness of implementation. As one commentator remarked: "It was big and you could see it" (Interview 8). Achieving visible progress on the ground provided strong additional motivation for all the participants.

9.6.3 Missed opportunities for dissemination at the time

The success of the Long Eau scheme begs the question of what impacts it had on floodplain or river restoration in the wider region in the longer term. Given the relative ease with which it was implemented, its low cost and the high satisfaction levels of all those involved, one might expect the Long Eau scheme to have become a model for similar projects in Lincolnshire and beyond. This, however, was not the case. The Long Eau scheme remained an isolated intervention largely unknown in the public realm.

One reason for this was the lack of any involvement of the local community in the project at any time during its implementation. Because planning permission was not deemed necessary, the parish council was not consulted. Although the project entailed providing public access to both sites as part of the Stewardship agreements, there were no efforts to engage the local public in a discussion of the project. Given the dearth of publicity about the project, both during and after its completion, there was little outside knowledge about the scheme. Today, the participants regard this as a lost opportunity for communicating the successes of the project, notably the benefits for flood defence and habitat quality and its exemplary function of multi-agency partnership.

Secondly, the achievements of the project have barely been analysed and documented, with the exception of a few isolated surveys on certain aspects of the scheme. A thorough, systematic post-project appraisal was never conducted. It is consequently difficult to determine what could have been done better with any degree of certainty. The absence of proper monitoring and evaluation is all the more surprising given that all the key actors regard the project in retrospect as a remarkable success, some even claiming it was one of the most fulfilling schemes they have ever experienced.

This can be linked to a third limitation of the project: the absence of follow-up activities. The Long Eau project was heavily dependent on a few key individuals. This was one of its strengths, but also one of its weaknesses. For when one of the key players moved on to another job and the focus of attention for the others moved to different types of projects interest in drawing lessons from the project declined. As one of the actors put it self-critically: "we took our eye off the ball" (Interview 8). As a result, there were no similar projects initiated in the region on the back of the success of the Long Eau scheme.

9.6.4 Limitations to duplicating the scheme today

In the wake of new administrative requirements for project managers to consult widely, enrol multiple stakeholders and justify their actions, one might imagine that the conditions today are more favourable for overcoming these weaknesses in the Long Eau scheme. Yet, looking back on the Long Eau project, all the key players agree that it would be highly difficult to implement a similar project under the institutional conditions now prevailing in the UK: "Today it would be a far different process that we'd go through" (Interview 1). The procedure of planning, funding and managing a restoration project of this kind has become far more complex, time-consuming and expensive.

In the first instance, the stricter stipulations regarding the planning, funding, consultation procedures and impact assessment of projects like this are having the combined effect of preventing them being implemented as quickly and as effectively as in the past. Projects of this nature are today subject to detailed environmental impact assessments, landscape assessments and cost-benefit analyses conducted prior to commencement (cf. Ledoux *et al.* 2005: 141). Many of these procedures require formalised consultation processes with a wide range of affected government agencies, NGOs as well as the general public. Since the Stewardship scheme has grown in size it has become more bureaucratic, relying on a point-scoring scheme rather than the local officer's discretion for assessing an agreement.

Whilst it is widely accepted that these institutional requirements have their justification as a means of improving accountability, policy integration and cost-effectiveness, the overall impact is to make the work of those responsible much harder. As one conservation officer comments:

> "The main problem now with all this is that some things have just got a lot more complicated. It's not necessarily a bad thing because a lot of the things we have to do mean that we get better schemes. But the trouble is the cost." (Interview 1)

Conservation and flood defence officers find themselves spending an increasing amount of their time managing the paperwork at their desks rather than exercising their skills out in the field. The scale of the new requirements means that it has become very difficult for a single organisation to manage a restoration project alone. Match-funding and co-management from other organisations is often needed, if not as a formal requirement then at least as a practical necessity. Yet, in the words of one conservation officer: "Managing collaborative projects is a bit of a nightmare in terms of the time that it takes" (Interview 1).

Stricter costing procedures have also been introduced in most agencies. Engineering work carried out by agency staff or equipment loaned out is now costed to the department using the service. The practice on the Long Eau project, where labour and equipment of the NRA were put at the disposal of the project managers, would not be feasible today (Interview 3). The common practice of key agencies contracting out work to consultants or other bodies has further complicated implementation procedures. Conservation officers at the Environment Agency (formerly NRA) no longer possess a fund which they can distribute at their discretion. It has become hard to raise even the smallest amounts of money to support habitat enhancement projects:

> "We don't have the scale of funding with the conservation budget to do a scheme like this anymore. So we'd have to be looking for other people to work with to do anything." (Interview 1)

Finally, the practice of using targets to guide and supervise the work tasks of agency officers has had the (unintended) effect of reducing interest in more high-risk projects, such as restoring floodplains. An officer at the Environment Agency explains how this happens:

> "You know, there are a lot of …demands but the ones that …make a difference are the ones …you are set as targets and you are not set …a target to create a wetland area. So it doesn't …bring me any money in, it doesn't help my salary. Whereas if I miss my targets then it does affect me, so, you know, I'm not going to jeopardise my position. If …a site comes along, even though it might be really well worth doing for a lot of reasons, if that means that I have to jump through a lot of hoops, …if it means I've got to spend hours and hours and hours to try and make it happen, then it's not going to happen." (Interview 3)

The time and resources spent on additional consultation exercises, planning procedures, preparatory studies and inter-agency collaboration is discouraging project managers from coming forward with project ideas. In particular, it is having the effect of reducing sharply the prospects of medium-sized projects, as a conservation officer at the Environment Agency explains:

> "It does mean, because of the complications, it's not worth doing unless it's a big scheme. So, effectively, you end up with very small schemes that are very, very simple, as in putting in a riffle or something, or you do big schemes where you create something like 80 hectares of salt marsh… You basically lose all your medium-scale schemes like this." (Interview 1)

The project manager is today often faced with the choice of either going for a project that is so small that it does not cross the thresholds which require the complex consultative and planning procedures or creating a major scheme which draws on the combined resources of multiple stakeholders.

9.7 CONCLUSIONS AND RECOMMENDATIONS

The Long Eau project set out to improve habitats for wildlife in the river channel, banks and floodplain whilst at the same time enhancing flood protection in the area by creating greater flood storage at a critical point on the river. The key players from the lead organisations hoped to demonstrate with the project how it was possible to enhance a low value habitat whilst at the same time achieving cost-effective flood protection benefits and maintaining the viability of the farms affected. This ambitious task was at the time – in the early

1990s – highly unusual, even on such a small scale as at the Great Carlton and Manby sites.

In retrospect, and in the opinion of those closely involved, the Long Eau project succeeded in meeting these objectives on the whole. The relocation of existing floodbanks back from both sides of the river created an additional flood storage space of 600,000 m^3, which increased the standard of flood protection on that reach of the river from a 1-in-10 to a 1-in-30 year flood risk. The total cost of this achievement – at ca. £70,000 – was extremely low. Significant habitat improvements to the area, notably the restoration of wet grassland in conjunction with the naturally flooding floodplain, have contributed to an increase in biodiversity, attracting large numbers of wintering and migrating birds. Given this record of success, what lessons can be drawn from the experience of the Long Eau project for floodplain restoration policy and practice?

9.7.1 Project management

The Long Eau project, whilst being distinctive in many ways, displays many features of an early generation of floodplain restoration schemes across Europe. Like other projects enacted in the early to mid-1990s it was small in scale and site-based. It involved a small number of actors and relatively straightforward funding mechanisms and implementation procedures. This enabled the project to be completed rapidly. The focus of interest was primarily on nature conservation issues, although flood defence concerns came to play an increasing part as the project progressed. In this respect, the degree of sectoral integration was relatively high. Typical of other early generation schemes, though, the level of interaction with national policy was very low. Only in the case of the Stewardship scheme can a direct impact of national or regional policy on the project design be detected. Finally, the project – like others of this period – had a low public profile; no attempt was made to enrol the support of the wider community or strategically important players.

Several factors contributed to the successful management of the Long Eau project. Firstly, the few key actors involved in the project were highly committed and complemented each other with their skills and resources. They had the backing of their superiors and enjoyed relative freedom to innovate. The two farmers on whose land the floodplain was restored were very cooperative. Three different funding sources for both the capital costs and revenue payments were made available at short notice. Moreover, there were no difficulties in using these funds for achieving benefits conventionally financed by other sources, notably a conservation fund used *inter alia* for improving flood protection. In the absence of any requirements for planning permission or impact assessments, the project was implemented on each site within the space of a few months.

Against this must be set the failure of the Long Eau project to generate similar, follow-up projects in the region. Although universally regarded as successful by all the participants, including the farmers, the project was not replicated on other sites along the Long Eau. Given the relative success and ease of implementation of the Long Eau scheme it is surprising that there was not a more concerted effort to persuade other farmers to follow suit. This reveals a structural weakness in relying too much on individuals. The freedom which the key actors required to implement the scheme so effectively was equally a freedom to move on to other issues following completion. The free rein which their organisations gave them was very valuable in encouraging innovation, but proved a weakness when it came to dissemination and follow-up activities. This suggests that some form of sensitive and un-bureaucratic monitoring of progress and potential by key public agencies is an essential complement to providing trusted experts with the leeway they need to be innovative.

9.7.2 Policy development

It is instructive to reflect on the lessons that a restoration project of this earlier generation hold for the challenges facing similar projects today. The institutional environment of floodplain restoration in the UK has changed substantially since the completion of the Long Eau project. Projects of this kind today need to attract co-funding from multiple sources, conduct extensive consultation procedures and prove cost-efficiency to senior management and the general public. Each of these requirements is, in its own right, often highly justifiable as a means of improving accountability, policy integration and cost-effectiveness. Their cumulative effect, however, is to severely complicate the tasks of implementing a project of floodplain restoration. The principal actors on the Long Eau project – all engaged in comparable tasks today – are in agreement that conducting a project of identical design would nowadays cost at least three times as much and take considerably longer.

Consequently, project ideas are often shelved at an early stage or run into serious difficulties during implementation. Experienced project managers, when they pursue projects at all, react by keeping their projects so small that they avoid the procedural requirements or by building major collaborative projects with multiple funding sources. We can observe, therefore, the emergence of a policy delivery gap between the intentions and the impact of government policy initiatives. Measures designed to improve the overall quality of interventions to enhance the environment are combining to create an institutional setting that is in several respects not conducive to delivering the policy targets at the operational level.

What recommendations can be drawn from this case for the development of policy pertinent to restoring floodplains? At a time when considerable effort is being given to refining the design of policy instruments for flood protection, nature conservation, sustainable agriculture and rural development so that they more readily deliver integrated policy objectives in a cost-effective way, it is important to reflect on how these instruments affect the scope for action of those entrusted with delivering policy targets at the operational level. This case study of the Long Eau project suggests that inadequate consideration is currently being given to the cumulative effect of these instruments and the policy signals they embody. A shift in perspective is needed so that experiences of project management at the operational level can feed back into the policy-making process. This will also require greater appreciation of the role of individuals – as addressees of institutional arrangements – in determining how policy initiatives work in practice. More concern needs to be given, in other words, to the games **real** actors play, and not just to assumed responses to given policy signals. This itself requires greater consideration of the contexts of action of individual actors and how these can create – or close – windows of opportunity for the restoration of floodplains or other valuable habitats.

At the same time as policy development needs sensitising to the problems encountered at the operational level, project management needs to be accorded greater self-responsibility. The Long Eau project illustrates very well what constructive, cooperative forces can be unleashed when the participants have the scope to shape substantively the content and style of a project. Particularly in the case of innovative projects where experience is limited, greater flexibility is needed in coping with the problems that inevitably arise in such complex undertakings. This requires a substantial degree of support for project managers from their respective organisations. It also requires that the project managers themselves cultivate support from within and beyond their organisations by publicising the project. They too could contribute to bridging the existing policy delivery gap by providing stronger feedback on the impact of policy initiatives and by seeking ways of exploiting new policy initiatives to the advantage of projects under their tutelage.

9.8 REFERENCES

Adams, W.M. Perrow, M.R. and Carpenter, A. (2004) Conservatives and champions: river managers and the river restoration discourse in the United Kingdom. *Environment and Planning A* **36**, 1929-1942.

EA (Environment Agency), Anglian Region (1996) *Saltfleet Haven Study. Final Report.*

EA (ed.) (1999) *Case studies: Great and Long Eau.* Leaflet. Bristol.

EA (2007) *The Steeping, Great Eau and Long Eau Catchment Abstraction Management Strategy. Consultation Document, January 2007.* Environment Agency, Lincoln.

Hall, J. and Tartellin, M. (1994) The Great Carlton / Manby Project 1993. In *Proceedings of the NRA Anglian Region Conservation Symposium, 14.04.1994* (ed. National Rivers Authority).

Ledoux, L., Cornell, S., O'Riordan, T., Harvey, R and Banyard, L. (2005) Towards sustainable flood and coastal management: identifying drivers of, and obstacles to, managed realignment. *Land Use Policy* **22**, 129-144.

Lincolnshire Wildlife Trust (ed.) (2000) *Lincolnshire Biodiversity Action Plan. Habitat action plan: Rivers, canals and drains.* www.lincsbap.org.uk/Rivers/rcd1.htm. Accessed on 06.12.2006.

Morris, J., Hess, T.M., Gowing, D.J.G., Leeds-Harrison, P.B., Bannister, N., Vivash, R.M.N. and Wade, M. (2005) A framework for integrating flood defence and biodiversity in washlands in England. *International Journal of River Basin Management* **3**(2), 105-115.

Moss, T. (2007) Institutional drivers and constraints of floodplain restoration in Europe. *International Journal of River Basin Management.* In press.

Sisson, D. (2001) Drainage channels – corridors for wildlife. In *Action for Wildlife in Lincolnshire. Rivers and Wetland Bulletin & Forum Conference 2001* (ed. M. Tartellin), pp.9-13.

Smith, P. (2001) Maintaining and restoring habitat on Lincolnshire rivers. In *Action for Wildlife in Lincolnshire. Rivers and Wetland Bulletin & Forum Conference 2001* (ed. M. Tartellin), p.14.

Wardle, R. (2001) Wetland opportunities with agri-environment, farmers and others. In *Action for Wildlife in Lincolnshire. Rivers and Wetland Bulletin & Forum Conference 2001* (ed. M. Tartellin), pp.17-19.

INTERVIEWS

Interview 1: Two conservation officers at the Environment Agency, representing the Anglia Region and the Northern Area, 4 December 2001.

Interview 2: Former conservation officer of the (then) National Rivers Authority, now Environment Agency, 15 May 2002.

Interview 3: Local flood defence officer of the Environment Agency, Anglia Region, 18 September 2002.

Interview 4: Farmer/landowner at the Manby site, 18 September 2002.

Interview 5: Farmer/landowner at the Great Carlton site, 19 September 2002.

Interview 6: Senior Engineer at the Lindsay Marsh Drainage Board, 19 September 2002.

Interview 7: Officer of the Farming and Wildlife Advisory Group (FWAG), Lincolnshire, 20 September 2002.

Interview 8: Stewardship officer for Lincolnshire of the (then) Countryside Commission, now Countryside Association, 20 September 2002.

Part D

NEW GENERATION SCHEMES OF FLOODPLAIN RESTORATION

10

The relocation of a dyke on the River Elbe: floodplain management as a challenge for intersectoral and multilevel coordination

Jochen Monstadt

10.1 INTRODUCTION

The relocation of dykes has developed into a key issue in the public discourse on ecological river restoration and flood prevention in Germany. This is particularly true since the disastrous flooding of the river Elbe in 2002, following which the restoration of floodplains has developed into one of the main policy issues of flood prevention (cf. Monstadt and Moss in Chapter 4 of this volume). One of the largest and institutionally most complex projects in Germany is a dyke relocation project on the middle stretch of the River Elbe, located in the "Flusslandschaft Elbe" biosphere reserve (Elbe river landscape). Around Lenzen, a small town in the federal state of Brandenburg, the restoration of 420 ha of floodplain area is currently being implemented. Dyke construction started in autumn 2005. Being

 ISBN: 1843390906. Published by IWA Publishing, London, UK.

one of the largest, most expensive and elaborate floodplain restoration projects in Germany, the dyke relocation project has attracted considerable political and technical attention and received much publicity as a consequence of the Elbe flooding in 2002.

This dyke relocation project represents a new approach towards the management of river environments and integrates modern approaches of both nature conservation and flood protection. Unlike traditional restoration projects it is not primarily intended to conserve a river ecosystem, which is already rich in rare and valuable species and biocoenosis, but to develop and restore a cultivated landscape with high ecological potential. Furthermore, the project does not focus on technical flood protection alone but also aims to minimise flood risks by giving more space to the river and by increasing flood storage capacity in the floodplain. In addition, the project targets the sustainable economic development of a rural region by promoting soft tourism, organic agriculture and environmental education.

In the following study this integrated project is analysed from an institutional perspective. Starting from an analysis of the physical and policy contexts of this restoration initiative, we investigate the development and the main objectives of the project, the instruments used and the stakeholders involved. Building on this, institutional drivers and constraints and emblematic issues of the project are discussed. The case study reveals a combination of driving forces that are of interest for deducing success factors for restoration projects. Strategic project management, professional fundraising, regional networking, extensive research activities, early stakeholder participation and personal involvement constitute driving forces which might also be relevant for other restoration projects. However, the analysis reveals that although the conditions for project development seemed to be conducive – considering the support of various actor groups, national policy shifts in different policy areas and the favourable funding conditions – such integrated projects are confronted with severe institutional hurdles. Project managers have had to cope with considerable institutional complexity in pursuing various policy objectives, applying several policy instruments and in meeting the interests, expectations and formal requirements of different funding organisations, of local farmers and land users and of numerous stakeholders from different policy communities and policy levels. This has complicated and delayed project development considerably: the decision-making process took nearly 15 years before the project was finally adopted.

10.2 THE PHYSICAL CONTEXT AND USE OF THE RIVER

10.2.1 Physical characteristics

The course of the Elbe flows from the north-eastern corner of the Czech Republic south-westerly almost to Prague and then continues north-westerly, diagonally crossing a large portion of eastern and northern Germany before emptying into the North Sea. As one of the longest rivers of central Europe, the Elbe spans a course of 1,091 km, commands a catchment of 148,268 km^2 (IKSE 2001: 7) and moves volumes of water typically ranging from 150 m^3/s in its upper length to around 300 m^3/s at the Czech-German border and 880 m^3/s near its mouth (Stubbs and Bonjour 2002: 1). Even though 1,384 m of altitude difference are covered along its course, the Elbe flows for most of its length as a slowly meandering waterway and most of the altitude differences are encountered within the first few kilometres. The major tributaries are the Vltava (Moldau), the Mulde, Saale and the Havel, which contribute substantially to the Elbe's current sediment load (ibid.). The catchment of the river Elbe encompasses the territory of four countries; however, the river itself and 99% of its basin are confined to only two: the Czech Republic (approx. 1/3) and Germany (approx. 2/3).

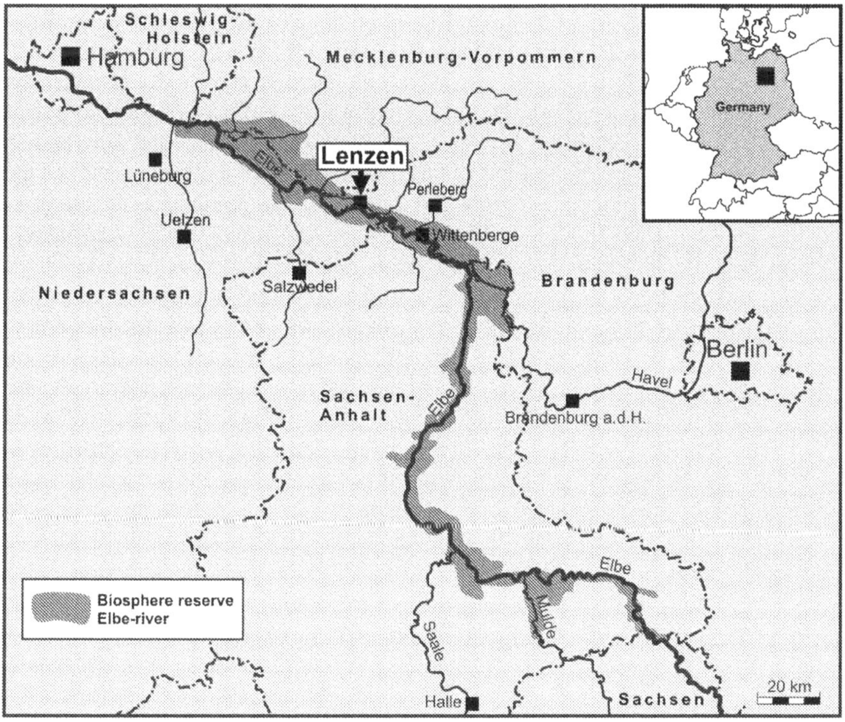

Figure 10.1: Map locating the Lenzen project area.

Three main landscape types can be distinguished along the upper, middle and lower reaches of the Elbe (cf. Stubbs and Bonjour 2002: 1). At its high altitude origin in the Czech Republic, tall peaks and cliffs of weathered sandstone dominate the region. The character of the lower reach of the Elbe is formed by the beginnings of the estuary and coastal landscape. The city of Hamburg and especially its container shipping port dominate the structure of the river for much of this lower stretch. The middle stretch of the river Elbe, where the project area of this case study is located (cf. Figure 10.1), is characterised by expanses of low-level landscape. The population density is much higher than in the upper reach of the Elbe. Agriculture plays an important role and industrial locations are prevalent.

Since the middle of the 19th Century the river Elbe has been transformed to make it navigable. As a result, the natural dynamics of the river have been restricted and its biodiversity has been diminished. The former natural vegetation of the hard- and softwood floodplain forests remained only in fragmentary form. Today, only 13.6% of the former floodplains of the Elbe can be used as a retention area for flooding (Jährling 1998). However, compared with other European rivers (e.g. the Rhine), engineered structures to control the waterway are less dominant along the middle stretch of the river. The river landscape and the hydro-morphological structure of the Elbe retains most of its natural characteristics as the former German Democratic Republic (GDR) did not implement its plans to canalise the river Elbe and to develop it as a waterway which would be navigable all year round. The landscape of the Elbe is characterised by extensively used willows and meadows, some residual hard- and softwood floodplain forests and different types of water bodies, but also dunes and dry grassland. Floodplain forests used to be widespread in the Elbe valley, but intense pressure by settlements and different land uses limited the extent of these forests dramatically over past centuries. Today only little of the original floodplain forests are left as small remnants on the Elbe. One of the main areas, which have hardly been altered by development, is located in the border region of the German Länder Lower Saxony, Saxony-Anhalt and Brandenburg. In this region the floodplain forests and flood meadows are the habitat of endangered species like beavers, storks and cranes.

10.2.2 Use and management of the river catchment

Although the river Elbe is considered move ecologically natural than other European rivers, the condition of the river has been changed tremendously. The regulation of the river course and interventions in the river environment by human settlement, shipping, agriculture, industry and flood protection have visibly altered the natural environment. It is only recently that nature conservation initiatives have tried to preserve and ecologically restore the river environment.

Shipping

Shipping has historically been very important within the Elbe region, both in connection with fishing and as a means of transportation. As the Czech Republic has no direct access to coastal waterways the use of the rivers for international trade has been, and still is, of economic importance. In order to make navigation possible all year round, three engineering projects were planned for the river Elbe.[1] Especially in the nature conservation community, these projects were considered to pose a serious threat to natural river ecosystems. It was suspected that the deepening of the river bed and the acceleration of the natural flow by the engineering measures would severely affect ecosystems and the landscape by contributing to the lowering of groundwater levels and the drying up of floodplain woodlands and other wetlands, as well as the loss of sedimentation etc.[2] As a result of the Elbe flood in the summer of 2002, the Federal government announced a temporary suspension of these projects on the Elbe. However, five years after the Elbe flooding the future of the engineering projects is still undecided and powerful advocates like the Czech Republic and trade associations have recently put the construction works back on the political agenda.

Agriculture

One of the major reasons for the continuous loss of retention areas on the river Elbe – aside from the increase of settlements in floodplain areas – is the intensification of the agricultural use of floodplains, which was made possible by building dykes. As long as the floodplains were periodically flooded, they were used extensively as grazed meadows and not as arable land, so they had a higher retention capacity. Land consolidation measures for the enlargement of the parcels of agricultural areas since the middle of the 20th Century eliminated swales, hedges, groves etc. As a consequence, precipitable water drains superficially and floods have become more probable.

Industry

Industrial pollution of the Elbe was a serious problem in the former GDR. Industrial sewage was barely treated until German reunification and water quality was very low. During the 1990s emissions from the main industries including pharmaceuticals, paper, metal, mining, leather and textiles were reduced significantly by new technologies, but also by the de-industrialisation of the East German region. However, a critical concentration of persistent metals in the River

---

1 For more details see www.wsv.de/Aktuelles/Projekte/Bau_Elbe/regelung_elbe/index.html, 01.12.2006.

2 For details see www.elbeinsel.de/basis/basis1.php, 01.12.2006. In addition to these ecological arguments, opponents of engineering measures suspect the constructions of intensifying the effects of flood events and call their economic soundness into question. Still, proponents of the projects point out that in comparison to road or air transport, water transport is the most ecological means to deliver heavy, mass loads of goods.

Elbe, such as cadmium and mercury, still remains. Even though water quality has been improved, there are still several industrial plants, especially along the upper part of the Elbe mainly in the Czech Republic, which do not yet meet the emission standards of the European Union (cf. Geller *et al.* 2002).

Flood protection

Flooding is a naturally recurring phenomenon which is to be expected given the very flat topography along the middle stretch of the Elbe River. Since 1850, about 86% of the natural retention areas of the Elbe River have been lost (IÖW 2002: 3) and only marginal parts of the original floodplain remain free for natural flood occurrences. In order to protect urban settlements, industrial zones, traffic routes, agricultural areas etc. dykes have been built along the whole stretch of the river, in Germany alone with a length of 1299 km (IKSE 2006: 34). Between 1991 and 2006 approx. 438 km have been restored. According to current studies, after the Elbe flood a further 528 km of dykes have to be reconstructed by 2015 for a total sum of €561.5 million (ibid.). In response to the disastrous flooding events in the summer of 2002, traditional flood management practices have come under increasing criticism and it is planned to supplement technical flood defence by measures to increase the water retention capacity in the river basin (IKSE 2003, 2006; cf. Monstadt and Moss in Chapter 4 of this volume).

Nature conservation

Even though the river ecosystems and the hydro-morphological structure of the Elbe were transformed by human activities, particularly the middle stretch of the River Elbe retained much of its ecologically natural form and condition. For that reason, UNESCO designated the first biosphere reserve in Germany on the middle Elbe in 1979, and expanded the Middle Elbe Biosphere Reserve to an area of 43,000 ha along a reach of 80 km after the reunification of Germany in 1990. Finally, UNESCO extended it to a new and much larger biosphere reserve in 1997. By creating this *"Flusslandschaft Elbe"* (Elbe river landscape) biosphere reserve, the length of the protected area grew to 375,000 ha along over 400 km of the river Elbe.[3] This new reserve extends across five German federal states (see Figure 10.1). It represents one of the biggest contiguous floodplain forests in Central Europe and the largest continuous river protection measure in the world. The biosphere reserve comprises a cultural landscape, designated as a World Heritage Site (The Garden Kingdom of Dessau-Wörlitz) and several Natura 2000 habitats. Current projects focus on the restoration of backwater lakes, beaver and willow management, floodplain forest development, environmental education and the promotion of tourism in the Elbe River region.

---

[3] However, in 2006 the extension in Saxony-Anhalt was partially cancelled and the reserve in this area was reduced by about one third from 190,000 to 126,000 ha.

10.3 THE POLICY CONTEXT

10.3.1 National policy context

Since the 1990s national policies in different policy fields such as spatial planning, nature conservation, agriculture and also flood protection have moderately shifted in favour of floodplain restoration (see for details Monstadt and Moss in Chapter 4 of this volume). In the summer of 2002 this promotion of floodplain restoration was intensified and the German government announced a far-reaching policy change in river basin management. Heavy rains in Central Europe had led to some of the worst flooding the Elbe region had witnessed in more than a century. The floods affected the cities of Prague and Dresden, and many villages. In Germany, 21 people lost their lives due to the floods and the direct damage to property amounted to more than €10 billion.[4] As a consequence, flood management practices have come under increased scrutiny. The Elbe flooding led to intense public discussion on the inadequacies of the current management scheme, the limits to technical methods of flood prevention and the negative effects of the ongoing development of the canalisation of the river for shipping. Furthermore, the floods have raised public awareness of global warming, which is likely to increase the occurrence of extreme weather conditions and flooding events.

Faced with a concurrent election campaign in Germany, the government gave high priority to the issue of flood management as a means of demonstrating political effectiveness. Immediately after the flood, it developed a five-point programme proclaiming a policy change towards precautionary flood protection. Moreover, the impending construction measures on the river Elbe were suspended. Based on this programme, the Federal government developed a Flood Control Act, with the aim of improving the management of flood risks, which was passed by parliament in May 2005 (cf. Chapter 4 in this volume). In addition, some weeks after the flood the German government announced the funding approval for the large protected area in Lenzen.

Another policy innovation relevant to the Lenzen project is the reform of the Federal Nature Conservation Act. Since 2002 it is possible to designate protected areas not only according to their present ecological value, but also regarding their future development and their ecological potential (cf. ibid.). As a result, the floodplain near Lenzen, which has been transformed by human intervention but has a high ecological potential, could be defined as a protected area in order to develop and restore it ecologically.

4 BMU 2006 (www.bmu.de/english/water_management/press_statements_speeches/pm/35486.php, 04.12.2006).

10.3.2 Initiatives in the catchment

Initiatives to restore floodplains on the river Elbe have existed since the beginning of the 1990s. As a result of the reunification of Germany the institutional basis for nature conservation in Eastern Germany has been improved and the Elbe river landscape biosphere reserve has been extended significantly. The conservation and restoration of the natural water retention capacity of floodplains was defined as a principal module of the international flood prevention action plan for the river Elbe, which was developed by the ICPE, the International Commission for the Protection of the Elbe (IKSE 1998).

Among the German Länder along the river Elbe the restoration of floodplains has been promoted predominantly in Saxony-Anhalt, while other German Länder along the river Elbe have not taken far-reaching initiatives. In Saxony-Anhalt, feasibility studies were carried out at an early stage proposing 32 projects for dyke relocation (Jährling 1994; Puhlmann 1994). Accordingly, these potential project areas were designated in the regional development plans as priority areas for flood protection.

As a result of the Elbe flood in 2002 the Federal government and the Länder announced a policy shift, departing from the traditional flood management approach of building dams and dykes to substituting or at least complementing it with more precautionary approaches (cf. Petrow *et al.* 2006). The main targets are to give space back to the river, to retain precipitation throughout the entire river basin in a decentralised way and to control urban development more effectively. Besides the coordination of immediate measures to repair dykes and the improvement of forecasting and warning systems, the ICPE was charged with revising the international flood prevention action plan according to these new priorities. The ICPE proposed that the Länder should take the initiative to relocate dykes and restore floodplains, and developed a list of appropriate areas for dyke relocation (IKSE 2003; cf. Figure 10.2). The Länder were requested to designate them as priority areas for flood protection in regional development plans and to promote the implementation of floodplain restoration. However, apart from in Saxony-Anhalt, most of the floodplain areas have not yet been designated in the preparatory zoning plans and binding development plans.

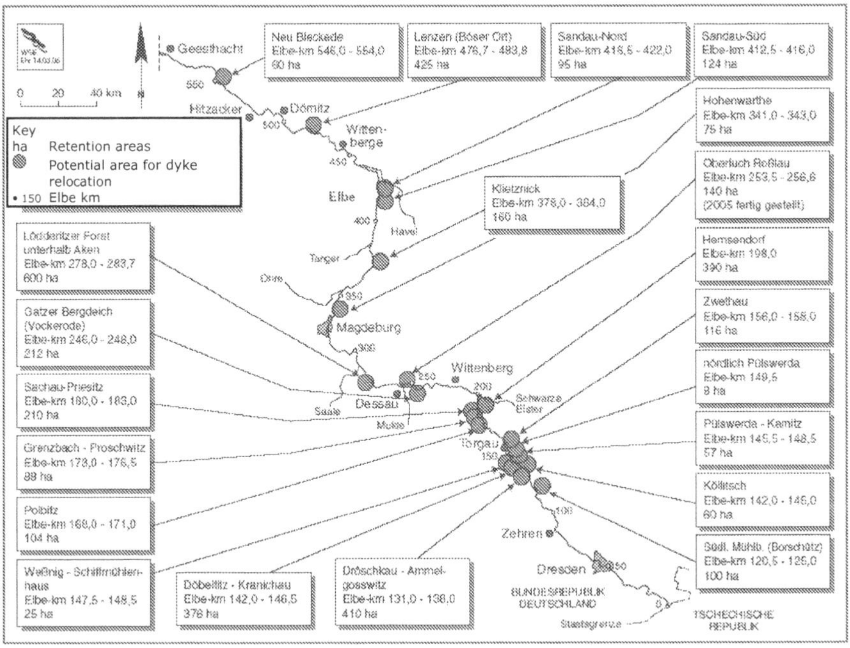

Figure 10.2: Potential areas for dyke relocation on the Elbe river. (Source: IKSE 2006: 25).

Four years after the Elbe flood the implementation of the proposed projects has failed to meet earlier expectations. In Saxony, the upstream German region on the river Elbe, implementation has not yet started and the flood protection plan only recommends potential areas for dyke relocation. In Saxony-Anhalt, where ambitious dyke relocation objectives were set up early, only one project near Roßlau had been implemented, as a compensatory measure for the building of a railway, by the year 2005. Another site in Saxony-Anhalt, the Lödderitzer Forst, has been designated as a large protection area for its natural floodplain forest and project implementation started in 2006. Four others of the proposed 32 projects are at an early planning stage (IKSE 2006: 24-26). Similarly, in the neighbouring Länder of Brandenburg and Lower Saxony only three projects for dyke relocation have been pursued. Among them, the dyke relocation project in Lenzen is the only project for which implementation is at an advanced stage (ibid). To summarize, although attempts towards coordinated and harmonized flood prevention have been made, there is still no consistent strategy at all planning levels and for all German riparian states (cf. Petrow 2006).

10.4 DESCRIPTION OF THE PROJECT

The relocation of dykes as a means of flood protection as well as for nature conservation is currently widely discussed among scientists and politicians. Despite a convincing scientific rationale in favour of such projects and various political proclamations, very few projects manage to get beyond the planning stage due to a large number of obstacles. One of the few projects being implemented is located near Lenzen, a small town in the Brandenburg part of the German "Flusslandschaft Elbe" Biosphere Reserve. Here, the restoration of 424 ha. of floodplain area is currently being carried out, with the intention of completing the construction works by the year 2008.

10.4.1 Purpose and objectives

The restoration of floodplain areas along the river Elbe has been discussed extensively since the beginning of the 1990s. Due to the high ecological value of the remnants of floodplain (forest) ecosystems as one of the most species-rich and most endangered habitat types, the preservation, but also the restoration, of floodplains has been promoted by conservationists. In the lower part of the Middle Elbe region near Lenzen, the river landscape has been well documented, it having constituted the last large floodplain forest on the north-eastern riverbank until the end of the 18th Century. The key objective of the project is thus to ecologically restore this historical floodplain and to create a biosphere network which reconnects the new floodplain forest to two existing ones nearby.

Even if nature conservation objectives have been the key driver of project development, the economic benefits of floodplain restoration as a strategy of flood prevention were another major argument. These benefits were confirmed by several feasibility studies on the relocation of dykes on the Elbe (Jährling 1993, 1998; Puhlmann 1994; BAW 2000). In particular at locations like the Lenzen project, where an abrupt river bend coincides with a narrowing of the river, flood prevention using technical measures is very costly. In these cases the objective of relocating dykes (and restoring floodplains as a side-effect) was even promoted by members of the technical flood protection department in order to minimise both the flood risks and the costs of an expensive refurbishment of the dykes. By increasing the river cross-section by means of the dyke relocation project in Lenzen the river level is expected to be lowered locally by 25 to 35 cm. This beneficial effect will continue upstream with a gradual decrease in the difference in the water level. The town of Wittenberge, situated about 25 km upstream, is expected to experience 5 cm of flood-peak decrease as a result of the project (Trägerverbund Burg Lenzen 2006).

Figure 10.3: Photo of the project area indicating the location of the relocated dyke. (Source: J. Purps, http://www.burg-lenzen.de/deichrueckverlegung, 08.01.07).

In addition to nature conservation and flood prevention objectives, an important argument for the restoration of the floodplain is its contribution to regional economic development. The project area is located in the structurally weak Prignitz region. This rural region is characterised by a lack of industry and other relevant regional economic activities except agriculture and by an unemployment rate of more than 20%. Faced with limited options for industrial development, regional development strategies focus on the endogenous potential of the region in sustainable agricultural production, soft tourism and environmental education, for instance with the commercial exploitation of regional products by cooperatives, with new services in soft tourism and recreation and by establishing a European centre for floodplain ecology. Moreover, the initial construction and plantation works as well as continuous maintenance and cultivation measures offer opportunities for local employment and income.

10.4.2 Origins and development

The difficult hydrological situation in the project area was recognised in the late 19th Century when the river Elbe flooded the adjacent land. Thus, some areas were declared flood-risk zones, where flood protection ought to be improved. More recently, the need for flood protection measures was demonstrated in the winter of 1988, when the ice of the river Elbe rose beyond the embankments and threatened to destroy the dyke. At the beginning of the 1990s scientists, representatives of the regional nature conservation authority and the largest local farmer reflected on ways of addressing this problem.

The specific idea of the dyke relocation project goes back to 1992, when experts from the State Agency for Large Protected Areas in Brandenburg (Landesanstalt für Großschutzgebiete – LAGS[5]) applied for EU LIFE funding in order to initiate a pilot dyke relocation project. After the EU and the state of Brandenburg approved funding, the project started in 1994 and was concluded at the end of 1998. Various scientific studies were conducted, e.g. on autochthonous soft and hard wood species, on the succession processes of floodplain woodlands and on different aspects of autochthonous vegetation, ground water hydrology and adapted agricultural land use as well as detailed analyses of the micro-relief. The project also encompassed the purchase of land (approx. 550 ha), the planting of alluvial forest (50 ha) and the establishment of a local nursery to grow native plant material for the initial plantation.

A focal point of criticism by sceptics from the technical flood defence and navigation policy communities was the uncertain effects of dyke relocation on the flow dynamics and flow rate of the River Elbe, on the water pressure at the future dyke openings, on local groundwater levels and on the navigability of the river. For this reason a further study was carried out by the Federal Waterways Engineering and Research Institute (Bundesanstalt für Wasserbau – BAW), which complemented the research of the LIFE project by examining the fundamental characteristics of the river's hydrology and morphology (BAW 2000). Two variants of dyke relocation were under discussion, the larger concerning the restoration of 670 ha of floodplain, the smaller only 400 ha. According to the hydrological calculations of the BAW, the smaller variant was more advantageous, because aggradations in the navigation channel of the river Elbe could be minimised. The hydraulic study recommended enlarging the flooding area to 400 ha by relocating the dyke. In order to prevent the river from meandering the study advises maintaining the existing dyke and creating openings in six places after the new dyke has been completed. Annual flooding of the area would then be expected due to the topographically low situation. By preserving the former dyke and ensuring the flooding area is not too close to the neighbouring settlements the study proved that the dyke relocation would have no notable negative effects on navigation, flow regimes and groundwater levels of the cultivated and built-up land in the Lenzen area.

In order to advance knowledge on ecological and socio-economic aspects of floodplains and their restoration the LAGS started an interdisciplinary project in 1996 funded by the Federal Ministry for Education and Research (BMBF). The

[5] In 2004 the LAGS was integrated, as the Department for Large Protected Areas, into the Environment Agency of Brandenburg (Landesumweltamt Brandenburg).

project was integrated into the national research programme "River Elbe Ecology", including 27 individual projects along the entire river Elbe. The project on floodplain restoration by dyke relocation focused on eight different topics, each of which was studied by specialised research institutes: groundwater hydrology, soil, forestry, vegetation, animal life, socio-economics, agriculture and environmental education. Different dyke variants were tested with reference to all these subjects. As a result, the project developed ecological and socio-economic concepts and guidelines for sustainable land use on the River Elbe (for details see Purps *et al.* 2003).

In order to continue the project activities, research and land purchases within the first EU LIFE project, the LAGS applied for a second project phase to implement the dyke relocation. After the European Commission had rejected the funding, the LAGS asked several national sponsors for funding. However, even though the Federal Agency for Nature Conservation indicated its interest in the project, the large amount of money needed for the restoration measures was not formally approved for a long while. One of the preconditions for receiving funding for large protection areas from the Federal Agency for Nature Conservation is that regional authorities co-finance the project with a minimum of 18% of the total amount and that private sponsors participate with at least 7% of the costs. Thus, the LAGS had to convince several actors to support the project and to bring together heterogeneous, actor-specific interests. In order to coordinate the relevant stakeholders an advisory committee integrating several local, regional and federal actors was founded within the "Trägerverbund Burg Lenzen (Elbe) e.V.", a local non-profit association (cf. Section 10.4.5). The Trägerverbund, which served as the formal applicant for funding, was designed to coordinate the further planning and implementation of the project, to bring together the funding organisations and other stakeholders and to manage the project's finances.

As a main precondition for this project the land required had to be transferred into public ownership. More than 520 private owners were affected (Drees and Sünderhauf 2006: 8). Because many of them refused to sell their land, a land consolidation project was started in 2000 parallel to the planning and fund-raising process. After negotiations with about 520 owners of in total 3,500 ha, about 160 owners throughout the whole project area offered to sell their land. By merging these plots and the already state-owned properties in the new flooding area (partially acquired during the EU LIFE project), the owners who did not wish to sell were compensated by the reallocation of their properties to the rear of the new dyke.

After a long period of uncertainty over whether public funding of the relocation would be approved, the Elbe flood in 2002 finally put floodplain restoration at the centre of public interest and accelerated the political decision-making process. A short time after the flood event and some weeks before the national elections, the federal environment minister agreed to the funding for the restoration of the floodplain near Lenzen as a large protection area of national interest. Additionally, the Federal State of Brandenburg and different private environmental associations and environmental foundations provided their obligatory share of the funding.

10.4.3 Current status

After funding was approved in 2002, two employees of the Trägerverbund started their work as project managers. In the first project phase until 2004, they coordinated the drawing up of a management plan for the project area and the procedures relating to acquiring building permission for the dyke relocation. In order to gain local acceptance and to develop common solutions for controversial issues (e.g. public access to the core zone of the biosphere reserve, the route plan, opportunities for local fishing), a facilitation procedure was organised by the Trägerverbund in 2003 including roundtable discussions and excursions with local stakeholders (Knierim 2004). The second phase from 2004 to 2008 includes the purchase of land, the termination of existing tenancy agreements, the relocation of the new dyke and measures to promote biotope development.

10.4.4 Physical interventions

The restoration of the floodplain comprises three main physical interventions:

Relocation of the dyke on the River Elbe

The main physical intervention consists of relocating the dyke along a distance of 7.4 km up to a maximum of 1.3 km inland from the river. For the construction of the new dyke materials from the floodplain will be used in order to (re-)create flat pools, old meanders and flood hollows by dredging the land along the main flood channels. This is considered necessary in order to develop the requirements for dynamic hydrological processes on a floodplain.

Despite the construction of this new, relocated dyke, the old dyke is largely to be maintained in order to preserve the main watercourse of the river Elbe for shipping and flood protection as recommended by the BAW study (BAW 2000; cf. Section 10.4.2). The existing dyke will be opened at six points (each of them being 200-400m in length) after the new dyke has been completed (see Figure 10.4). A limited number of lowered segments will be built into the existing dyke for the flooding and subsequent emptying of the floodplains (Trägerverbund Burg Lenzen 2006).

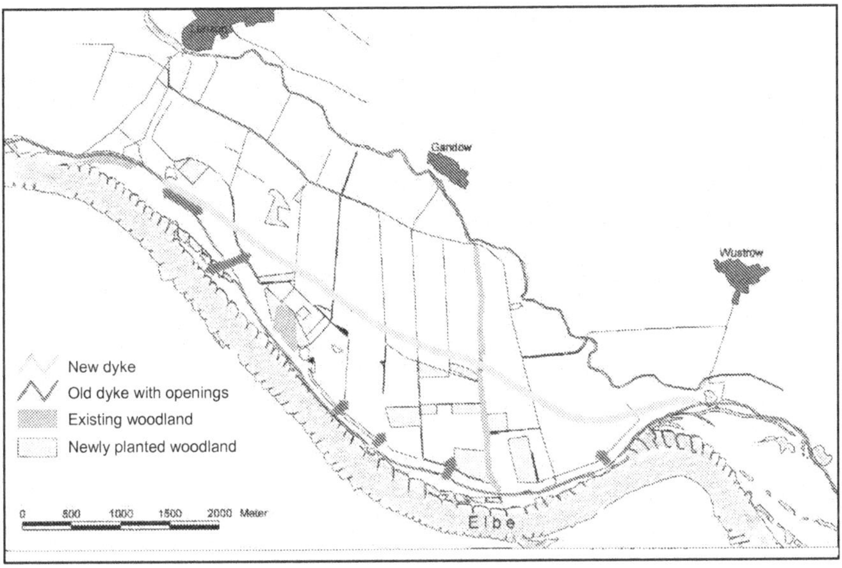

Figure 10.4: Course of the new dyke in the project area. (Source: Trägerverbund Burg Lenzen 2006).

Planting a floodplain forest

Since tree cover as a reproductive source of floodplain woodland is largely lacking in the project area, the natural regeneration by successional processes was expected to take extremely long periods of time. In order to enhance the autochthonous soft and hard wood species in the floodplain, initial plantings on 70 ha have already been completed under the EU LIFE project. Over the following years, another 100 to 120 ha patches of initial planting will be spread across the area in order to speed up the re-growth of a hardwood floodplain forest. Autochthonous black poplar and willow species are being planted along the riverbank in order to improve the connection between the remaining wooded patches near the river and across the river with the larger floodplain forests in Lower Saxony. They are complemented by hardwood species typical of the regional river environment. The trees have been raised in a local nursery cultivating autochthonous plant material.

In addition to floodplain woodlands wet grassland, reeds, dry grassland and hedges will be developed. The acceleration and conservation of different types of river biotopes typical of the floodplains of the River Elbe is considered an important precondition for the development of natural habitats for beavers, birds (such as storks, quail and osprey) and rare amphibians like the tree frog. Once established, the floodplain woodland and the other plant communities will develop from succession without forestry measures.

Establishment of a year-round grazing system

Until recently the area to be periodically flooded was mostly agricultural land, used extensively as grazed meadows. Aside from the floodplain forests with hardwood and willow forests, 80 ha of the new floodplain area will be turned into extensively horse-grazed floodplain meadows. The meadows will be the habitat of a robust horse breed called "Liebenthaler Wildlinge" during the whole year except for flooding periods (Trägerverbund Burg Lenzen 2006).

10.4.5 Organisation and actor involvement

One of the principal promoters and key actors in the project's development was the State Agency for Large Protected Areas in Brandenburg (LAGS, now Environment Agency of Brandenburg, cf. Footnote 5). The agency provided technical input to the whole development process, from the conception and planning phases to fund-raising and the management of the project. It successfully applied for funding from the EU LIFE programme, the national research programme "Elbe ecology" and for national funding for large protected areas. Furthermore, it coordinated different stakeholder groups and their interaction and had a central function in public relations measures.

From the beginning of the project there was close cooperation between the nature protection bodies and Lenzen Holding, which is the largest land-user in the region, cultivating 4,100 ha of agricultural land near Lenzen. The holding integrates the conventional agricultural holding VAL Agrar AG and the Landscape Management Agency Lenzen, which manages approx. 1,200 ha of grassland according to the guidelines for environmentally sensitive agriculture. The holding owns or rents most of the land in the project area, cultivates the trees for the initial plantings and will cultivate the meadows and manage conservation measures in the floodplain area. In view of the employment and steady income opportunities, the manager of the holding has been one of the main promoters of the project from the very beginning. Because of his reputation as a highly respected local figure, he has proved to be one of the key actors for gaining local acceptance and for persuading doubters.

> "Cooperation between nature conservationists and farmers was simply good from the very beginning, and they smoothed things out on the spot... The procedure [for the exchange of plots, J.M.] would not have been possible without the manager of the holding, as the land-owners here are often older people and of course they take care whether the exchange is O.K. for the leaseholder of their plot. And if the leaseholder said no, don't sell it, then we would be unable to buy it" (Interview 2).

The members of the research teams under the EU LIFE and Elbe ecology programmes were also crucial in promoting the project. As their research results proved the benefits – or at least the feasibility – of the project, even the sceptics could, in most cases, be convinced.

One of the largest German environmental groups, the BUND (German Federation of Environment and Nature Conservation) played an important role in the project's development and promotion, too. In the mid-1990s the BUND acquired ownership of a small castle in Lenzen, which is currently being turned into the "European Centre for Floodplain Ecology, Environmental Education and Visitor Information" focusing on floodplain ecology, river restoration, sustainable land use and conservation. As the river dynamics in the floodplain and the growing floodplain forest provide ideal circumstances to demonstrate floodplain ecology, the project area serves as the most important object of reference for educational purposes.

Even though the flood defence unit of the regional Brandenburg Ministry for the Environment had a rather passive role in the process of project initiation and development, its general support became important for project implementation not least because it is the body responsible for the dyke construction works.

> "The flood protection authority sees its responsibility in the construction of dykes, and they said they could master this bottleneck technically" (Interview 2).

Besides these representatives of regional flood defence, the Federal Waterways Engineering and Research Institute (BAW), as the supreme authority for research and consultation on hydraulic engineering in Germany, was integrated at an early stage of the planning process. This proved strategically helpful in undermining any potential criticism or doubts about the feasibility of the project by the subordinate flood defence authorities.

> "And with the Federal Waterways Engineering and Research Institute we also have the decisive partner because it is, after all, a Federal waterway. For all the discussions that we have to have over and again concerning the waterway, it is important that – already in the run-up phase – we have been able to involve the competent authority at the highest level as an advisor" (Interview 2).

The "Trägerverbund Burg Lenzen (Elbe) e.V.", a local non-profit alliance originally founded to renovate the castle of Lenzen, served as the formal applicant for funding and played an important role in project development and implementation. Its members are the town of Lenzen, the local district, the local agricultural holding as the largest land user of the region and the nature conservation association BUND as the owner of the castle since 1994. After funding was approved, two persons were employed by the alliance to develop the project's management plan, to coordinate the implementation process, to bring together the funding organisations and other stakeholders and to manage the project's finances. Furthermore a project steering group has been created, including the representatives of the Federal Agency for Nature Conservation, the State Agency for Large Protected Areas and the ministerial department for flood protection in Brandenburg, the local land user, the neighbouring municipalities and authorities involved, environmental groups and other relevant stakeholders. The steering group accompanies the whole implementation process with the objective of gaining local acceptance, coordinating the interests and proposals of different stakeholder groups and providing the Trägerverbund, as the project manager, with technical expertise.

10.4.6 Instruments used

Project development and implementation were shaped by a complex mix of policy instruments.

Funding

As in most of the floodplain restoration projects analysed, funding is a predominant instrument for project implementation. Specific to the funding of the Lenzen project is, firstly, that project managers depend on a multitude of different funding sources from the European Union, the federal research and environment ministries, the regional bodies of nature conservation and flood defence and private sponsors. Secondly, the amount of funding totalling more than €12 million (plus €6.3 million for regular flood protection) is comparatively large.

- 1994-98: About €3 million were funded by the EU LIFE programme and the Land Brandenburg (50% each), which was invested especially in buying land (approx. 550 ha) and in initial planning activities.
- 1996-2000: The Federal Ministry for Research financed a research project analysing floodplain restoration by dyke relocation on the Elbe (€1.95 million).
- 2002-2006: The Land Brandenburg transferred the regular flood protection budget, which had originally been planned to maintain and redevelop the existing dyke in the project area, to the relocation of the dyke (€6.3 million).
- 2002: Since the designation of the floodplain as a large protected area the ecological restoration and the opening of the old dyke have been financed by the Federal Nature Conservation Agency (€5.25 million), the Ministry for Agriculture, Environment and Land-Use Planning in Brandenburg (€1.25 million) and private environmental foundations and associations (€0.5 million).
- Since 1992: The technical support, project management, monitoring and public relations activities by the LAGS is difficult to monetarise.
- 2000-2007: The land trust scheme (Bodenneuordnungsverfahren), implemented for the reparcelling of agricultural land and the re-allocation of property rights by the Land Brandenburg, has not been quantified financially.

Protected areas

The instrument of protected areas has been a key factor behind the project's implementation. The designation of the Flusslandschaft Elbe Biosphere Reserve has been a key driver for nature conservation activities and the promotion of agri-environmental issues in the region. The biosphere reserve is subdivided into core areas, buffer zones and transition areas, which can be differentiated according to the specific extent of their conservation. The floodplain near Lenzen will become part of the core zone after completion of the project. The re-opened floodplain will be designated a nature conservation area (see Monstadt and Moss in Chapter 4 of this volume). In this core area agricultural land use is restricted and public visitors will only be given partial access. The neighbouring biotopes of the regularly flooded area on the riverside of the dyke are also part of this core area, which covers a total of 994 ha. Its surrounding buffer zones – like most of the biosphere reserve – are part of a landscape protection area in which restrictions on agricultural land use are less severe.

Technical flood protection measures

Studies carried out after German reunification proved the necessity of extensive dyke refurbishment measures along the river Elbe. Whereas the most crucial repair measures along the Brandenburg part of the river Elbe were carried out during the 1990s, this did not apply to the dyke segment near Lenzen. With the decision to restore the floodplain, the regular dyke refurbishment budget was transferred to relocating the dyke.

Rural development and agri-environmental instruments

EU regulations on rural development and the regional policies of the Land Brandenburg include agri-environmental programmes which compensate farmers for costs incurred in carrying out environmental improvements on their land, such as reducing pollution, improving biodiversity, adopting organic agriculture and managing non-farmed land. Economic profits for the Lenzen Holding thus derive not only from food production, but increasingly from managing the local landscape and environment in a broader sense. Apart from improving the environment, these measures also support other parts of the rural economy like tourism by improving the landscape's attractiveness and increased biodiversity. Furthermore, many activities, such as landscape management, create local jobs and safeguard traditional skills (cf. Dehnhardt and Petschow 2001).

Land consolidation procedure

Following the privatisation of the former co-operative and restitution claims of the former private landowners, ownership of the land has become highly fragmented. For this reason stakeholders from both nature conservation and agriculture applied for a land consolidation procedure to re-allocate and exchange property rights for 3,600 ha of land in the Lenzen region. This procedure, which was started in 2000 (Drees and Sünderhauf 2006), allowed for the acquisition of the floodplain area by the Trägerverbund.

10.5 PRINCIPAL DRIVERS AND CONSTRAINTS OF FLOODPLAIN RESTORATION ON THE PROJECT

Although project development and implementation were very complex and time-consuming, the Lenzen project had comparatively ideal preconditions: an integrated regional administration responsible for nature conservation, flood protection and agriculture, a proactive and supportive farmer owning and cultivating most of the floodplain area and a sparsely populated project area which already included protected areas. In detail the drivers and constraints can be described as follows.

10.5.1 Driving forces

One of the key drivers of project development was the constant *personal commitment of regional actors and their cooperative interaction.*

- First, the representatives of the LAGS gave constant technical support, organised professional funding, made the project a showcase of their activities and personally invested much time and enthusiasm.
- Secondly, the manager of Lenzen Holding was extraordinarily open-minded towards the project ideas. He did not object to the extensification of land use and the re-allocation of property rights, but regarded the project as a chance to ensure a durable income for the agricultural Holding and for regional economic development. He was a key actor not only in supporting project planning and implementation, but also as a catalyst for gaining acceptance among local residents and the municipality.
- Thirdly, stakeholders of the regional flood defence authority made the project idea acceptable. By carrying out several studies they demonstrated the economic viability of dyke relocation projects. The fact that the regional flood defence and the nature conservation departments were both part of the same ministry simplified the implementation of a project combining nature conservation and flood defence interests. Moreover, the flood defence administration in Brandenburg is quite centralised compared to other German Länder, where decentralised public bodies (so-called dyke associations – “Deichverbände”) carry out flood defence. This comparatively centralised structure has proved to be less vulnerable to NIMBY (not in my backyard) conflicts.
- Fourthly, environmental groups promoted the project by establishing a floodplain restoration centre and supporting soft tourism. Good personal contacts and constructive cooperation among these promoters of the project won local acceptance, maintained political pressure and improved the prospects for accessing research and project implementation funds.

- Fifthly, the Trägerverbund (partially in cooperation with the LAGS) organised excursions, discussions and a facilitation procedure to gain local acceptance and to consider local interests in project development.

Another main driver is the *growing public acceptance of new approaches to flood prevention*. On the one hand the value of floodplains and their environment has a prominent status in the public nature conservation discourse. On the other hand the demand for effective and sustainable techniques to minimise flood damage has been intensified. The higher probability of flood events is no longer just an uncertain prognosis of climate change research. The frequency and intensity of flood events are already – as the Elbe flood in 2002 and other flood events in Europe have shown – visibly increasing.

Further, *progressive European and national floodplain restoration policies* have continued to be key drivers for regional and local action. First, funding programmes like the EU LIFE programme, the German research programme "Elbe ecology" and the German funding programme for large protected areas gave considerable support. They stimulated basic and applied research on the ecological and socio-economic aspects of floodplain restoration and also made the development and implementation of the project possible. Secondly, the reformed Nature Conservation Act no longer exclusively designates protected zones in biotopes which are currently of high ecological value, but also in those which have high ecological potential. Thirdly, a new shift in agricultural policy has extended the scope of action for the local farmer.

When the project was at an advanced planning stage, the final funding approval of the restoration project was driven by the *occurrence of a flood disaster*. In the aftermath of the Elbe flood the project was used as a means of demonstrating political effectiveness (cf. Sections 10.3.1 and 10.4.2).

10.5.2 Constraints

Although a majority of the local population considered the project and the establishment of a floodplain restoration centre as opportunities for regional economic development, project managers had to cope with a lack of local acceptance among some local interest groups and the resistance of local land users. However, these contributory factors, often cited as the principal restriction for floodplain restoration, could be minimised at an early planning stage by excursions, exhibitions and project presentations at local events by the LAGS and by the personal commitment of the largest local landowner. The remaining controversial issues such as public access to the conservation area and permission for fishing could be reconciled by a facilitation procedure (Knierim 2004). Consequently, the problem of local acceptance decelerated the implementation process, but as external investors and powerful local and regional stakeholders promoted the project, it never called the project into question.

One main reason for the prolonged planning process was the innovative character of the project. Project managers could refer neither to empirical knowledge and experiences derived from the implementation of other integrated floodplain restoration projects in Germany, nor to approved policy concepts and problem-solving techniques. Planners and project managers have therefore been lacking reliable knowledge on (alternative) strategies for floodplain restoration. Consequently, a lot of research activities and expensive and time-consuming project management by trial and error have been necessary in order to fill this *gap in knowledge and practical experience.*

Additionally, project managers had to overcome considerable *hurdles in the fund-raising process.* Requests for funding had to cope with sophisticated and complex requirements. Especially the complex requirements for funding large protection areas relating to the co-financing from numerous public and private organisations made it difficult to handle the different substantive interests, technical claims and institutional requirements of each sponsor. Project managers had to persuade numerous public and private sponsors and to negotiate a compromise between their specific interests and requests.

Furthermore, project management has not only been a question of convincing and encouraging numerous sponsors, but also of *coordinating* the interests of the *several policy sectors and private stakeholders* involved. Spatial planning, nature conservation, flood protection and agricultural policies of the Land Brandenburg as well as local policies of the three neighbouring municipalities, environmental groups and other relevant stakeholders had to be considered. As successful project implementation depended on the crucial and often voluntary contribution of each of those stakeholders, the coordination capacity of the employees of the LAGS, and since 2002 those of the Trägerverbund, was sorely tried. They have been responsible not only for supporting the project with technical expertise, gaining local acceptance and organising professional fund-raising, but also for enrolling the different stakeholders in the planning process. Even though the LAGS provided the considerable personal commitment of several employees, the planning process took ten years from the first request for public funding to the funding approval and it will take six more years until the greater part of the project is implemented.

In addition, the pressure to create a *catchment-oriented approach to floodplain restoration* has been rather limited. Until the Elbe flooding in 2002 a strategy towards flood prevention involving the enhancement of retention capacity, for instance with floodplain restoration, was largely missing. Aside from the project in Lenzen, only a few other floodplain restoration projects have been implemented, all of which are essentially site-based (cf. Section 10.3.2). Each of them was primarily coordinated by the regional authorities of the German Länder. Catchment institutions like the ICPE or the administration of the biosphere reserve had a rather weak position limited to transferring knowledge, organising an exchange of experiences and calling on the Länder to initiate preventive flood management. Consequently, ecological synergy effects, achievable by creating a cross-border network of floodplain biotopes, or economic benefits gained by cross-border co-operation, remain underdeveloped so far.[6] Although the ICPE has recently developed a catchment-oriented flood action plan designating appropriate areas for dyke relocation and floodplain restoration, progress in planning and implementation could not live up to the proclamations following the Elbe flood (cf. Section 10.3.2).

10.6 SELECTED EMBLEMATIC ISSUES

Considering the emblematic issues of the case study – understood as particular, project-specific aspects of the institutional conditions for floodplain restoration – the Lenzen project is characterised by high complexity of project development and implementation. The project is, firstly, marked by a high degree of epistemological and normative uncertainty relating to the concept and practice of floodplain restoration. Secondly, the project managers have had to cope with various policy objectives and to coordinate a variety of actors from different policy fields and levels. Thirdly, the project managers have had to handle complex formal procedures in fundraising and planning which have made project development and implementation remarkably lengthy.

10.6.1 Managing uncertainty: enrolling sciences and reproducing nature

One of the main emblematic issues is the innovative character of the project and its high epistemological and normative uncertainty: project development and implementation could refer neither to substantiated ecological knowledge and practical experiences nor to standardised solutions and normative criteria derived from the planning and implementation of other large integrated floodplain restoration projects. The degree of scientific uncertainty on eligible ways to ecologically restore the floodplain, to relocate the dyke etc. was correspondingly high and different solutions were contested scientifically and normatively. In technical terms, the idea of restoring the floodplain was complex as "it is difficult to guarantee that ecosystems will respond as expected, with attendant risks of the evolution of landscapes and ecosystems in un-planned and socially undesirable ways" (Adams *et al.* 2004: 1930). This is particularly true for large floodplain restoration projects in which ecological and hydrological linkages between river and floodplain, both within one reach and longitudinally, are altered and patterns of river flows and flooding across the catchment are to be considered. Moreover,

6 For example, Lower Saxony is located on the opposite side of the river to the Lenzen project area in Brandenburg. However, neither the project area nor other floodplain conservation and development measures have been extended to the Lower Saxony side of the river.

the restoration activities raise normative issues, for example in determining what "natural" conditions should be the target of the restorer, in identifying what past state of nature to restore to, and in defining how "natural" the resulting facsimile should be (ibid: 1929f.)

The answer to this scientific and normative uncertainty was thus elaborate research and development (R&D) activities conducted throughout the whole planning process. In 1994 the LAGS initiated an EU LIFE project that aimed at proving the ecological potential of the floodplain, which up to that point was used extensively as grazed meadows and where floodplain woodlands had all vanished. Various scientific studies on the woodland components of floodplain restoration and initial plantings of floodplain forests were conducted. Based on these activities an interdisciplinary research team financed by the German Research Ministry conducted more in-depth research on floodplain restoration and on the development of floodplain forests by dyke relocation on the River Elbe (cf. Section 10.4.2).

The comprehensive scientific studies not only recommended the relocation of the dyke and the conservation of the floodplain biocoenosis, but also the creation of a floodplain environment that reproduces the "natural" state of the floodplain. Natural habitats are to be created by large initial plantings of floodplain trees, groves and shrubs, typical of the floodplains of the river Elbe. Additional interventions were recommended in order to promote the dynamic hydrological processes typical of a natural floodplain, involving extensive earth-moving works, the (re-)creation of shallow watercourses and flood hollows.

Although the project has been promoted *inter alia* for its benefits to flood protection, the expensive restoration measures accompanying the dyke relocation increase the flood storage capacity only slightly by retaining water and by slowing down the velocity of river flows by trees and other barriers. From a flood prevention viewpoint, it would be much more cost-efficient to leave the river environment to a greater extent to develop of its own accord and to invest the available amount of money in re-opening further floodplain areas instead of investing in expensive plantation measures. At least from a long-term perspective the effects on flood prevention (and nature conservation) could be much more substantial if investments were concentrated on physical interventions extending the area in which rivers could move without restraint.

Despite the high expenses for "re-naturalisation" of the floodplain and its limited benefits for flood defence, the R&D activities were conducive to the man-made re-naturalisation of plant communities. The studies conceptualised the potential ultimate state of the river ecosystems and developed practical guidelines for reproducing an original state of nature. To achieve this, the floodplain environment in Lenzen is not primarily left to natural dynamics and successional ecological processes, but physical interventions like plantation, construction works etc. are carried out to accelerate and regulate the restoration of the river environment. The studies thus provided an intellectual framework for an approach to nature conservation which is actively developing an ecosystem in a direction which is considered to be the former ("natural") state of the biotope or an ultimate state of natural succession.

Although such restoration approaches have been highly contested in the German discourse on nature conservation and ecosystem restoration since the 1990s, the competing approach of wilderness was not discussed or applied by the variety of R&D activities conducted or coordinated by the LAGS. In this conflicting approach, the restoration and conservation of ecosystems is understood as leaving nature to its succession processes and minimizing human interventions. As the environmental groups and the responsible organisation for the floodplain restoration centre were interested in having access to an impressive demonstration site for floodplain restoration, they appreciated the scientific recommendations. Moreover, the municipality was dependent on local investments, and the Lenzen Holding was interested in a steady income. As the institutional interests of diverse actors were in favour of the measures planned and the allocation of funding, this expensive approach to nature conservation and flood prevention was promoted by most of the stakeholders.

10.6.2 Managing complex networks: horizontal and vertical coordination as a challenge for floodplain restoration

A striking feature of the Lenzen project is the high complexity of project development and management, owing to the involvement of a multitude of actors from different policy fields, policy levels and private sectors. Only their constructive co-operation, encouragement and sponsorship and the private initiative of some key promoters have made the project possible.

From a *horizontal perspective* several policy sectors have been involved in project planning and management. The key actors who initiated and promoted the project belong to the policy community of nature conservation, like the LAGS, the environmental groups, representatives of the floodplain-ecology centre, the Federal Agency for Nature Protection and several environmental foundations. These key actors had to convince and cooperate with the regional policy community of flood defence, crucial for the acceptance of the project and funding the relocation of the dyke. Their support was assured with the help of research activities and studies by regional flood defence authorities, the Federal Waterways Engineering and Research Institute (BAW) and the research network "Elbe-ecology". Additionally the Lenzen Holding, the largest regional land user, has been a vigorous promoter of the project. By embedding the project in a regional strategy to develop the region for soft tourism and organic farming (or at least less polluting agriculture) it was viewed as a chance to draw economic benefits and to create opportunities for employment in this structurally weak region.

From a *vertical perspective* the project was influenced by different policy levels. The main initiatives derived from the co-operation of the local land user, the regional administration for nature conservation in Brandenburg and the regional flood defence agency in Saxony-Anhalt. Apart from this interaction between the local and regional levels, EU policy contributed with the EU LIFE project on floodplain restoration and gave significant support to the project management. By contrast, national policy did not engage in the project or offer support at an early stage with funding programmes, legal instruments or policy initiatives. For a long time federal policy was only involved via its research programme "Elbe-ecology". When the European Commission refused the funding of a subsequent EU LIFE project, the national policy level finally joined more actively in project development. Only as a result of the Elbe flood in 2002 did federal initiatives on preventive flood management introduce a policy shift, giving higher priority to soft engineering measures, and finally the Federal Agency for Nature Conservation funded a major part of the project's implementation.

As a result of the large number of stakeholders, project development and implementation generated high expenditure on intra- and inter-organisational coordination. The LAGS, in particular, with its limited staff and financial resources, was seriously challenged to coordinate the fund-raising and the designation of protected areas and to raise support in the flood defence community and the local population. Bringing together the heterogeneous interests and various activities of each stakeholder demanded considerable management resources. Due to the limited institutional capacity to coordinate the multitude of actors, project management required the private and often voluntary initiative and enthusiasm of some key actors. The project was only made possible thanks to close personal relationships between key actors of the LAGS, the manager of Lenzen Holding, experts in flood defence, the biosphere reserve administration and activists in environmental groups.

Only after the approval for funding was a special project management office established within the Trägerverbund. Its function was to gain public acceptance and coordinate the interests of different stakeholders with a facilitation procedure, to develop a project management plan, to prepare the project approval procedures (Planfeststellungsverfahren) and to coordinate the contracting and implementation of the project. In addition, an advisory committee was founded to discuss and evaluate the progress of the project and to consult with the project management. As several local, regional and federal actors participate in this committee, it is used as an institution to coordinate policies across several policy levels and sectors.

10.6.3 Managing complex formal procedures: fundraising and planning as a challenge for floodplain restoration

The Lenzen project could be implemented as a result of various driving forces, with severe constraints largely absent. For example, there has been a high level of public acceptance and energetic involvement of the local farmer, the regional bodies of nature conservation and flood protection and considerable financial support from the federal and European levels. In view of this, it is at first sight rather astonishing that the planning process took ten years and that several more years will be needed to implement the project.

> "Dyke relocations of this size are a special type of project. They require time, patience and a highly integrative approach. For quick fixes following flood disasters they are unfortunately not very appropriate" (Interview 1).

This very time-consuming process is not only the result of the innovative character of the project and the need for R&D activities, or the expensive process of coordinating the interests and activities of the various stakeholders involved. The main reasons for this were various procedural and bureaucratic hurdles confronting the key promoters. First of all, the formal requirements for public funding were very complicated: in the case of the EU LIFE project and particularly in the case of the final funding of project implementation, diverse organisations from different policy levels and private sectors had to participate in co-financing the project. The specific demands and interests of various funding organisations had to be met. Secondly, the project managers were challenged by the complicated formal procedures of the German planning system. The instruments of the German regional planning and water management laws for granting the floodplains legal status in order to bind regional and land-use planning is a very time-consuming, costly and complex administrative procedure, in which powerful opponents in agriculture and forestry, technical flood defence and municipal land-use planning have the opportunity to intervene or at least to decelerate the project. Thirdly, the land consolidation procedure has been very time-consuming, as the property rights were fragmented and various actors had to be persuaded and enrolled (Drees and Sünderhauf 2006). Last but not least, the project managers are also challenged by project implementation, as they have to coordinate the technical performance of construction works and the interests and demands of different sponsors and various stakeholders.

10.7 CONCLUSIONS AND RECOMMENDATIONS

In contrast to early floodplain restoration projects, which in most cases concentrate on single sites and which can be characterised by a limited goal structure (predominantly focusing on nature conservation), limited stakeholder involvement and the application of only a few instruments, the management of larger projects is much more complex. At this larger scale, scientific and normative uncertainties increase, the projects have direct implications for the traditional concerns of flood defence, agriculture, forestry and various other land users and stakeholders at different scales and they have to cope with complex procedural standards. The requirements of policy development and project management in the case of these larger projects are thus enormously high.

10.7.1 Policy development

Deducing policy recommendations from the project in Lenzen, we can first point to the vital need for policy reforms capable of reducing the bureaucratic hurdles confronting project development and implementation. One of the key problems to be corrected is the complex funding structure. The case study has illustrated that the complicated funding system, requiring the co-financing of various public and private actors with different expectations and interests, causes overcomplexity and the inefficient use of time and staff resources. In the fund-raising process, the LAGS had to cope with heterogeneous requirements for funding, different time-scales, complex formal procedures etc. Furthermore, the beneficiaries downstream of the floodplain in no way participated in funding the project. In order to reinforce better coordinated and less complicated funding of floodplain restoration projects, it would be more efficient to concentrate the allocation of public funding on a single organisation responsible for the whole river basin and financed by the different riparian German Länder, the Federal government and the Czech Republic. This organisation could organise multilateral negotiations on the cost-sharing of flood protection and be responsible for funding and coordinating measures for preventive flood management and the ecological restoration of rivers.

Apart from this institutional bundling of funding, the case study brings to mind the problem of the high costs of large restoration projects. Even if the costs of €12 million (plus the budget from regular flood protection) were to a certain extent due to the innovative character of this pilot project and various R&D activities, cost minimisation should be considered more seriously for future projects in order to distribute the available funding for the implementation of further floodplain restoration projects (see also Section 10.7.2).

The case study provides evidence that the problem consists not only of the complicated funding structure, but of an overall deficiency concerning a river basin approach to promoting floodplain restoration projects. In the past, floodplain restoration activities between bordering federal states have not been well coordinated. In the aftermath of the Elbe flooding, the role of the ICPE in flood prevention has become more proactive. Since then, initiatives for the development of a flood protection plan for the river Elbe have been introduced which include the restoration of floodplains as one of the main objectives. Based on feasibility studies, the ICPE recommended giving space back to the river

Elbe by implementing 15 dyke relocation projects along the German stretch of the river (IKSE 2003: 24f. and 2006: 25). However, due to the lengthy, costly and complex funding and planning procedures, only a few large projects are at an advanced planning or implementation stage. As project development and management is organised in a rather decentralised way, the exchange of decentralised knowledge and project experience has, until now, been rather weak. Given, in addition, the relatively limited influence of river basin organisations in promoting floodplain restoration projects it is unlikely that far-reaching changes will occur in the near future. This is particularly true since the German Länder have been authorised by the reform of federalism to diverge from national water, regional planning and nature conservation legislation (see Monstadt and Moss in Chapter 4 of this volume). In view of this highly decentralised structure and the complex formal procedures, it would be advantageous if the activities of the ICPE were not limited to the definition of the potential floodplains to be restored. The organisational concentration of project management or, alternatively, an institutional framework to guarantee the intensive co-operation and exchange of experience and knowledge of project managers could provide for important synergy effects and greater professionalism in floodplain restoration.

Recent legislative initiatives by the Federal Ministry for the Environment, committing the federal states to develop flood protection plans, coordinate the regional plans inter-regionally and internationally and reconcile the interests and concerns of the upstream and downstream residents, should be advanced further. In addition, it seems advisable to establish a new river basin agency, which supports projects in floodplain restoration and other strategies integrating flood prevention and nature conservation, accumulating and bundling knowledge, expertise and management capacities. Alternatively, the systematic extension of the competencies of the ICPE towards flood prevention and nature conservation planning and management could be equally effective.

10.7.2 Project management

The floodplain restoration project at Lenzen was, as a whole, managed in a successful manner. Strategic network management, project marketing, fundraising and knowledge management can be used as an encouraging model for project management. From the very beginning, the project was developed in a strategic actor network. On the one hand, the project has never been exclusively planned by a single policy community, such as public nature conservation authorities, environmental groups, environmental research or other actors of the nature conservation policy community. Instead, representatives from flood protection, agriculture, local municipalities and regional planning participated from an early planning stage and took part in the specification of the project design and the promotion of the project. On the other hand, this extensive and inclusive network was built and coordinated by a rather exclusive "core network", consisting of very few people from the fields of nature conservation and flood protection as well as the local manager of Lenzen Holding. Within this core network the key actors worked together closely, built up amicable relationships and invested considerable personal effort into promoting the project. In addition to this informal network, an advisory committee integrating several local, regional and federal stakeholders was institutionalised at an advanced planning stage to coordinate project implementation across several policy levels and sectors. In addition, project publicity, including local exhibitions on the project, the employment of two professional project managers and a facilitation procedure with local stakeholders provided for public acceptance. Furthermore, the project was presented and promoted at local events and festivals, and respected local personalities, especially the manager of Lenzen Holding, acted as intermediaries between the local population and public nature conservation and flood protection authorities. As regards funding, project managers filed their applications professionally and gained considerable financial support, which allowed them to implement remarkable restoration measures and generate extensive knowledge, which could be of considerable value for subsequent projects.

Even if the management of the Lenzen project has been successful and very professional, and the high operating and financial expenses can be justified to a certain extent by the innovative and explorative character of this model project, subsequent projects will probably be challenged by having less funds at their disposal. Less expensive nature conservation measures, such as leaving the floodplain more to processes of natural succession, could contribute to lower expenditure and improve the balance between costs and benefits in nature conservation and flood protection. Presumably, the cost-benefit balance will improve the more projects are implemented and the more knowledge and experience is generated. However, better coordination between the single floodplain restoration projects along one river, the systematic extension of the tasks of the river basin organisations towards floodplain restoration and the establishment of a new floodplain restoration agency could contribute to policy diffusion and more efficient project management.

10.8 REFERENCES

Adams, W.M., Perrow, M.R. and Carpenter, A. (2004) Conservatives and champions: river managers and the river restoration discourse in the United Kingdom. *Environment and Planning A* **36**(11), 1929-1942.

BAW (Bundesanstalt für Wasserbau) (2000) *Untersuchung der Auswirkung von Maßnahmen im Elbevorland auf die Strömungssituation und die Flussmorphologie am Beispiel der Erosionsstrecke und der Rückdeichungsbereiche zwischen Wittenberge und Lenzen – Schlussbericht.* Karlsruhe.

BMU (Bundesministerium für Umwelt, Naturschutz und Reaktorsicherheit) (2003) *Eckpunkte für ein Artikelgesetz zur Verbesserung des vorbeugenden Hochwasserschutzes*. Bonn.

Böhm, H.R., Heiland, P., Dapp, K. and Haupter, B. (2001) *Development of spatial planning instruments to improve river flood prevention and awareness through regional planning, interregional co-operation, economical instruments, information management and hazard management.* Executive summary, Darmstadt University of Technology in co-operation with University of Berne. Darmstadt/Berne.

Dehnhardt, A. and Petschow, U. (2001) *Rückgewinnung von Retentionsflächen und Altauenreaktivierung an der Mittleren Elbe in Sachsen-Anhalt.* Forschungsbericht des Instituts für ökologische Wirtschaftsforschung, Berlin.

Drees, A. and Sünderhauf, R. (2006): *Land Consolidation as a Tool for Flood Prevention.* Paper presented on the XXIII FIG Congress "Shaping the Change", Munich, 8-13.10.2006.

Förster, S., Kneis, D., Gocht, M. and Bronstert, A. (2005) Flood risk reduction by the use of retention areas at the Elbe River. *International Journal of River Basin Management* **3**(1), 21–29.

Geller, W., Puncochár, P., Guhr, H., v. Tümpling, W., Medek, J., Smrt'ak, J., Feldmann, H. and Uhlmann, O. (eds.) (2002) *Die Elbe – neue Horizonte des Flussgebietsmanagements.* 10. Magdeburger Gewässerschutzseminar. Teubner, Stuttgart.

IKSE (Internationale Kommission zum Schutz der Elbe) (1998) *Strategie zum Hochwasserschutz im Einzugsgebiet der Elbe.* Magdeburg.

IKSE (2001) *Bestandsaufnahme des vorhandenen Hochwasserschutzniveaus im Einzugsgebiet der Elbe.* Magdeburg.

IKSE (2003) *Aktionsplan Hochwasserschutz Elbe.* Magdeburg, 24.10.2003.

IKSE (2005) *Internationale Flussgebietseinheit Elbe. Merkmale der Flussgebietseinheit, Überprüfung der Umweltauswirkungen menschlicher Tätigkeiten und wirtschaftliche Analyse der Wassernutzung.* Bericht an die Europäische Kommission gemäß Art. 15 Abs. 2 der Richtlinie 2000/60/EG, Dresden.

IKSE (2006a) *Erster Bericht über die Erfüllung des „Aktionsplans Hochwasserschutz Elbe" im Zeitraum 2003 bis 2005.* Magdeburg.

IÖW (Institut für ökologische Wirtschaftsforschung Berlin e.V.) (2002) *Für eine Wende in der Flusspolitik: „Nachhaltige Elbe 2020" jetzt auf den Weg bringen.* Ausführliche Stellungnahme des Instituts für ökologische Wirtschaftsforschung (IÖW) zur Nationalen Flusskonferenz. Berlin, 12.09.2003.

Jährling, K.-H. (1994) Bereiche möglicher Deichrückverlegungen in der Elbaue im Bereich der Mittelelbe: Vorschläge aus ökologischer Sicht als Beitrag zu einer interdisziplinären Diskussion. In *Die Elbe im Spannungsfeld zwischen Ökologie und Ökonomie* (ed. H. Guhr *et al.*) 6. Magdeburger Gewässerschutzseminar. Internationale Fachtagung in Cuxhaven vom 8. bis 12. November 1994:, pp. 326-355.

Jährling, K.-H. (1998) *Deichrückverlegungen: Eine Strategie zur Renaturierung und Erhaltung wertvoller Flußlandschaften?* Magdeburg.

Knierim, A. (2004) *Bericht zum Moderationsverfahren im Naturschutzgroßprojekt "Lenzener Elbtalaue" über den Zeitraum Februar bis Dezember 2003.* Müncheberg.

Puhlmann, G. (1994) *Bereiche möglicher Deichrückverlegungen im Gebiet der Mittleren Elbe zwischen Hirschmühle / Prettin und Dornburg (Elbe-km 168-301) als Grundlage für eine interdisziplinäre Diskussion.* Staatliches Amt für Umweltschutz Dessau/Wittenberg, Abteilung 2 Wasserbau (ed.), Wittenberg.

Puhlmann, G. and Jährling, K.-H. (2003) Erfahrungen mit "nachhaltigem Auenmanagement" im Biosphärenreservat "Flusslandschaft Mittlere Elbe". *Natur und Landschaft* **4**, 143-150.

Purps, J., Damm, J. and Neuschulz, F. (2005) Re-establishing floodplain forests and dike relocation: a model project carried out at the Elbe River, Germany. *Large Rivers (Archiv für Hydrobiologie)* **15**, 561-568.

Petrow, T., Thieken, A.H., Kreibich, H., Merz, B. and Bahlburg, C.H. (2006) Improvements on Flood Alleviation in Germany: Lessons Learned from the Elbe Flood in August 2002. *Environmental Management* **38**(5), 717-732.

Stubbs, M and Bonjour, S. (2002) *Transboundary Management Efforts of the Elbe River.* Working-Paper submitted to the Doctoral Research Seminar "The Science and Politics of Freshwater Management", ETH Zürich / Eawag, Dübendorf. (www.eawag.ch/research_e/apec/seminars/Case%20studies/elbe.pdf, 15.02.2003).

Trägerverbund Burg Lenzen (2006) *The large conservation project "Lenzener Elbtalaue" – a project outline.* (www.burg-lenzen.de/deichrueckverlegung/welcome.html, 05.12.2006).

Weichselgartner, J. (2003) Nach der Elbeflut 2002: Und danach? *GAIA – Ecological Perspectives for Science and Society* **12**(4), 245-248.

INTERVIEWS

Interview 1: Director of the Elbe River Landscape Biosphere Reserve, Dessau, 27 February 2002.

Interview 2: Two conservation officers at the State Agency for Large Protected Areas, Brandenburg, Rühstädt, 24 January 2003.

Interview 3: Manager of a local agricultural holding, Lenzen, 24 January 2003.

Interview 4: Two conservation officers at the Federal Agency for Nature Conservation, Bonn, 29 January 2003.

Interview 5: Officer of the International Commission for the Protection of the Elbe, Magdeburg, 5 February 2003

Interview 6: Officer of the Water and Shipping Directorate East, Magdeburg, 5 February 2003

Interview 7: Conservation officer at the Trägerverbund Lenzen, Lenzen, 6 February 2003.

Interview 8: City councillor of the city of Lenzen, Lenzen, 6 February 2003.

Interview 9: Flood defence officer at the State Agency for Environment, unit for flood protection and hydraulic engineering, Potsdam, 7 February 2003.

Interview 10: Research associate at the State Agency for Consumer Protection and Agriculture, Teltow-Ruhlsdorf, 7 February 2003.

11

Restoring floodplains on the River Seine: combining flood prevention with regional development

Maria Hagemeier and Axel Klaphake

11.1 INTRODUCTION

The Bassée plain on the River Seine is one of the largest floodplains in France and is located in the Paris region (Région Ile-de-France), where one in six French people live. It is part of the Seine river basin (40,000 sq. km^2) at the confluence of the Seine and Yonne rivers. The restoration project at La Bassée envisages measures to restore ecological functions combining flood control and the protection of biological diversity. This case study of the project, conducted primarily in 2002/2003, therefore allows the investigation of important institutional and instrumental features which characterise the planning and implementation of large-scale floodplain restoration projects in France.

In contrast to the other French case study on the Bourret project discussed in this volume (see Chapter 8), the main objective of the Bassée project is flood prevention. More precisely, the project aims in particular to reduce flood risks and potential flood damage in the Paris conurbation. The case study will therefore be used to address some of the most striking features which characterise the development of projects relating to flood control policy in a large river basin. In particular, the institutional aspects of multi-level and multi-actor decision-making are illustrated, as well as upstream/downstream coordination.

The project's technical approach to floodplain restoration involves the creation of an artificially inundated area in the form of a controlled polder system. Although the effectiveness of a controlled polder system for the lowering of the flood crest is largely uncontested, it is not clear whether and, if so, to what extent polders might also contribute to biodiversity targets. Consequently, the case of the Bassée addresses an important and controversial issue concerning the compatibility of flood control and nature protection targets in floodplains with controlled flooding. Today, the Bassée area is integrated into the network of SAND projects with the objectives of increasing flood defence levels in the Paris region and making the scheme part of a spatial development project.

11.2 THE PHYSICAL CONTEXT AND USE OF THE RIVER

11.2.1 Physical characteristics

The River Seine is located in northern France, rising on the Plateau of Langres, near Dijon, and flowing northwest past Troyes, Fontainebleau, Paris and Rouen into the English Channel. The river is about 776 km long and is navigable for riverboats from Bray-sur-Seine, 560 km from its mouth, and for ocean shipping from Rouen, about 121 km from its mouth. The Seine drains an area of more than 77,700 km^2. The river has a high water level in winter and a low water level in summer and most of the permanent rivers of the 3,000 km network meander through gentle hills and wide plains on a foundation of sedimentary rocks (see United Nations 2003).

Like many European rivers, the Seine was nearly entirely channelled at the beginning of the 20th Century in order to enhance the agricultural use of the rich alluvial soils, and to secure river navigation and flood prevention. Today, the flood disturbance regime of the Seine floodplains is no longer maintained, and the riparian habitats are facing strong pressures due to human activities. Severe habitat degradation along the Seine has been caused by changes in land use, modifications of river dynamics and the decline in water quality (see Meybeck 1998).

In history, the Seine has experienced various disastrous floods (for the details see ACRI 1997). In January 1910 an extreme flood occurred and the river rose more than 7 metres above its normal level in Paris. Serious flooding also occurred in 1955, 1993 and 1995. The worst flood flow known in Paris is 2500 m^3/s while the most extreme low flow in dry periods is 15 m^3/s. The average flow of the Seine is 250 m^3/s. The annual fluvial dynamics were modified by upstream reservoirs that modulate the river flow (see below). The most dangerous tributary in terms of flood risks is the Yonne, which meets the Seine approx. 100 km upstream from Paris (see Hydratec 1998).

The floodplain of the Bassée project is located to the south-east of the Paris conurbation, 70 km upstream from Paris, and covers a total surface of around 16,000 ha. (see Figure 11.1). The floodplain is approx. 40 km long, running from the confluence of the Aube, several kilometres upstream from Nogent/Seine, to the town of Monterau-Fault-Yonne, where the Yonne meets the Seine. The mean width of the floodplain varies from 3 to 5 km.

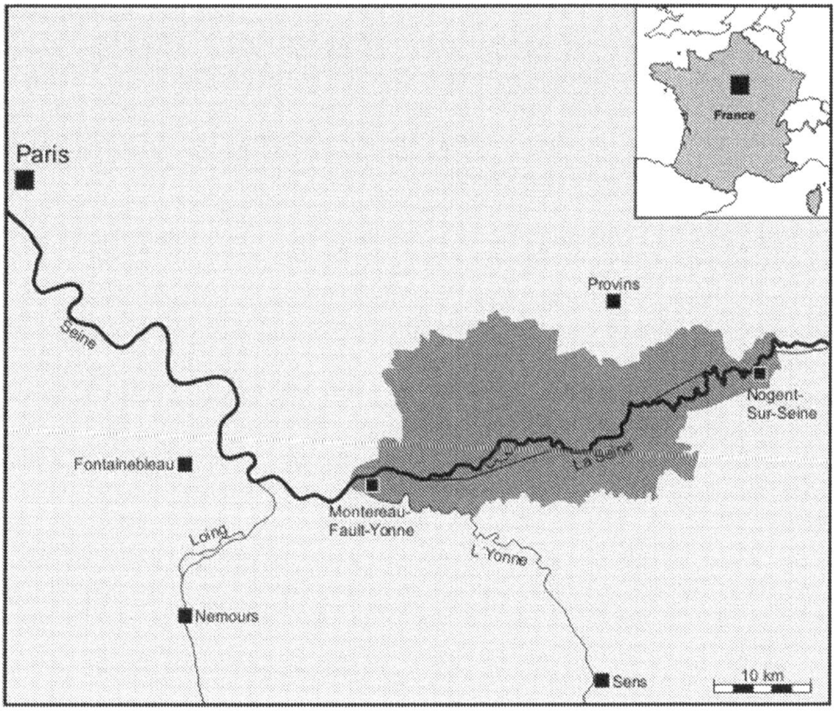

Figure 11.1: Map locating the La Bassée project area.

11.2.2 Use and management of the river

By French standards the Seine basin is densely populated and hosts one third of French industrial and agricultural production. The Paris urban area in the centre of the river basin divides the watershed into two very unequal parts, a fairly "clean" low polluted upstream area and a "dirty" downstream area where environmental problems mainly occur because of waste water from Paris.

Both physically and socioeconomically the floodplain at La Bassée can be divided into two parts (see Fustec *et al.* 2000). The upstream part is located to the west of the small town of Bray-sur-Seine and has a rural character with a low population density; industrial production is of little relevance and agriculture plays a major role in the economy. The area is still regularly inundated in the case of larger floods and can therefore be judged the "natural" part of the floodplain. Due to the regular floods, there are several ecologically important areas (natural alluvial forests, wetlands, open land, meanders etc.), some of which are under legal protection, e.g. a natural alluvial forest that covers around 800 ha. With an area of approx. 10,000 ha which is still floodable, the upstream part of the Bassée is the largest natural water expansion zone in the Seine basin and, therefore, its preservation is perceived as a primary task of flood risk management. The main use of the land in this part of the floodplain is agriculture and farmers demand additional protection because floods in the springtime occasionally cause considerable damage (Interviews 2, 5).

The physical context of the downstream part of the floodplain, between Bray-sur-Seine and the Yonne confluence, differs greatly from the upstream part because land use is not threatened by flooding, and the intensive exploitation of aggregates has already shaped the landscape for some decades. Due to the channelisation of the Seine and the exploitation of sand and aggregates, the area has experienced a significant drop in the water level and the drying out of formerly inundated wetlands (Fustec *et al.* 2000). The quarry industry is by far the most important regional economic factor and significantly contributes to local incomes (see IIBRBS 2002a). However, the area still has a rural and peripheral character in which, apart from the quarry industry, few economic activities have developed. Due to the various man-made pressures, the ecological value of the downstream part of the Bassée is rated as comparatively poor. The area shows a high potential for tourist and leisure activities, and hunting and angling are already important land-use-related activities that are mostly practised by the regional population (Interview 4).

The management of the fluvial dynamics is largely influenced by several upstream dams which are mostly managed by the interdepartmental organisation IIBRBS (Institution Interdépartementale des Barrages Réservoires du Bassin de la Seine). The IIBRBS is a specialised state institution for flood and drought protection and manages four sets of reservoirs upstream of Paris which provide flood and drought protection, increasing leisure activities (see ASCA 1996) and wildlife habitats. In order to protect the Paris conurbation against floods similar to that of 1910, the need for water storage has been estimated at 2.2 billion m^3 for the entire Seine basin (see Barraqué *et al.* 1998). Only 38% of this requirement is met by the four dam reservoirs, the total capacity of which is 830 million m^3, plus 24 million m^3 resulting from the Crescent et Chaumecon dam, which is managed by the power company EDF.

11.3 THE POLICY CONTEXT

11.3.1 National policy context

National programmes and initiatives which provide the background for the Bassée restoration project are concerned with wetland preservation, nature protection policy and flood risk management.

National wetland policy is an important contextual factor, the national government having adopted a national wetlands action plan in 1995. As a result the Bassée plain has been chosen as one of the 87 wetlands of national importance by the Ministry of the Environment. Further, the national implementation of the EU Habitats Directive provides the framework for the project as around 900 ha (forest, marshes) in the floodplain were designated a Natura 2000 site by the European Commission. Large parts of the project area are protected by the EU's Birds Directive.

The national relevance of the project site can be ascribed not only to its importance for nature protection but also to the economic relevance of the floodplain, which is due to the significance of the area for the exploitation of alluvial aggregates (Baron and Piketty 2001; Dasnias *et al.* 2000). The production of alluvial materials on the site account for 45% of the region's entire economic production and the quarry industry is still on the increase (see IAURIF 2002). There are therefore significant conflicts between economic and ecological interests.

In addition, national flood control policy indirectly influences the project. As a reaction to the disastrous flood events in the 1950s and 1960s, the national government promoted the implementation of flood control measures in the basin and the setting up of the specialised institution IIBRBS, which is the initiator of the Bassée project, for flood prevention in the Seine basin. The creation of the IIBRBS and the strong involvement of the national government in the construction of early flood control dams demonstrate that flood prevention on the Seine attracts nationwide attention due to the prominent role of the Paris conurbation and the considerable damage potential there. More recently, and because of previous natural catastrophes, the national government launched a multi-annual programme for protection from natural risks in the early 1990s and developed new measures to protect the Paris conurbation. In particular, because French laws oblige planners to take into account the worst flood ever experienced over a 10-year, 30-year and 100-year period when calculating potential damage and planning protection measures, the decision-makers were required to consider additional flood control measures that might reduce flood risks in the future. According to calculations of potential damage in the Seine basin which were carried out in the 1990s (Hydratec 1998; SIEE-SOCOTEC 1994), a case of river over-flow on the same scale as that of 1910 (100-year flood) would today cause enormous damage. This assessment explains the general national political support for additional flood risk management measures. During the national debates in the 1990s the deficiencies of large-scale technical solutions and, in particular, the potentially negative ecological and economic effects of large dams were emphasised, which has meant that interest in the flood retention capacity of larger floodplains has grown considerably (see Monstadt in Chapter 5 of this volume).

11.3.2 Initiatives in the catchment

Different instruments and initiatives play a role within the Bassée project. In this context, various planning documents are of particular importance. In accordance with the Water Law of 1992, the Master Plan for Water Development and Management (SDAGE) for the Seine-Normandy basin was approved on 20 September 1996 after four years of extended dialogue (see Water Agency Seine-Normandie 1996). The SDAGE lays down qualitative and quantitative objectives, provides for measures aiming at the restoration of wetlands and riparian ecosystems and calls for the implementation of the relevant European directives (e.g. the Habitats Directive) and national regulations. From 1997 the Agency started buying or restoring wetlands, in particular for groundwater protection; in 2001, a permanent wetland unit was created. With the new 5-year programme of the Agency, which started in 2002, an expansion of river restoration activities is envisaged.

At the regional level, the Regional Urban Development Plan (Schéma Directeur de la Région Ile-de-France – SDRIF) is an important planning document because it requires the preservation of open space in the Bassée area and gives attention to flood prevention (see IAURIF 2003). In addition, the implementation of the new flood risk planning instrument, the Foreseeable Natural Risk Prevention Plan (Plan de prévention des risques naturels prévisibles – PPR), which was first applied in the Seine basin after 1996, raised the awareness of decision-makers at the local and regional levels and supported the Bassée project in general terms (Interviews 2, 4).

During the 1990s, the IIBRBS had considered the construction of three additional dams in the Yonne basin but these projects have largely lost momentum due to doubts concerning the ecological and economic effects as well as strong local political opposition (Interview 2).

11.4 DESCRIPTION OF THE PROJECT

11.4.1 Purpose and objectives

The principal objective of the project is the enhancement of flood retention capacity and the reduction of potential flood damage in the Paris conurbation via the construction of polders in the downstream part of the floodplain. According to available estimates and assuming a flood scenario comparable to the 100-year flood of 1910, the project could in practice lower the flood crest in the downstream areas by around 15-30 cm, an effect which might significantly reduce the vulnerability of the conurbation in economic terms. This original flood defence target still plays a significant role within the currently ongoing SAND project, as the hydraulic part aims to increase the flood defence level for the Paris region by retaining peak floods via a polder system.

Apart from this primary target, the creation of artificial water reserves is intended to enhance biodiversity in parts of the area. In the SAND project, ecological flooding is still envisaged but it is not yet clear whether this will take place over the entire polder area or only in ancient dried-up river arms. In addition, it is intended to link the physical objectives with a socioeconomic development perspective. Accordingly, the project should serve as a promoter of the further socioeconomic development of the floodplain area and trigger a new dynamic for the development of an ecologically sound long-term strategy. Specifically, the intention is to develop "a project within the project" with a clear socioeconomic perspective (spatial development project) in which the initiation of environmentally friendly tourism and the encouragement of inter-municipal development strategies are possible elements. The general goal of this socioeconomic approach is to place the hydraulic project in a broader spatial development perspective and to create ecological as well as socioeconomic benefits for the floodplain region. Within the ongoing SAND project, the spatial development project plays an important part, combining flood protection targets with the economic development of the affected region. This part of the project aims to ensure the compatibility of flood retention with gravel extraction and fresh water supply. Moreover, a coherent transport infrastructure is to be established which is in harmony with the future dyke system. Major targets of the Bassée SAND project are also nature protection and restoration, as well as the rehabilitation of ancient Seine meanders and dried-up wetlands (SAND Project 2003).

11.4.2 Origins and development

The first initiatives for an improvement in the floodplain environment came from ecological inventories and studies carried out in the 1980s. At that time the ecological debate was dominated by the wish to push back the quarry industry and to preserve still existing habitats and species. In particular, the flood retention function of the alluvial wetland and its economic value were neglected and such a utilitarian perspective was not used to support nature protection arguments.

This situation changed somewhat after the publication of a report on the possible future of the Bassée in 1992 which was carried out by the Reflection Group on the Management of Dams (Groupe de Réflexion sur la Gestion des Barrages-Réservoirs – GREBAR) in the Seine basin (GREBAR 1992). The GREBAR report stressed the economic value of the maintenance of the flood retention capacity of the floodplain and compared it with other opportunities to reduce damage in the Paris conurbation. The report outlined a possible nature protection strategy for the floodplain which would include the acquisition of 10,000 ha by the state environmental administration and the creation of large protection zones. Although these recommendations were unrealistic in political terms and provoked a storm of protest at the local level, the results increased the awareness of regional decision-makers and provided the regional environmental administration and non-governmental organisations (e.g. the regional department of the national environmental ministry DIREN, the regional council and various environmental groups) with useful arguments for the development of a site-specific environmental policy strategy. In line with the results of the GREBAR report are the recommendations from the Dambre report which suggested a long-term environmental strategy to conserve the natural areas on the site, in particular the alluvial forests and the wet meadows (Dambre and Seven 1996).

In the late 1980s, the IIBRBS and other flood control or water policy actors began to notice the possibility of creating artificial water reserves for flood prevention purposes on the downstream floodplain. In 1994, a first report on potential damage in the Seine basin was published which roughly assessed the different options. One of the main results of this study was that damage caused by a flood comparable to the disastrous flood of 1910 could, today, amount to €12 billion and that, as already mentioned above, only 30% of this damage could be reduced by the flood prevention measures already in place (see Hydratec 1998). In addition, in 1994-1995 a technical study on the feasibility of a polder system on the downstream part of the Bassée was conducted (IIBRBS 1995).

A further large study of flood risks and different risk management options at the basin level was commissioned by IIBRBS, the Ile-de-France region, the DIREN and the Water Agency in the mid-1990s. One of its key results was the general assessment that a mix of floodplain restoration projects, the preservation of still existing flood retention zones, and the construction of new dams would be advisable (see Hydratec 1998). The initiation of the project was also supported by the increasing cooperation between the quarry industry and environmental policy-makers (Interviews 2, 4).

11.4.3 Current status

The study presented here is based on research conducted mostly in 2002/2003 during the early planning stage of the project. Today, the Bassée project is in the process of being implemented. The Bassée area is integrated into the European SAND projects network, which is supported within the programme INTERREG IIIB.[1] It is divided into two main parts, a hydrological and a spatial one. The hydrological project involves 12 municipalities and a total area of 4,300 ha with a retention area extending over 2,500 ha. The spatial development project involves 47 municipalities (along 100 km of the river). This is an integrated spatial development project in a rural area involving regional, national and international interests that go beyond the local level. The coordinating and main partner within the current project is the IIBRBS. Other partners and funding organisations are the DIREN Ile-de-France, the Water Agency Seine-Normandie and the Région Ile-de-France.[2]

11.4.4 Physical interventions

The technical approach of the project is the creation of an endyked polder area (approx. 2,500 ha) in the form of seven "natural" water containers (see IIBRBS 2002b). The idea behind the project is to prevent the accumulation of floods from the Seine and the Yonne basins at the confluence downstream of the Bassée. Such an accumulation is mainly responsible for the occurrence of larger floods in the Seine basin and shows the most significant negative economic impacts. Therefore, in case of a coincidence of floods from the Yonne tributary and the River Seine, water should be pumped out of the Seine or channelled via inlets into the downstream floodplain, which is no longer naturally floodable, and water should be stocked for a couple of days, thereby delaying the outlet of the Seine in the polder area and in this way lowering the flood crest. Because of the hydraulic situation, technical experts are of the opinion that it will be necessary to revert to pumps to fill the water retention areas. Furthermore, the construction of an

---

1 See SAND project 2007: Spatial quality enhancement, Alleviation of flood damage and nature enlargement by development and redevelopment of mineral extraction sites along rivers (http://www.sandproject.nl/, accessed on 25.07.2007)

2 See: Institution Interdépartementale des Barrages-Réservoirs du Bassin de la Seine 2007: L'étude Globale d'aménagement de la Bassée (http://www.la-bassee.com, accessed on 25.05.2007).

adequate drainage infrastructure (drainage channels etc.) will be necessary in order to guarantee the outlet of the water when the flood risk is over.

To ensure the effectiveness of the physical intervention, the dykes need to be 3-4 meters high, so that the total estimated water storage capacity would amount to 55 million m^3 (IIBRBS 2005). The frequency of the use of the water reserves depends on the frequency of large floods. Experts are of the opinion that, on average, the polder system might need to be used once per annum (IIBRBS 2002b). The duration of the use of the water retention areas would be 4 to 6 days while the full outlet of the water will take around 8 days. Consequently, the total duration of use of the polders might amount to 14 days, or even longer depending on the occurrence and the characteristics of the floods. Ecologists are demanding the regular use of the infrastructure in order to provide for ecosystems typical for a floodplain (ecological flooding) (Interviews 2, 4). Consequently, the polder system might be used regularly independently of the actual flood risks for this reason.

11.4.5 Organisation and actor involvement

Government ministries

Within the national government there are at least four ministries which are directly or indirectly (via their regional administrations) involved in the decision-making process. The Ministry for the Environment and Development (Ministère de l'Écologie et du Dévelopement) is the ministry responsible for flood control management and water policy and is the coordinating administration for both environmental protection/water management and flood control policy. Therefore this administration is involved in the flood risk management of the Seine basin, which is certainly the river attracting the most national political attention. Regarding the initiation and the development of the Bassée project, the DIREN Ile-de-France, which is the regional department of the Ministry for the Environment and Development, is more directly involved and, due to its local expertise, generally supports the implementation of the planned retention reservoirs because it acknowledges the potential to use the project for a more comprehensive ecological restoration of the floodplain.

Secondly, the Economics Ministry is a relevant actor because of financing options for projects such as the Bassée. The third relevant national ministry is the Ministry of Public Works and Transport. This ministry is not only responsible for the regulation of land use (local and master plans) and the definition of building codes concerning natural risk management but also for various water policy related issues and the navigability of the waterways. The ministry is represented in each region (Diréction Régional de l'Equipement –

DREF) and at the departmental level through the Public Works Departmental Headquarters (Diréction Départementale de l'Équipement – DDE) (Interview 6). A further ministry involved is the National Culture Ministry with its regional administration, DRAC, as there are important archaeological sites in the Bassée which should be protected until the end of the excavations.

Departmental services

The departmental level, in the Bassée case the Département Seine-et-Marne, could gain in importance because of its official responsibilities and its power to (co-)subsidise flood management measures. In the Bassée project, the departmental institutions DDE (Direction Départementale de l'Equipement) and DDAF (Direction Départementale de l'Agriculture et de la Forêt) are of pivotal importance because both are strong administrative services (in terms of financial resources) and exercise considerable power over project implementation (flood control, infrastructure and traffic routes on the one hand, agriculture and forest policy and rural development on the other). The DDAF is in charge of agricultural and forestry policy as well as water-related issues such as water pollution caused by agricultural practices. In the Bassée project, the DDAF particularly aims to ensure the consideration of the farmers' economic interests and the protection of the groundwater supply.

Interdepartmental organisations

Interdepartmental organisations have gained importance in the project's preparatory stage to coordinate the various administrative (and non-administrative) actors. Here, in particular, the Water Agency and interdepartmental flood control organisations show the potential to act as coordinators and key actors (Berger 1996). The Water Agency is not prepared to take on an active role because of its general lack of specific authority over flood control, its small budget for river restoration, and – regarding the Bassée project – its focus on groundwater protection. This has contributed to the reserved position of the organisation towards the planned polder project (Interviews 1, 2, 4).

In contrast to the low involvement of the Water Agency, the IIBRBS provides the most important interdepartmental contribution to the project's initiation and development. Without the involvement of IIBRBS the project would certainly not have been initiated, and its active and coordinating role is a key factor in the recent progress in the project's preparation and planning. In general, the main focus of the IIBRBS is on flood control, which – besides low flow management – is its main mission, while the other targets of the planned Bassée project (biodiversity, regional economic development) appear to be secondary from the organisation's point of view and do not *per se* form an integral element of the IIBRBS's problem perception. The IIBRBS attempts to integrate the relevant stakeholders in the decision-making process and appears to be prepared to consider other project targets such as nature conservation and regional development if, through this, the likelihood of the project's implementation can be increased.

Région Ile-de-France

In addition to the interdepartmental organisations and given the partial decline in national policy interventions in the implementation of flood control measures, the regional level is gaining in importance. The Ile-de-France region represents an additional political actor which is also characterised by complex internal decision-making. The role of the government of the Ile-de-France region is important because it includes the Paris conurbation with its four departments and it covers all the potential beneficiaries of the project as well as the Bassée region and the department Seine-et-Marne. The Ile-de-France region is responsible for the broad spatial cohesion and coordination of spatial development policies and is consequently an important actor in the development of floodplain restoration measures. It might also contribute to the necessary upstream/downstream coordination by promoting thinking across borders.

Municipalities and landowners

The active involvement of local municipalities in developing and negotiating the project proved very time-consuming. In addition, compensating the individual landowners and local communities for the losses expected from the project is expensive. The local actors were the main supporters of the demand for linking the hydraulic project to a broader economic and social development perspective in the form of a spatial development project.

11.4.6 Instruments used

Different instruments are used to implement the project. From 2001 to 2004, a feasibility study was carried out by the IIBRBS which considered the comparability of the envisaged hydraulic project with the current use of the area and with economic development and nature protection targets (IIBRBS 2005). Not surprisingly, financial resources are highly relevant for the subsequent implementation of the project. With the integration of the Bassée project into the European INTERREG IIIB Programme, a major source of finance has been acquired. Currently, the financing of the SAND project is mixed, with resources from the European Union (INTERREG IIIB), from the IIBRBS, from the Ile-de-France region and from the Water Agency Seine-et-Marne. The EU funding amounts to 50% of the total project costs (SAND Project 2003). A financial mechanism will be necessary to compensate landholders for possible economic losses due to the use of the polder infrastructure. In particular, farmers expect damage due to the future flooding of their land and are already claiming compensation.

When this study was carried out, in 2002/2003, it had not yet been decided in detail which planning instruments should be used. The use of nature protection instruments (e.g. protection zones) will depend on the future land use in the project area, which will depend, in turn, on the extent of the implementation of the ecological flooding proposed by the DIREN. If this strategy is implemented in larger parts of the project area, granting legal protection status (e.g. as a nature reserve) is to be expected.

Besides the financial and planning instruments, in the earlier preparation stage of the project, information and communication instruments were widely used, e.g. public hearings, working groups, public exhibitions, regular project newsletters and bilateral consultations with local decision-makers. In this way, it was possible to achieve a high level of transparency, which appears to have been crucial in minimising the initially strong local opposition.

11.5 PRINCIPAL DRIVERS AND CONSTRAINTS OF FLOODPLAIN RESTORATION ON THE PROJECT

11.5.1 Drivers

Risk awareness and risk analysis

The most important driver of floodplain restoration in this project is growing risk awareness as a consequence of flood events in recent decades. The Bassée case study thus corroborates the findings of the general analysis of the policy context of floodplain restoration in France (cf. Monstadt on Chapter 5 of this volume). Estimates made by the scientific studies mentioned above assume that the large-scale hydraulic project will be very effective in terms of flood control and can achieve a significant reduction of flood damage in the Paris urban area.

The case study also demonstrates that public-financed research and risk analysis were important drivers of project development (Interviews 1, 2, 4). Progress in flood modelling and economic assessments of potential flood damage was achieved during the 1990s when several large scientific studies were carried out. Apart from the actors in the Seine basin mentioned above (IIBRBS, Water Agency etc.), the national research ministry developed research programmes on flood risks and prevention measures, the results of which directly or indirectly contributed to the design of the project. The research results on potential flood damage in the Paris conurbation not only gave rise to an improved scientific basis for decision-making but also to an intensified media discussion, both contributing to higher risk awareness.

Nature protection policy

Besides the flood risk research, various studies concerning the ecological and socioeconomic status of the Bassée floodplain contributed to growing knowledge on the importance of the area regarding biodiversity. Therefore, nature protection policy is a driver of the project as it (partly) supports not only the preservation of the existing floodable areas on the upstream floodplain but also the "artificial" flooding of its downstream sections. Since the project does not intend to restore the natural river dynamic but, rather, to create a controlled environment and artificial alluvial wetland, there are also some reservations on the nature protection side. Some NGOs, for example, doubt the technical feasibility and, in particular, the ecological effectiveness of the project, and expect negative impacts on certain species (Interview 2).

Quarry industry

Furthermore, and somewhat surprisingly, the fact that large parts of the Bassée are used for the exploitation of aggregates supports the project. Since there is in any case a need for the restoration of the landscape heavily transformed by the quarry industry, there is general support for the development of a long-term strategy for the Bassée, in particular for the time after the phasing-out of the quarries in the next 15-20 years. In addition, because of the changed legal framework, the quarry industry is increasingly involved in the general restoration of the river landscape. The expanse of the quarry industry on the floodplain also reduces conflicts between agriculture and the proposed project (Interview 2). The fact that the Bassée area has been shaped by gravel extraction is one of the main reasons that the project was integrated into the EU-funded SAND projects, as these projects aim to combine flood protection with socio-economic development in European mineral extraction sites along rivers. This was an important factor behind the implementation of the Bassée project.

Political support

Additionally, the project is certainly favoured by the support of the regional government (regional council). In general, the Ile-de-France region has recently considerably increased its involvement in river restoration and flood control policy via planning instruments (e.g. the Ile-de-France Regional Urban Development Plan – SDRIF) which include river restoration and flood control objectives and support appropriate projects. Furthermore, the regional government aims to upgrade the Bassée floodplain not only in environmental terms but also with regards to social and economic development. In general, the high level of political support for the Bassée project can mostly be traced back to the political importance of flood protection for the Paris conurbation.

Research

In the European context, research on flood prevention and floodplain restoration and research on the restoration of gravel extraction areas is gaining more importance. This is shown for example by the INTERREG IIIB programme, which supports projects working on these topics. The SAND projects in different regions of Europe, supported by this programme, all deal with "spatial quality enhancement, alleviation of flood damage, and nature enlargement by development and redevelopment of mineral extraction sites along rivers" and thus demonstrate the scientific and political interest in these issues.

11.5.2 Constraints

Funding and maintenance

The restoration of the Bassée floodplain has emerged on the political agenda because technical solutions such as the construction of large dams in the Yonne basin do not appear to be cost-efficient and are locally controversial. However, the planned project cannot be considered to be a low-tech and low-budget approach comparable to other floodplain restoration projects because the flooding of the downstream section of the Bassée necessitates not only major physical interventions but also major financial resources for the construction and maintenance of the infrastructure, for land acquisition and for the compensation of landowners. For a long time in the early stages, financing was unclear and was therefore a limitation during the early phases of the project. This restriction has now been eliminated, as the financing modalities have been resolved.

Low involvement of the Water Agency

One restriction on the project is the rather low level of involvement of the Water Agency. The Water Agency contributed to the financing of the preparatory study but in the planning phase of the project it was not yet clear whether it would support its implementation (Interview 1). The main reason for its reservations is the existence of important groundwater resources of the Bassée, which must be protected in accordance with the relevant water planning documents (SDAGE). According to the Water Agency, the floodplain restoration project has potentially negative impacts on groundwater quality because the flooding of the area with pollutants from the Seine might contribute to decreasing groundwater quality. However, in the current SAND project, the Water Agency is one of the participating project partners and has apparently abandoned its reservations towards the project.

Archaeological sites

A further conflict regarding the feasibility of the project concerns some archaeologically important sites in the project area and the relevant administrative bodies. For example, the agreement of the national culture ministry and its regional administration (Direction régionale des affaires culturelles – DRAC) depends on the precondition that damage to the archaeological sites can be avoided.

Local objections

In addition, there are more traditional restrictions to the project resulting from opponents at the local level. Some local communities do not fully accept the project because they fear restrictions to local economic development and, in particular, a decrease in real property values in or near the planned polder area (IIBRBS 2003; Interview 5). Similarly, some mayors and farmers had general reservations because they perceived the project as an idea from Paris and were not willing to agree for that reason. Following the trust-building exercise, active involvement of the local stakeholders during the early stages of the preparatory study, a comparatively high level of transparency in the planning process and the announcement of financial compensation, objections at the local level could be significantly reduced.

11.6 SELECTED EMBLEMATIC ISSUES

The planned Bassée project has been chosen as a case study not only because of the great political attention that the project receives but also due to its illustrative character for typical institutional features of flood control policy in France. In order to bundle and systematise the most important institutional features three emblematic issues have been selected. The first concerns aspects of multi-level decision-making and the complex system of interests and powers involved. Secondly, the measures debated to compensate the affected Bassée region for the acceptance of the project will be discussed. Finally, the Bassée project is presented as an example of typical conflicts between flood control and biodiversity targets which can arise in planning processes.

11.6.1 Multi-level decision-making and complex institutional juxtapositions

One of the most striking features of the project is the variety of actors and governmental levels involved (see Section 11.4.5) and, compared to other floodplain restoration projects, the high level of the political actors involved.

This circumstance can be traced to the characteristics of the project site, its importance for the Paris conurbation in terms of flood control and, at the institutional level, the complex institutional juxtaposition which is a general feature of flood control policy in France. The multitude of actors in the development of the project goes hand in hand with a multitude of problem perceptions and diverging priorities regarding the development of the site.

Two main aspects are worth underlining. First, the coordination of the national administrations and their regional agencies is not easy to achieve and, as the Bassée project illustrates, some contradictory or competing interests are involved. Second, in spite of the considerable political attention which flood control policy in general, and the Bassée project in particular, attracts, it is important to note that France has generally experienced a certain withdrawal of the various central state administrations from spatial planning and the implementation of flood control measures. As a general rule, the central state is increasingly disengaging itself from a direct role regionally and since the early 1980s considerable policy-making and implementation powers have been transferred to local government at the regional, departmental and municipal levels. The dominating paradigm tends to mean that national administrations attempt to act as facilitators of floodplain restoration but concrete activities and projects should be generated locally or regionally and in a partnership-based manner in which the national government might play the role of partner but not act in a hierarchical way (see Berger 1996; Duran and Thoenig 1996).

The Bassée project illustrates the changing climate of negotiations between the IIBRBS, an interdepartmental organisation, and the region or the local communities which are affected by measures that primarily provide services for the Paris region. Formerly, the IIBRBS and the national ministries acted rather hierarchically, 'imposed' the measures on the relevant municipalities and involved the local stakeholders only late in the decision-making process, if at all. But since the national government has begun to withdraw from the planning and implementation of large flood protection projects and tends to encourage regional and local actors to carry out activities themselves, the political and regulatory back-up of the IIBRBS's actions has changed and the organisation is increasingly obliged to act in a consensus-oriented manner. In order to circumvent local opposition, the IIBRBS puts its main emphasis on trust-building and transparency, and declares the decision-making process to be completely open in order to underline the fact that the construction of the polders depends particularly on the willingness of the municipalities affected to accept the project. According to the most recent position of the IIBRBS, an agreement with the local municipalities is a prerequisite for the project's implementation, which means that, in particular in comparison to earlier flood control measures, decisive power regarding the project realisation has shifted to the local municipalities.

The majority of the municipalities on the Bassée floodplain are organised in the intercommunal organisation Pays Bassée-Montois. In general, the *pays* are territories that do not necessarily conform to existing administrative units but to functional contexts (for details concerning the *pays* see Buller 2000 and Monstadt in Chapter 5 of this volume). The Pays Bassée-Montois was established in 1996 with the task of developing a locally generated development perspective. Due to its intermunicipal character and the organisational and financial weakness of the individual municipalities, the Pays Bassée-Montois is clearly an important actor for overcoming political barriers at the local level. In particular, as a local governance structure it should facilitate the channelling of financial support to the region, adjust the envisaged spatial development project to the already envisaged development of the floodplain for recreation and tourism, and serve as a contact organisation for the departmental and regional administration. In particular, the organisational form of the *pays* can be characterised as an institutional innovation that increases the negotiating power of the small local municipalities. But there are also some doubts regarding the quality of the present organisational capacity of the Pays Bassée-Montois and, furthermore, the interests of the various municipalities in extending the activities of the Pays appear to be diverse (Interview 2). Even if implementation deficits and diverging interests of the individual municipalities are to be expected, the Pays Bassée-Montois represents a shift from a top-down, state-led national rural development policy to an analysis of regional needs and the mobilisation of local actors which offers opportunities for negotiations and should, at least in theory, fit well into the local development agenda and the intended spatial development project. As a result, local government might be more prepared to use the planned polder project for the creation of additional economic benefits for the floodplain region.

The Bassée project illustrates various important features of the current institutional setting in France and its strengths and weaknesses regarding the feasibility of large floodplain restoration projects. In the allocation of state powers we see a complicated juxtaposition of responsibilities at various governmental levels. Relevant powers are allocated to the national, regional and departmental levels, which makes coordination between state bodies and the development of a coherent approach for the implementation of the Bassée project highly demanding. Therefore, the analysis of the Bassée case confirms the results of other actor-oriented research on flood control policy. These studies show that a large number of different functions and participants makes decision-making time-consuming and complex (Böhm *et al.* 2002).

Moreover, compared to the former top-down planning of flood control dams, the planning of the Bassée project is characterised by negotiations and bargaining targeted at achieving a broad consensus. Therefore, the project is in line with a general shift in French water policy from a confrontational relationship between the state and the local communities to increasing cooperation, the integration of both top-down and bottom-up procedures, contractual agreements and project management (see Duran and Thoenig 1996).

11.6.2 Geographies of compensation

One of the most important factors hampering floodplain restoration and the extension of retention areas is the fact that, in general, economic incentives for implementing the relevant measures are lacking. Normally, the positive externalities of restoration projects occur not only in the region in which the projects are implemented but to a large extent elsewhere, typically downstream of the respective river section. The costs of floodplain restoration (acquisition of land, planning costs, compensation to farmers etc.) and possible losses (e.g. the reduction of recreational value) mostly occur in the region where the projects are implemented. Therefore, there is regularly a spatial mismatch of costs and benefits in the sense that those who bear the costs benefit only partly, if at all, from the project itself. Typically, floodplain restoration projects primarily aiming at flood control are confronted with the classic upstream-downstream dilemma. Consequently, without any upstream-downstream compensation mechanism for sharing costs evenly the construction of additional extension areas like the planned Bassée project is unlikely (see Durth 1996).

Against this general background, the Bassée project is an interesting example of how different funding models for overcoming the problems resulting from the mismatch of costs and benefits might be combined. Put simply, there are three basic compensation models which are applicable to the financing of floodplain restoration (see Böhm *et al.* 2002). First, national or regional funding might be used. A higher level of government, covering both the positively and adversely affected regions, would here be responsible for financing the project and compensating the affected municipalities. In this case the geographical scope of the governmental level providing the funding could match, or fully cover, the scope of the physical interconnections, and the spatial scope of the funding institutions would thus be congruent with the geographical scale of costs and benefits. Second, funds specific to the river basin might be distributed based around centralised allocation of funds across the catchment in the form of subsidies corresponding to the effects of the measures. Ideally, those who cause an increase in flood risks (e.g. through the construction of dykes) and the beneficiary parties would contribute to the funds, while those who bear the costs of retention measures would receive financial support. Third, direct negotiations between the regions benefiting downstream and the region carrying the main costs upstream might take place and the beneficiary region would contribute to the financing of the project's implementation or compensate the region where the project is located. Because of their respective strengths and weaknesses, a certain mix of these three economic approaches might be suitable to promote the implementation of flood prevention measures and to speed up the completion of suitable projects (see Böhm *et al.* 2002).

It is obvious that the current financing of the preparatory study and the likely funding structure for the Bassée project could come close to this general recommendation even if some elements would probably go missing. Corresponding to the first funding model, a national and regional financial contribution is expected. The national budget will probably be involved via the Ministry for the Environment and Development or the regional administration DIREN. Concerning the considerable costs of the ongoing preparatory study, the DIREN already finances up to 28% of the costs while the general budget of the Ile-de-France region contributes 21%. In addition, it is expected that the national and the regional budget will contribute up to 50% of the entire project costs.

Furthermore, with the financial involvement of the Water Agency a river-basin-wide fund would be used which roughly corresponds to the second model mentioned above. The Water Agency took responsibility for approx. 30% of the preparatory costs.

Thirdly, and probably the most interesting aspect, the beneficiary region, in the form of the four departments of the Paris conurbation, would directly bear a considerable share of the project's costs, because the IIBRBS would be the operator of the polder system and take responsibility for (at least a considerable share of) the direct costs (e.g. in the form of expenses for the dyke construction, pumps and land acquisition) and, potentially, a share of the indirect costs (e.g. water damage, indemnity for damage to landowners, reduction of agricultural land) which might be caused by the project. Therefore, the current preparatory process of the Bassée project can partly be interpreted as a form of direct negotiation between the beneficiary departments that are represented by the IIBRBS and the department and municipalities adversely affected. The strategy of the IIBRBS is to design the entire project (hydraulic project plus spatial development project) in such a way not only that the costs of the project are minimised but also that internal benefits are generated for the adversely affected region in the form of new opportunities in tourism and recreation.

The negotiations between the Paris region benefiting from the project and the Bassée region are hampered not only by the complicated institutional framework and the interplay of various state administrations and the different levels of local government but also by the fact that various cost and benefit components of the project are not easy to assess. There is no consensus regarding the planned local development projects intended to compensate the region for possible negative effects caused by the polders. The linkage of the spatial development project to the current establishment of a local development agenda by the Pays Bassée-

Montois might be a suitable approach insofar as additional stimuli for local development could be achieved. In any case, it is probable that the Ile-de-France region – probably co-financed by the national government – will offer financial support for the spatial development project and support local development projects in exchange for the acceptance of the polders by the local stakeholders.

11.6.3 Connecting biodiversity benefits to flood control policy

There are different perceptions regarding the ecological impacts of the Bassée project. In general, there is no broad consensus regarding the ecological desirability of the polder system because not all the relevant actors support the view that the artificial creation of additional retention areas via pumps and polders can be equated with an ecologically oriented restoration of the floodplain. In this respect the Bassée project is a representative example of current European-wide debates on the ecological dimension of polders.

Critical views are expressed particularly by some environmental NGOs (e.g. Association Seine-et-Marnaise pour la Sauvegarde de la Nature) but some doubts also remain in the environmental administration (Interviews 2, 6). These criticisms relate to the artificial character of the planned project, the possible negative impacts on certain species and habitats already present on the project site, and the consequences for the natural scenery arising from the necessary dykes, which could not only divide ecologically important habitats but also have a negative impact on landscape aesthetics. In addition, some critics doubt the effectiveness of the project regarding flood control and do not consider that the flood control effects compensate for the possible negative environmental impacts.

Additionally, some environmental actors worry that the effect of the polders could be to reduce the inundation of the upstream part of the floodplain. The reason is that, under certain hydraulic circumstances, the use of the polders could weaken the flooding of the upstream floodplain, and ecosystems depending on regular natural flooding could potentially be harmed. Interestingly, farmers suffering from flood damage in recent years have already discovered this possibility and are demanding that the creation of the polders be linked to the reduction of flood risks in the upstream floodplain (see IIBRBS 2003; Interview 2).

In order to overcome obstacles regarding the ecological dimension of the polders, the IIBRBS recently made an effort to adapt its position to the views of the DIREN and other environmental actors. Accordingly, the IIBRBS proposed devoting at least smaller parts of the project area to ecological flooding even although the organisation fears that local political opposition might be provoked by the regular use of the polder system for ecological purposes.

In the Bassée case the potential implementation of the polders appears justifiable from an environmental point of view because of a lack of alternatives and because of the natural characteristics of the site. In particular, the DIREN's demand for the ecological flooding of, at least, larger parts of the project area might gain the support of environmental groups. Furthermore, mitigation measures appear necessary to avoid possible adverse effects on the environment.

11.7 CONCLUSIONS AND RECOMMENDATIONS

11.7.1 Policy development

The decision-making process, as illustrated here, is very complex and there is a great need for broad cooperation between the various state actors, local governments and non-state actors. The Paris conurbation is highly vulnerable to disastrous flooding in economic and social terms, leading to the national relevance of flood defence and stimulating significant policy involvement in the implementation of flood defence measures in the Seine basin. Although the IIBRBS and some water policy actors (e.g. DIREN) have succeeded in the past in communicating the vulnerability of the area and presented detailed calculations of the potential flood damage, flood defence policy has only recently advocated additional flood retention zones.

In contrast to other French river basins, there is no basin-wide organisation in charge of flood defence policy in the Seine basin. For historical reasons, various regional organisations such as the IIBRBS still exist but the flood defence administration clearly lacks a truly basin-wide perspective. Consequently, the establishment of a basin-wide institution might support the development of a coherent flood defence approach and facilitate the coordination of the various departments and municipalities. In particular, such an institutional innovation might facilitate interregional negotiations and contribute to an early integration of the affected upstream regions in the planning of flood defence measures.

In addition, a comprehensive river restoration and flood defence action plan for the Seine basin should be developed. There are various important planning documents which cannot be regarded as representing a well-integrated and holistic policy approach. Equally, there is no broad consensus regarding spatial priorities in flood defence policy. Moreover, nature protection, water policy and flood defence policy are not well combined and integrated into overall development concepts. In general, intersectoral coordination is still a major task for the improvement of the present policy approach. As the analysis of the Bassée project has illustrated, however, intersectoral coordination is challenged by significant conflicts between economic, social, environmental and water policy targets.

Finally, the internal coordination of the state administration is not sufficiently guaranteed and, because of the complex juxtaposition of powers at the different administrative and governmental levels, there might be a need to upgrade the role of the regional environmental administration (DIREN) in order to facilitate the development of an action plan and the implementation of projects.

11.7.2 Project management

The overall organisational approach of the Bassée project appears well suited to facilitate the involvement and participation of the various actors. The implementation of various communication measures and the organisation of the preparation of the project in the form of a project committee and different working groups have resulted in considerable transparency in the decision-making process. Despite this positive assessment of the project's organisation, the following aspects might be worth further consideration.

Firstly, the involvement of the local municipalities still appears weak, although progress could be made as a result of the involvement of the Pays Bassée-Montois and individual mayors. Secondly, the reconciliation of environmental and flood defence targets remains a primary task for the further development of the project. The proposed ecological flooding of the polder system might be a consensus-oriented solution but there are still doubts regarding its ecological effectiveness. Thirdly, the involvement of the Water Agency in the project's preparation can be regarded as comparatively limited and its rather unclear position is obviously a restricting factor for the implementation of the planned project.

Finally, divergent opinions still exist regarding the necessity of a broader land acquisition strategy. Whilst the IIBRBS tends to support a conflict avoidance strategy and attempts to purchase the land necessary for dyke construction from the owners, other actors (e.g. DIREN) argue for more enforced acquisition of large parts of the project area in order to avoid future conflicts with individual landowners and to facilitate the implementation of a comprehensive floodplain restoration programme. Here, a comprehensive strategy has to be found in order to avoid conflicts and to increase local acceptance for the planned project measures.

11.8 REFERENCES

ACRI - Mécanique appliquée – Observation de la Terre Sciences de l'Environement (1997) *Etude hydrologique des crues historiques du bassin de la Seine*, Paris.

Baron, P. and Piketty, G. (2001) *Plaine Alluviale de la Bassée*. Report for the Ministry for the Environment and the Ministry for the Economy, Paris.

ASCA – Application des Sciences de l'Action (1996) *Retombées économiques locales des barrages réservoirs du bassin de la Seine*. Report commissioned by the IIBRBS, Paris.

Barraqué, B., Berland, J.-M. and Cambon, S. (1998) Country report France, in *Institutions for Water Resources Management in Europe* (ed. F. N. Correira), Vol. 1, pp. 85-182, Rotterdam.

Berger, B. (1996) *Les institutions interdépartementales face aux inondations des partenaires privilegiés pour une gestion globale*. Research report commissioned by the Water Agency Seine-Normandie, Nanterre.

Böhm, H.R., Heiland P., Dapp, K. and Haupter, B.. (2002) *Spatial planning and supporting instruments for preventive flood management*. Project IRBMA – Sponge 5, Final Report, Darmstadt / Berne.

Buller, H. (2000) *Actors, Institutions and Attitudes to Rural Development : The French National Report*. Research Report to the World Wide Fund for Nature and the Statutory Countryside Agencies of Great Britain, Department of Geography, University of Paris.

Dambre, J.L. and Seven, P. (1996) *État de la situation et perspective d'évolution de la plaine alluviale de la Bassée*. Report 94-166, Ministère de l'Environnement, Agence de l'Eau Seine Normandie, Paris.

Dasnias, P., Kovacs, J.C., Leloup, V. and Said Hassani, M. (2000) *Evolution de l'utilisation récente des sols*. PNRZH – project report 07, Paris

DREIF – Direction Régionale de l'Equipement d'Ile de France (1990) *Faisabilité d'aménagement des gravières en bassing anti-crue*, Paris.

Duran, P. and Thoenig, J.C. (1996) L'état et la gestion publique spatial. In *Revue Francaise de Science Politiques* **46**, 580-623.

Durth, R. (1996) *Grenzüberschreitende Umweltprobleme und regionale Integration*, Baden-Baden.

Fustec, E. Bendjoudi, H., Gaillard, S., Weng, P., Boyer, F., Boet, P., Schannen, O., Crogennec, S and Billen, G. (2000) *Fonctionnement du territoire 'Bassée', PIREN-Seine*. Research Report, Paris.

GREBAR –Groupe de Réflexion sur la Gestion des Barrages-Réservoirs (1992) *Quel avenir pour la Bassée?*, Paris.

Henrichfreise, A. (2003) Denen Polder dem naturverträglichen Hochwasserschutz. In *Natur und Landschaft* **78**, 150-153.

Hydratec (1998) *Evaluation des dommages lies aux crues en Region Ile-de-France*, Paris.

IAURIF – Institute d'aménagément et d'urbanisme de la région d'Ile-de-France (2002) *Exploitations de Materiaux de Carrieres en Ile-de-France. Evolution 1990-2000. Mutation de l'Occupation des Sols*, Paris.

IAURIF – Institute d'aménagément et d'urbanisme de la région d'Ile-de-France (2003): Le SDRIF et la ressource en eau : les nécessité d'une action concertée. In *SDRIF - Bilan stratégique, No. 29*, Paris.

IIBRBS – Institution Interdépartementale des Barrages Réservoires du Bassin de la Seine (1995) *Étude de faisabilité d'a,énagements hydrauliques dans le bassing de l'Yonne et dans la one de la Bassée*. Dossier 2 Étude des aménagements de la Bassée, Paris.

IIBRBS – Institution Interdépartementale des Barrages Réservoires du Bassin de la Seine (2002a) *Les connaissances relatives à l'occupation des sols sur le secteur du projet hydraulique*, Paris.

IIBRBS – Institution Interdépartementale des Barrages Réservoires du Bassin de la Seine (2002b) *Bilan des connaissances relatives au projet hydraulique*, Paris.

IIBRBS – Institution Interdépartementale des Barrages Réservoires du Bassin de la Seine (2003) *Exposition de l'étude globale pour l'aménagement de la Bassée. Catalogue des exposants*, Paris.

IIBRBS – Institution Interdépartementale des Barrages Réservoires du Bassin de la Seine (2005) *Étude Globale de l'aménagement de la Bassée. Un projet hydraulique d'écretement des crues au sein d'un projet de territoire*, Paris.

Meybeck, M., de Marsily, G. and Fustec, E. (1998) *La Seine en son basssin. Fonctionnement écologique d'un système fluvial anthropisé,* Paris

SAND Project (2003) SAND Project brochure.

SIEE-SOCOTEC (1994) *Analyse empirique des désordres et des dommages résultants des différentes crues de la Seine et de la Marne dans l'agglomération continue d'Ile-de-France*, Paris.

Thibault, C. (2000) Paris / Ile-de-France : For a sustainable environment. In *Informationen zur Raumentwicklung* **11/12**, 679-690.

United Nations (2003) *Water for People, Water for Life.* World Water Development Report, Paris, chapter 19.

Water Agency Seine-Normandie (1996) *Schéma Directeur d'Aménagement et de Gestion des Eaux (SDAGE) du bassin Seine Normandie*, http://www.eau-seine-normandie.fr/

INTERVIEWS

1. Water Agency Seine-Normandie, Regional Departement Ile-de-France.
2. DIREN, Regional Environmental Administration, Service for Landscapes and Nature.
3. DIREN, Service for Flood Control.
4. IIBRBS, Interdepartmental Organisation for Flood and Drought Protection; Project Management la Bassée.
5. Pays Bassée-Montois.
6. Departemental Administration for Public Works (DDE) Seine-et-Marne.
7. Departemental Administration for Agriculture and Forests (DDAF) Seine-et-Marne.
8. Fédération de Seine-et-Marne pour la Peche et la Protection du Milieu Aquatique.

12

The Parrett Catchment Project: between rhetoric and reality

Ross Beveridge and Timothy Moss

12.1 INTRODUCTION

The Parrett Catchment Project (PCP) in Somerset, England, was one of the flagship schemes of integrated catchment management in the UK. In a variety of ways it embodied recent shifts in opinion about how to manage rivers and landscapes. Launched in 2000 against the backdrop of bitter disputes between farmers and conservationists and severe flooding events in 1999/2000, the PCP's remit was to resolve conflicts over land and water management activities by building a broad partnership with which to develop innovative forms of sustainable flood management across the whole catchment of the River Parrett. A 50-year Action Strategy outlined the project's integrated approach to meeting flood defence needs in ways that would benefit wildlife, support the local economy and offer new opportunities for rural development. With the support of

 Restoring Floodplains in Europe: policy contexts and project experiences edited by Timothy Moss and Jochen Monstadt. ISBN: 1843390906. Published by IWA Publishing, London, UK.

local and national policy communities it succeeded in accessing substantial funding, stimulating joint projects and initiating a dialogue between actors of the upper and lower catchment. What distinguished it from early generation projects of floodplain restoration elsewhere was (a) its multiple objectives, (b) its broad spatial orientation to the catchment, (c) its long-term vision, (d) the emphasis on partnership-building, (e) its linkage to national policy initiatives, and (f) the combination of instruments used. For this reason the PCP reaped considerable praise in professional circles, winning national awards and being selected as an example of best practice for flood management nationally. Expectations of the PCP – both locally and nationally – were high.

However, in April 2007, within just seven years of its creation, the PCP was shut down and replaced by a Somerset-wide organisation, the Somerset Water Management Partnership. Though there were broader, regional issues behind this change, it is clear that the PCP came under increasing criticism for its perceived failure to deliver. Beyond its impressive strategy documents, extensive consultation exercises and substantial external funding the PCP achieved comparatively little in terms of improvements to flood protection, wildlife and the rural economy. Many of those closely involved grew impatient at the lack of progress on the ground and questioned the viability of maintaining the broad project partnership in the absence of tangible benefits. One important organisation, English Nature, even withdrew from project meetings, though it maintained its funding.

This chapter investigates the story of the Parrett Catchment Project, from its promising beginnings and ambitious objectives to the difficulties which contributed to its termination.[1] Crucially, these difficulties are shown to have emerged as early as 2002, when interviews with leading stakeholders in the PCP revealed a variety of frustrations with the PCP and doubts over its long-term viability. That such concerns were being expressed about the PCP so soon after its establishment raises some important questions and, furthermore, invites a look behind the rhetoric of the vision statements and press releases to investigate the realities of participating in projects of this type. By contrasting the expectations with the realities of such an ambitious scheme we do not set out to criticise the PCP. The objective is, rather, to explore why it is proving so difficult to re-order the complex relationship between water and land use in accordance with recent national policy guidance on integrated catchment management. Some of the observations made are, of course, specific to the PCP. The principal message, however, is targeted at the relationship between the strategic level of policy design and the operational level of project management. On the basis of documentary analysis and interviews with nine leading stakeholders the lessons that can be learned from the attempt by the PCP to translate policy objectives into practice are explored. As with the other case studies, the chapter begins with a description of the

1 The authors would like to thank Manuela Siener for her assistance in researching this chapter.

project, then summarises the principal drivers and constraints of floodplain restoration before providing more in-depth analysis of selected emblematic issues on the institutional challenges of pursuing integrated catchment management.

12.2 THE PHYSICAL CONTEXT AND USE OF THE CATCHMENT

12.2.1 Physical characteristics

The Parrett catchment covers a total surface area of 1690 km^2 primarily in the County of Somerset in the South West of England. The catchment incorporates five major rivers – the Parrett, Tone, Isle, Yeo and Cary – which rise in West Dorset and Exmoor and all flow into the Parrett Estuary near Bridgwater and subsequently into the Bristol Channel (see Figure 12.1).

Hydrologically, it is important to distinguish between the upper and mid catchments and the lower catchment floodplain, known as the Levels and Moors. The Levels and Moors, covering over 60,000 hectares, represent a huge floodplain surrounded by a crescent of hills formed at the end of the last Ice Age some 10,000 years ago. The run-off from the upper and mid catchments of the River Parrett and its four tributaries passes rapidly through the relatively narrow river valleys. At times of heavy rainfall this water collects in the downstream floodplain, where the gradient of the watercourses is very low. The flood risk which this poses for the Levels and Moors is compounded by the exceptional tidal range of the Severn Estuary and the fact that the floodplain has a lower ground level than the coastal zone, due to siltation from the estuary (EA 2001: 2-1f.; EN 1997: 1). Because the ground levels of the Moors lie approximately six metres below peak tide levels, the main river channels are embanked so as to carry the fluvial flow of the upland waters through the lower Levels and Moors to the coast. As the capacities of the Parrett and Tone channels are insufficient to discharge peak flood flows to the sea, the excess water flows over the riverbanks and spillways into the lower lying Moors (EA 2002: 12f.). In order to drain floodwater, but also to provide for livestock drinking in the summer, a complex arterial network of rivers, drains, rhynes and ditches regulated by pumping stations, dykes and sluices has been developed over the centuries which now characterise this unique cultural landscape (EN 1997: 2; Countryside & Community Research Unit 2000: 4). However, this impressive legacy of hydrological engineering is unable to prevent severe damage and disruption caused by increasingly frequent flooding events. Furthermore, the flood risk is likely to increase substantially as a consequence of global warming and climate change. It is estimated that the sea level on the Somerset coastline will rise by 250 to 500 mm by 2050 (WWF 2002: 3).

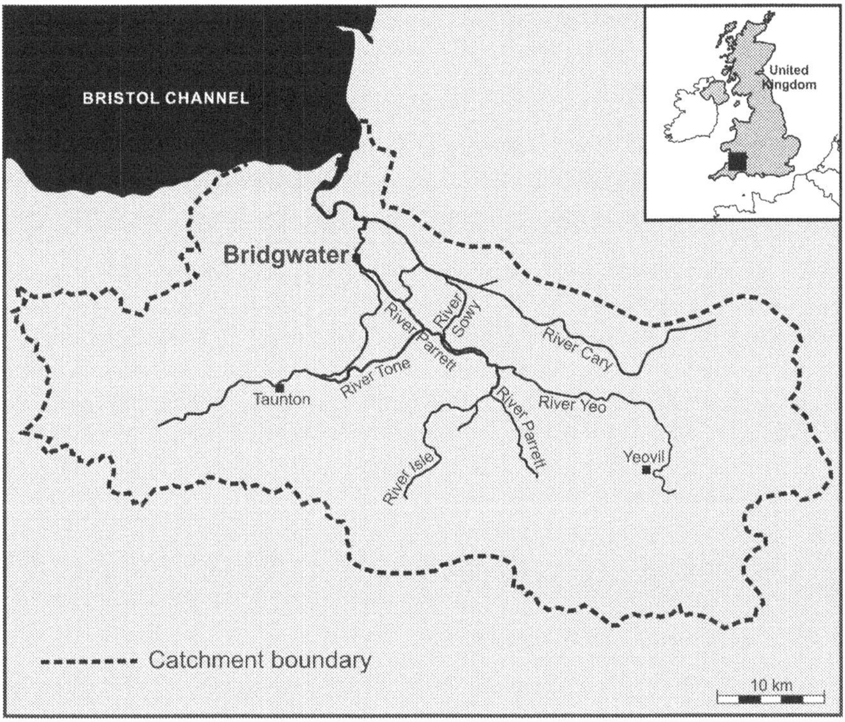

Figure 12.1: Map locating the Parret project area.

12.2.2 Use and management of the catchment

The Parrett catchment incorporates 166 parish councils, and has three major towns: Bridgwater, Taunton and Yeovil (see Figure 12.1). The predominant form of land use in the catchment is agriculture. Due to the varying topography, relief and the range of climatic and soil conditions between the upper and lower catchments in particular, the types of agriculture vary considerably from sheep rearing, cider orchards and potato growing to dairy farming (Land Use Consultants 2001a: 22). Increasingly intensive forms of agricultural production over past decades have exacerbated the management of water resources in general and the provision of flood protection in particular. Woodland clearance, the removal of hedgerows, improved land drainage and modern ploughing techniques have contributed to much higher rates of rainwater run-off and sediment transportation from the middle and upper part of the catchment. Any attempts to reverse some of these agricultural practices, and thereby minimise the flood risk in the Levels and Moors, are very difficult to achieve (Interview 3; for a detailed discussion of the issues and alternatives see Morris *et al.* 2002). Beyond the sheer complexity of the water management problem itself, the ownership structure of agricultural land is highly fragmented, posing a special challenge to integrated catchment management.

The nature conservation value of the Parrett catchment is huge. The catchment includes 47 Sites of Special Scientific Interest (SSSIs) covering a total area of some 9,400 hectares, as well as large-scale Environmentally Sensitive Areas (ESAs) and some Special Landscape Areas. The Levels and Moors is the largest area of wet grassland remaining in England, representing 20% of this national resource. They are of national importance to nature conservation, in particular for their populations of breeding waders and wetland birds (EA 1999: 9). Some 914 hectares of the Levels and Moors have been designated as a Special Protection Area (SPA), which provides specific protection under EU legislation to birds and their habitats (Land Use Consultants 2005: 31). Finally, the Brackets Coppice has been designated as a Special Area of Conservation (SAC), meaning its rich biodiversity is protected under EU legislation.

12.3 THE POLICY CONTEXT

12.3.1 National policy context

The emergence of the PCP needs to be understood against the backdrop of growing national policy interest in the beneficial role of floodplains for flood protection, biodiversity, water resource management and sustainable rural development. As elaborated in Chapter 6 on the shifting policy context of floodplain restoration in England and Wales, a variety of policy and planning initiatives from diverse government agencies are combining to create a powerful discourse on the importance of protecting and restoring floodplains. The Department for the Environment, Food and Rural Affairs (Defra) has launched a strategy for flood risk management entitled “Making Space for Water”, calling on the widening of river corridors and the promotion of multi-functional wetlands (Defra 2005). Defra’s vision statement addresses the need to consider social, economic and environmental aspects, long-term drivers such as climate change, policy connectivity (e.g. to planning, rural development, agriculture and nature conservation), co-funding options, local participation and transparent performance targets (Defra 2005: 14-15). The Environment Agency’s recent position statement “Effective Flood Management” advocates the re-establishment of natural floodplains, with benefits for wildlife as well as flood protection, and the designation of strategic floodplain storage areas, with incentives to farmers to allow land to be used for this purpose (EA 2006). From the field of land-use planning, the Planning Policy Guidance Note 25 highlights the value of floodplains and washlands for flood defence, habitat creation and landscape amenities and requires the consideration of catchment-wide issues when preparing development plans or determining planning applications (DTLR 2001).

The PCP sat nicely within this policy framework for a more catchment-orientated, multi-functional and stakeholder-based approach to the management of floodplains. Its strategy documents, as we shall see, echo the language used by national agencies cited above. The PCP was, however, not simply a reflection of the ongoing shift in national policy on flood protection, nature conservation and land-use planning towards floodplains and river catchments. It was also a pioneer in seeking to put these new policies into practice. The close correlation between the line of argument for integrated catchment management at national and project level was no accident, but part of the PCP strategy. The PCP presented itself as an operational pilot for numerous national policy initiatives. It explicitly aspired, for instance, to contribute to UK Biodiversity Action Plan targets relating to wet grassland and wet woodland. It was selected as one of five pilot studies for implementing Catchment Flood Management Plans (CFMP) in England and Wales, an instrument designed to identify long-term sustainable policies for managing flood risks across a whole catchment. It was designated by Defra as a target area for Countryside Stewardship agreements with farmers over more environmentally sensitive water level management. Through these and other activities the PCP won a reputation as a flagship project of integrated catchment management. It reaped the rewards in the form of successful grant applications and prizes, such as the Award for Planning for Inland Water at the RTPI National Annual Awards for Planning Achievement in 2001.

12.3.2 Initiatives in the catchment

The PCP was a product not only of a shift in national policy towards floodplains but also of earlier regional initiatives to improve the management of land and water use in the catchment, particularly in the Levels and Moors. Indeed, the PCP developed out of a series of projects, plans and partnerships for collaborative environmental management in the region from the mid-1990s onwards. These are, in chronological order:

(1) **The Levels and Moors Partnership (LAMP)**. LAMP was formed in 1995 in response to calls from local communities and interest groups for a new way of working with the local authorities and government agencies. Serving 86 parish councils with wetland, it functioned as a forum for discussion between communities and organisations, and replaced the Countryside Forum and the Somerset Peatlands Reclamation Advisory Board in 1996. It was itself replaced – along with the PCP – in April 2007 by the Somerset Water Management Partnership.

(2) **The Somerset Levels and Moors Water Level Management Action Plan (WLMAP).** This action plan was launched in 1999 to review flood management practices – in particular the management of pumping stations – with a view to reversing recent losses of wetlands and biodiversity in the area (EA 1999: 14). Run by the Environment Agency, the action plan sets out a strategy for day-to-day management of water levels in the Levels and Moors, taking into consideration aspects of water quality, biodiversity, archaeology and recreation.

(3) **The Wise Use of Floodplains project (WUF).** WUF was a project conducted by the World Wild Fund for Nature (WWF) in collaboration with LAMP and supported by the EU LIFE-Environment fund between 2000 and 2002. The project's purpose was to demonstrate how wetlands in floodplains can contribute to improved catchment management (cf. Morris *et al.* 2002). Within the spatial remit of the floodplain of the rivers Parrett, Isle and Tone only, an intense programme of dialogue between community and agency representatives was conducted. This process resulted in the formulation of five water management principles for the floodplain, support for an EA review of flood management practices and an important impulse for the PCP.

(4) **The Parrett Catchment Water Management Strategy Action Plan (PCWMSAP).** This action plan, launched in 2002 by the Environment Agency, looks beyond the issue of water levels covered by the WLMAP to broader concerns of flooding, nature conservation, community development and climate change in the Parrett catchment (EA 2002). It presents a strategy for pursuing the Environment Agency's vision of the environmentally sustainable development of the Levels and Moors and practical steps for implementing this strategy.

(5) **The Rural Environmental Facilitation Service (REFS).** REFS is the successor project to WUF established in 2002 under the leadership of LAMP and with financial support from Defra, the Environment Agency, English Nature and the PCP itself (LAMP 2002). Its overall aim is to support stakeholders in their work towards robust solutions for the sustainable management of water and land in the region. Its two priorities have been to manage a stakeholder dialogue to develop a catchment flood management plan for the Brue and Axe catchments and to help stakeholders appraise the role of washlands in the future management of flood and water levels across the whole Levels and Moors.

A few general points emerge from this overview of parallel initiatives to the PCP. First, there were several attempts to address the complex issues of land and water management in the catchment – both prior to, and concurrent with, the PCP – involving a wide range of local stakeholders. Second, most of these initiatives tackled only a part of the issues addressed by the PCP, having either a smaller spatial remit (LAMP, WUF, WLMAP), a narrower focus of activities (WLMAP, REFS) or fewer lead actors (WLMAP, PCWMSAP). Furthermore, they resulted in little substantive change on the ground, being more concerned with creating better forums for discussion (LAMP, WUG and REFS) or creating broader strategies within which work should be conducted (WLMAP and PCWMSAP).

12.4 BRIEF DESCRIPTION OF THE PROJECT

12.4.1 Purpose and objectives

The overall aim of the Parrett Catchment Project was "to achieve sustainable and integrated land and water management in the Parrett Catchment area to meet the needs of the local community, the environment and the rural economy" (PCP 2001c: 1). To achieve this over a period of 50 years, four objectives were identified:

(1) developing an Integrated Catchment Management Plan (ICMP) for the Parrett Catchment;
(2) developing a sustainable approach to integrated flood management across the whole catchment;
(3) providing a range of measures for modifying land use across the catchment; and
(4) developing an integrated approach to rural development.

These objectives were to be pursued by means of intensive collaboration within a broad partnership of local community organisations, land managers, interest groups and the responsible authorities. The PCP, from the beginning, set itself up to be a model for best practices in floodplain and catchment management with the potential to influence government thinking and policy at regional, national and international (EU) levels (PCP 2001c: 3). Interestingly, there is no explicit mention of nature conservation in these key objectives. This is not to suggest that it was ignored. Indeed, its 'Action Strategy' stressed that "nature conservation is an important consideration throughout the Parrett Catchment" (Land Use Consultants 2001: 3). It does, however, suggest that the project was from the start more focused on agricultural activities and flood protection than on actively promoting nature conservation.

12.4.2 Origins, development and current status

Given the range of initiatives to enrol local stakeholders in the future management of the Parrett catchment, it is easy to forget the depths of division between the farming and conservationist communities which characterised past actor relations. Up until the late 1990s attempts to develop a coordinated approach to flood management and nature conservation in the region were fraught by a bitter conflict of interest over water levels in the floodplain. This came to a head, hitting the national headlines, with an infamous incident when effigies of leading nature conservationists were ceremoniously burned in the car park of a pub frequented by local farmers. If this event launched a phase of sober reflection on the state of stakeholder relations in the catchment, then a series of flooding disasters around the same time triggered widespread criticism that the existing flood management system was inadequate to protect settlements, livelihoods and infrastructure. In the five years prior to the formation of the PCP in 2000, there had been six major instances of flooding in the catchment. Most important in terms of the response, it was generally perceived that the flooding problem was getting worse (Land Use Consultants 2001: 21).

The PCP was created as the principal collective response to this dual problem of poor stakeholder relations in managing the land/water interface and the inadequacies of existing flood protection practices. As the principal butt of criticism for flood damage, the Environment Agency was particularly keen to develop a more collaborative approach to flood protection in the region. In March 2000 the PCP was launched by Somerset County Council, LAMP, the Environment Agency (EA), the Internal Drainage Boards (IDBs), English Nature (EN), the National Farmers' Union (NFU), the Wildlife Trust and the Royal Society for the Protection of Birds (RSPB), among other organisations. There followed an intensive participatory process to develop an integrated approach to managing the catchment, involving events organised by the Wise Use of Floodplains project (WUF), the Levels and Moors Partnership (LAMP) and the PCP itself (Land Use Consultants 2001: 32).

A major breakthrough was achieved at a public forum held in October 2000, where a common vision was developed between farmers, nature conservationists, local communities, the county council, the EA and other agencies. At the public forum the representatives agreed on a "potential list of components needed to develop an integrated flood management strategy with multiple benefits" (Land Use Consultants 2001: 31). On the basis of this list an "action strategy for integrated catchment management" was prepared by consultants. This strategy was substantially informed by a series of workshops held in 2000 and managed by the WUF project and LAMP.

The Action Strategy of the PCP was launched on 5 February 2001. It set out the principles, framework and timetable for developing "a fifty-year Action Strategy for guiding integrated decision making about land use and water management in the catchment" (Land Use Consultants 2001: 1). The strategy was founded on the conviction that actions to relieve the effects of flooding would be most beneficial when they were integrated with social, economic and environmental interests. From the perspective of a 50-year period of gradual transformation, the Action Strategy named a number of project objectives related to integrated land and water management. It also prioritised initiatives and actions that should be addressed in the near future (1-5 years) and which were of medium- or long-term relevance (5-10 years and more than 10 years).

Following the foundation of the PCP and the launch of its Action Strategy, a Business Plan was developed and the project was institutionalised, with a project coordinator, steering group and forum (see below). On the basis of the Action Strategy and the clear demonstration of agreement over the need and purpose of the PCP by the stakeholders, the project benefited subsequently from considerable political and administrative support, both regionally and nationally (Candy and Jenner 2002: 9). The national Minister for Fisheries and the Countryside, Elliot Morley, expressed his support of the PCP. The high public profile of the PCP proved instrumental in gaining access to external funding, in being selected as a pilot for various policy initiatives for integrated catchment management and in easing the passage of administrative procedures.

As a result the PCP was able to achieve the following:

- expand the Farming and Wildlife Advisory Group's (FWAG) programme of arable to grassland reversion and provision of soil management advice and plans;
- complete 6 pilot on-farm flood storage schemes;
- increase woodland development within the catchment by around 30 hectares;
- increase funding for the Environment Agency's flood management activities; and
- raise public awareness of river and floodplain management issues in the catchment with educational campaigns and river festivals (Somerset County Council 2005).

However, in 2005 the future status of the PCP came under review. With EU-funding coming to an end in January 2006, Somerset County Council, working alongside other organisations, funded an external assessment of both the PCP and the LAMP (Land Use Consultants 2005) which suggested that both organisations should be wound down and eventually replaced. After a period of consultation with stakeholders the 'Somerset Water Management Partnership' was established in April 2007.[2] This new organisation will attempt to deal with all water-related issues within the county's main river catchments – the Parrett, Axe and Brue. Its 14 partners – many of whom were in the PCP – will be

[2] See Somerset County Council, the Parrett Catchment Project *News*. Accessible under http://www.somerset.gov.uk/somerset/ete/pcp/news/index.cfm. Accessed on 13.04.2007.

coordinated by Somerset County Council, and together they will attempt to "build upon the success of the Parrett Catchment Project".[3]

12.4.3 Physical interventions

It is difficult to ascribe specific physical interventions to the PCP because the PCP was not a conventional scheme to restore a particular floodplain but rather an initiative to promote integrated catchment management, including floodplain restoration options. The PCP did not undertake physical interventions itself but persuaded others to do so. For instance, flood management measures and improvements to flood defence infrastructures are carried out by the IDBs, the EA or the individual landowners, not by the PCP itself (Interviews 1 and 6).

One of the major projects fostered by the PCP was to facilitate a discussion amongst stakeholders about building a tidal sluice that is regarded by many as a prerequisite for keeping the flood and water management system working stably. In addition, the PCP managed to acquire funding for the EA to conduct a feasibility study addressing the issues of where and how to establish such a sluice. However, the sluice project – being so central to water level management in the floodplain – proved a divisive issue within the PCP and a decision has not yet been made.[4]

Whilst the sluice project, if eventually approved, is planned within a long-term perspective (10-15 years), the PCP pursued more short-term measures to create washlands for storing (flood)water within the catchment. Over a period of 3 to 7 years, schemes were to be implemented addressing the following themes:

- temporary retention of floodwater in floodplain areas;
- small reservoir flood storage by interception in mid-catchment;
- arable reversion to woodland.

For various reasons – not least the foot-and-mouth outbreak in 2001 – the planned projects were seriously delayed. Some pilot projects to "farm water" in the upper catchment, where farmers were compensated financially for accepting wetter conditions on their land, have since been completed. However, being on

[3] See Somerset County Council, News Release: *new water management partnership ready to make a splash*. Accessible under http://www.somerset.gov.uk//somerset/latestnews/pressreleases/details.cfm?releaseID=1217. Accessed on 29.07.07.

[4] See Environment Agency "Choice for our future: Managing flooding on the Somerset Levels and Moors – a discussion document", accessible under http://www.environment-agency.gov.uk/regions/southwest/1450090/?lang=_e. Accessed on 28.08.2006.

such a small scale, their value for flood protection is more symbolic than substantive.

12.4.4 Organisation and actor involvement

One of the distinctive features and strengths of the PCP was the broad partnership of actor groups it embraced. The principal partners were Somerset County Council (SCC), the Environment Agency (EA), Defra (formerly the Ministry for Agriculture, Fisheries and Food), the Internal Drainage Boards (IDBs) and the local authorities. The SCC, responsible for strategic planning in the county and widely viewed as an unbiased voice on catchment management issues, was the lead partner in the PCP. The EA plays a key role as the lead authority for flood defence measures, exercising its powers through the Wessex Regional Flood Defence Committee and the Somerset Local Flood Defence Committees, which decide on flood defence measures. The EA is influential via its flood defence and water management planning authority and expenditure on flood defence works. Defra, responsible for flood defence policy in England and Wales, was important to the PCP for providing grant aid for flood defence investments and for administering agri-environmental schemes which offer subsidies for environmental enhancement to farmland, such as wet grassland. The IDBs – many of which now operate within the Parrett Consortium of Drainage Boards – are responsible for flood defence works on those smaller rivers not covered by the EA and secure funding for this from landowners and local authorities. The four local authorities of Taunton Deane, Sedgemoor, South Somerset and Mendip cover most of the Levels and Moors and play an important role in implementing schemes for small-scale flood retention and advocating change to land management practices.

Beyond these core organisations several others were very active and instrumental in the PCP. Nature conservation interests were represented by the national agency Natural England, formely English Nature (EN), the Somerset Wildlife Trust, the environmental protection section of the EA and the Royal Society for the Protection of Birds (RSPB). Farming interests were represented by the local authorities and to some extent by the IDBs, but also, more widely, by the National Farmers Union (NFU) and the Country Landowners Association. Other supporting organisations, often spanning diverse interests, included the Council for the Protection of Rural England (CPRE), the Countryside Agency, the Farming and Wildlife Advisory Group (FWAG), the South West of England Regional Development Agency, WUF and LAMP (Land Use Consultants 2001).

Such a wide array of diverse and influential partner organisations demanded an effective organisational structure. The principal decision-making body of the PCP was the **Steering Group.** The Steering Group comprised all the key organisations active in the catchment and met quarterly to decide on all strategic issues relating to the PCP. It agreed on spending and was responsible for reviewing and agreeing the Business Plan annually. The Chair of the Steering Group was voted in annually (PCP 2001c: 11). The **Parrett Catchment Forum** met at least bi-annually to keep all the stakeholders involved in the PCP fully informed about the progress of the project (Land Use Consultants 2001: 64). It comprised two groups: the Management Group, consisting of the core funders, and the Stakeholder Group of all member stakeholders. For the day-to-day management of the project a project secretariat with a full-time **Project Co-ordinator** was formed in 2001. The coordinator was responsible for communicating internally and externally, developing the integrated catchment management plan and promoting action on the ground in collaboration with the participating agencies, local communities and interest groups.

12.4.5 Instruments used

The PCP set out from the beginning to enrol a variety of instruments to support the diverse activities involved in promoting integrated catchment management, ranging from planning and strategy instruments to funding schemes and networking activities. The idea was to apply a mix of instruments reflecting the multiple dimensions and interests – but also opportunities – involved in managing water and land in ways beneficial to local communities, wildlife and agricultural production. In some cases the PCP itself was the organisation responsible for applying these instruments. More commonly, though, the PCP's role lay in supporting one or more partner organisations in taking advantage of a particular instrument.

The PCP set great store on strategy development and planning. Its principal products were, indeed, its strategy documents, notably its own Action Strategy and the Parrett Catchment Water Management Strategy Action Plan published by the EA but strongly influenced by the PCP process. One of the PCP's original objectives was to develop an integrated catchment management plan in order to "identify an integrated programme of projects and activities within the context of the water (and land) management issues and opportunities faced by the area and its economic, social and environmental situation" (PCP 2001c: 4). The project subsequently identified 12 objectives for its work:

- Rural: changes to agricultural land management; creating temporary flood storage areas on farmland; creating new wetland habitats; woodland development.
- Water-oriented: dredging and maintaining river channels; raising riverbanks; upgrading pumping stations; spreading floodwater across

the moors; building a tidal sluice or barrier downstream of Bridgwater; upgrading channels to enhance gravity drainage.

- Urban: controlling runoff from development; restricting new development on the floodplain.[5]

The PCP also acted as a pilot for the national Catchment Flood Management Plans (see above).[6]

In terms of funding instruments, the PCP and its members used their influence to great effect, accessing a number of lucrative sources to finance feasibility studies, flood defence improvements and follow-up projects. For instance, in 2002 it secured £650,000 for flood mitigation work within the Joint Approach for managing Flooding (JAF) project, funded by the European Regional Development Fund. Most significantly, the PCP managed to achieve a huge increase in funding for flood management measures in the catchment from the EA and local government. The flood defence committees of the region, which in the past had to manage with an annual budget of £3-4 million, saw this increase to around £14 million (Interview 3). The PCP was also drawing on funding from agri-environment schemes to promote changes to forms of land management conducive to nature conservation and flood protection. This included the creation of floodplain woodland under the 'Woodland Grant Scheme' and 'Farm Woodland Premium Scheme' run by the Forestry Commission.[7]

Arguably the most important 'instrument' of the PCP was its networking activities. These went far beyond simple communication and consultation exercises with local stakeholders. The PCP defined itself in terms of a broad stakeholder partnership. This was substantiated by the multiple activities used to generate debate on the issues it addressed and common ownership of the strategies agreed, ranging from project meetings and workshops to negotiations with interest groups. Collaboration and coordination were not confined to the Parrett catchment area but extended also to relevant organisations and individuals nationally and on the European stage, who were effectively lobbied for support (see Section 12.6.2).

---

[5] See Somerset County Council: The Parret Catchment Project, accessible under http://www.somerset.gov.uk/somerset/ete/pcp/whatwearedoing/. Accessed on 13.04.07.

[6] See Environment Agency: Catchment Flood Management plans, Vol. 1, Policy Guidance, accessible under http://www.environment-agency.gov.uk/commondata/acrobat/cfmp_1426671.pdf. Accessed on 13.04.2007.

[7] See Somerset County Council http://www.somerset.gov.uk/somerset/ete/pcp/whatwearedoing/rural/woodland/. Accessed on 13.04.2007.

12.5 PRINCIPAL DRIVERS AND CONSTRAINTS

This section summarises the principal drivers and constraints of the Parrett Catchment Project. The summary acts as a bridge between the project description above and the in-depth analysis of selected emblematic issues below.

12.5.1 Drivers

Given the orientation of the PCP towards national policy initiatives for integrated catchment management, it is not surprising that the drivers and constraints affecting the PCP resonated closely with those pertinent to the national context of floodplain restoration (see Moss in Chapter 6 of this volume). Thus the most powerful drivers of the PCP – as nationally – lie in the field of flood management. The immediate trigger for the PCP were the series of severe flooding events in 1999/2000 which, in retrospect, created a window of opportunity for re-thinking past flood mitigation strategies. Since then, growing concern over the local impacts of global warming – especially a significant long-term rise in sea levels – has heightened and sustained fears of serious flooding events in the future. Criticism levelled at the EA for failing to provide the level of flood protection expected in 1999/2000 was important in convincing the organisation to seek a broader consensus for a more integrated approach to flood management (Interviews 3, 4, 5 and 8).

A further driver was the recognition that an approach of this kind could only be pursued by resolving the long-standing, bitter dispute over water levels involving primarily the farming community and nature conservationists. Appreciating the connectivity between land and water use across the catchment and exploring new ways of managing water in the framework of its multiple functions became a central motive for those most closely involved. Foremost among these, and the acknowledged mastermind of the PCP, was an influential former councillor who used his extensive contacts – both locally and nationally – and social skills to push the PCP from an idea to a reality (Interviews 1, 4, 5, 6 and 7). In addition, though of lesser importance, nature conservation tools, such as stipulations governing designated protected areas, have played some role in the PCP's efforts to regulate water levels.

The idea of seeking ways of allowing water to flow into floodplains came to play a central role in the rationale of the PCP. The controlled flooding of sections of the Levels and Moors at times of severe floods and the retention of flood waters in the upper catchment were important because they offered potential benefits to all the partners. They appealed to local communities as a way of mitigating flood damage, to nature conservationists as a means to enhance wildlife and to farmers (to some extent) as a way of controlling flooding and as a potential source of additional income via the agri-environment schemes. Given the uncertain future of agricultural production in the UK and the diminishing appeal of farming to young people, alternative approaches to managing land and water are gaining credence, in Somerset as elsewhere.

12.5.2 Constraints

The difficulties which the PCP experienced in putting its strategy for integrated catchment management into practice are indicative of the existence of powerful constraints. First and foremost is the diversity of interests of the participating actors. The impressive advances made by the PCP in bringing actors together and generating agreement on key management principles should not disguise the continued existence of substantial differences of opinion within the project or the inequalities of influence between partner organisations. This is explored in more detail below.

Secondly, in its efforts to seek integrated ways forward, the PCP, like other initiatives of this kind, was running up against the problem of fragmented responsibilities for flood, land and water management, both within the catchment and in the country as a whole. This was, of course, to some extent structurally preconditioned by the diversity of the issues addressed. However, the complex division of responsibility for flood defence between Defra, the EA, the IDBs and local authorities – to take one particular case – made progress particularly difficult. Consequently, implementation was restricted largely to small-scale pilot projects, as in the upper catchment, with only limited impact. Even here it has proved difficult to persuade farmers to retain water on their land. The relevant funding schemes are not geared to long-term changes in land use but offer compensation for loss of revenue only in the medium term (Interviews 3 and 5).

The final significant constraint the PCP faced was its own complexity. The PCP provides a salutary lesson on the difficulties of putting into practice a long-term, catchment-based and stakeholder-oriented strategy for managing land and water use to the benefit of the local environment and economy. Translating the rhetoric into reality proved extremely hard.

12.6 SELECTED EMBLEMATIC ISSUES

This section explores what lessons can be learnt from the experience of the Parrett Catchment Project in translating policy objectives for integrated catchment management into practice. This is done through the perspective of three emblematic issues selected to expose some of the key challenges encountered in pursuing a catchment-wide, long-term, inter-sectoral and cooperative approach to managing water and land on a floodplain. First, there is an examination of the limits of the consensus created by the PCP, a non-statutory, voluntary organisation with a limited remit operating in an unstable, disagreement-prone catchment; it is argued here that the price of securing consensus was action on the ground. Second, the challenges of applying such an approach within the broader policy and regulatory context are explored. It will be shown how even a highly proactive and strategic organisation such as the PCP can struggle to achieve its objectives within existing policy-making and funding frameworks which, despite promoting such an approach, are not always equipped to assist implementation. Third, just how far the joint decision-making processes established by the partnership resulted in the joint pursuit of a common interest is questioned. In doing so, it will be shown how power asymmetries in some ways became reinforced in the PCP, casting doubts over how far there was genuine co-operation in, and common ownership of, policy delivery. To conclude, an assessment is offered of what the PCP's experiences can tell us about the prospects for truly integrated catchment management.[8]

12.6.1 The price of consensus

One of the defining aspirations of the PCP upon its establishment was to introduce a new collaborative form of decision-making to prevent a return to the conflicts between "hereditary adversaries" (Interview 3) which had characterised relations in the past. The foundation for this was the 50-year vision and Action Strategy for the integrated management of the catchment, devised in consultation with stakeholders. By giving everyone a voice in decision making on the future of the catchment the PCP aspired to generate a consensus around key issues of common interest:

> "We are very much about trying to understand what locals think, what they consider is the way forward, and then – together – developing an agreed approach that works" (Interview 2).

The assumption was that if interests could be united around a long-term strategy, it might provide "sufficient time to win the confidence of people on the ground" (Interview 5). The potential benefits of this partnership approach for the participants and the project as a whole were obvious. For participants, the pooling of resources (knowledge, finances and contacts) opened new avenues of influence beyond their traditional remit or resources. They acquired shared competence

---

[8] This section is based upon documentary analysis conducted in 2006 and interviews with key participants in the project in 2002: the PCP management team, Somerset County Council, the Environment Agency, the Consortium of Parrett IDBs, the Royal Society for the Protection of Birds, the Council for the Protection of Rural England and the Somerset Wildlife Trust (see list in Section 12.8).

across the catchment. Clearly, the unity of the PCP engendered a new optimism: "in my view it is a partnership beyond many of those that occurred before" (Interview 3).

However, we have to ask what the costs of this unity were. It appears that what we might term a new institutional culture emerged with the PCP, one which concentrated on unity but avoided the hard decisions necessary to make progress on the ground. Even within the space of two years certain participants were voicing concerns over the preference for general vision statements over more specific actions. Frustration began to emerge over the importance accorded to signing up to agreements, rather than putting them into practice (Interview 7). As a result, after two years, the PCP had not "really achieved anything else apart from the partnership. We've got lots of nice documents and we all meet up and have worthy thoughts and say worthy words" (Interview 5). In 2005 the Land Use Consultants confirmed this limitation by noting the "possible criticism that PCP has not sought to lead opinion (*and* has sought to retain support of the majority)" (Land Use Consultants 2005: 20). This neatly sums up the dilemma for the PCP and consensual-based approaches in general: the coalition of interests it created was an undoubted achievement and therefore something to be valued in itself. However, being a coalition, and only coalition, was also a fundamental limitation:

> "While that coalition is all pulling together, it's hugely powerful... If that consensus starts to break down, then the project will fail. That's an intrinsic weakness as well. The strength is a double-edged sword" (Interview 8).

This statement points to the tricky balancing act required of the PCP. Preserving the consensus was essential to the vision of integrated catchment management, to attracting funding and preventing a return to conflict amongst actors. However, preserving it at all costs and avoiding divisive issues became self-defeating as the lack of visible progress on the ground generated disillusionment amongst some participants. Through adopting this path of least resistance it never "quite found its own particular role in the delivery/implementation side of things" (Interview 3). There was a sense that the partnership around the long-term vision was just covering up underlying conflicts between participants rather than genuinely overcoming them. This became most apparent in the PCP's reluctance or inability to tackle seriously what many see as the biggest challenge facing the Parrott catchment: controlling the water levels (Land Use Consultants 2005: 21). This points to a certain dilemma in attempts to manage collective action, one which has been highlighted by researchers working in the field of policy and public administration (Mayntz and Scharpf 1995; Scharpf 1997; Benz and Fürst 2002). They have noted that institutions in which actors are autonomous and competitive are often able to find innovative solutions to problems despite being prone to conflict. Conversely, such a capacity for action is often stifled in "an institutional setting which fosters integration, compromises, co-operation, control and stability" (Benz and Fürst 2002: 30).

In search of explanations for this development one can point firstly to the relatively low level of commitment by participating organisations required to agree to a long-term vision of the catchment. Such an agreement was attractive and did not impose immediate demands upon participants or require them to move a great deal. The second point to make is that the PCP, as a voluntary partnership and a non-statutory organisation, had no remit to force through implementation of the often very difficult but necessary measures for delivering improvements within the catchment. Such authority ultimately remained dispersed amongst its key members, most importantly the EA, local government and the IDBs. Consequently, doubts have to be raised about the PCP's ability to ever have prompted changes to the entrenched practices and procedures of its member organisations. This concern was reflected in the comment of a former member of the PCP management team that:

> "someone's actually got to do it [implement the 50 year plan]. That takes time and resources… If we're not careful we're going to be appearing to be saying a lot and not doing much" (Interview 1).

There appears to have been a growing recognition amongst key participants that this balancing act could not be sustained forever and that, in the absence of progress on the ground, old tensions might re-emerge. A key member of the PCP management team said the main aim was to:

> "not allow potential jealousies to develop if I can. I don't think that will be forever… there will be a point where we get to be less fashionable" (Interview 9).

This sense of the PCP having a shelf-life, of losing its lustre, was underscored by an "inevitable" tailing off in enthusiasm amongst participants, not simply because of a perceived lack of progress but because of the actual effort required to keep the PCP going:

> "There is a bit of fatigue. I feel it myself. Devoted quite a lot of energy to it in three years, beginning to flatten off. You could spot that amongst the key people in there" (Interview 3).

This question-mark over the long-term future of the PCP prompted the key funding organisations – Somerset County Council, the Environment Agency, English Nature and the Rural Development Service – to commission a consultancy, Land Use Consultants, in 2005 to appraise the performance of the PCP and the Levels and Moors Partnership (LAMP). The report concluded that the Somerset Levels and Moors would be better served by a change in organisational structure which would see both organisations "wound up" (Land Use Consultants 2005: v). It should be stressed that this recommendation was based on broader issues relating both to the performance of LAMP and the needs of the area as a whole, and not merely the performance of the PCP. However, despite praise for much of its work, the report revealed a much less harmonious partnership struggling to maintain both its consensus and catchment-wide approach as some organisations sought to advance their own interests more forcefully:

> "It is argued that certain interest groups have become increasingly focussed on specific issues like the case of the Parrett Sluice, and are pursuing single objectives rather than considering the wider needs of the area" (Land Use Consultants 2005: 2).

The debate over the PCP's support for the creation of a tidal sluice across the Parrett, seen by some as the best way of managing the floodplains, represented an attempt to finally tackle a big issue, take on a major role in delivering change and risk losing the consensus. The fall-out from this decision did indeed prompt a break in the partnership, with English Nature withdrawing from meetings (though not withdrawing funding) because it felt that the "current focus and direction of the partnership is not assisting its own objectives of achieving favourable conditions in the statutory nature conservation areas" (Land Use Consultants 2005: 2). To some extent the disruption to the partnership precipitated by the tidal sluice dispute was a consequence of the limited remit and lack of power of the PCP, the long shadow cast by earlier conflicts and the differences in expectation of the PCP's members, factors which combined to strain the PCP's laudable efforts to use consensus as a platform for action. It can also be attributed, however, to this very strategy of pursuing and protecting consensus at all costs, in that this left the PCP inexperienced in tackling and resolving contested issues. When faced with its first major case of contention the partnership proved brittle.

12.6.2 Playing the policy game

With its vision of a 50-year plan for the integrated management of the catchment and its promise of a truly new form of collaborative decision-making, the key members of the PCP were, from the outset, very aware of the need to 'push the right buttons' not just within the catchment itself but, significantly, within central government and the European Union. It recognised the importance of creating something that those observing from the outside could both believe and invest in: that of the united partnership working towards common goals:

> "If it comes from the Parrett Catchment Project and it's all signed up by 75 local organisations and it goes to Whitehall, it would be far better than one organisation whinging away or the odd letter coming in" (Interview 7).

An analogy with the marketing world might say that the PCP's sensitivity to its image represented a process akin to 'brand' creation. This by no means implies a lack of sincerity within the PCP nor does it question the critical importance of creating the 'brand'. Unity, at least in theory, provided the PCP not merely with the means to engage with, and manage, most aspects of the catchment, but also the less tangible strengths of kudos and access to networks of funding at regional, national and international levels which participants like the RSPB could provide. This merging of "very visible things and very invisible things", a "whole set of slightly different experiences and resources" (Interview 3) was a crucial factor in making the PCP a "powerful lobby for influencing policy" (Interview 8).

The emphasis on brand creation underlines the extent to which the PCP was necessarily conscious of, and driven by, the need to preserve and actively promote a positive image in order to make some form of genuine progress. It was recognised from the beginning that only a clear and united strategy would win over the interest and support of the Environment Minister, Elliot Morley:

> "We had been trying to engage the minister in local issues for some time. But by and large he had said: 'while you're all arguing amongst yourselves – conservationists, farmers, EA, drainage boards – all arguing the toss, I'm not going to come down there and bang heads together. But if you can actually come up with some sort of strategy that you all agree with, then I'll have a look at it and give it my backing if possible'" (Interview 3).

So with this thinking in mind the PCP set about tailoring, or branding, its actions (and proposals for actions) to fit with the policy priorities of central government and European levels in order to receive funding and broader political backing. As one participant said, "we're operating on all sorts of levels" (Interview 9). This interplay with higher scales of action was a major strength of the PCP, with the key figures acknowledged as being "extremely good as a group to respond to opportunities. That is down to particular personalities, characteristics of players" (Interview 1). Perhaps the most important of these players was a former chairman of the PCP, who is praised for being entrepreneurial in his pursuit of funding and, vitally, very well connected: "he's got good networking skills and he's got good contacts up in Whitehall" (Interview 2).

This ability to jump scales and to utilise unconventional, informal methods gave the PCP two strategic advantages when searching for funding. First, "we know what's going on before a lot of other people do" and second, "in knowing what is being aimed at, you can actually gear your response…to meet what you think and we think is the sort of thing that, say, the government departments are looking for" (Interview 9). The PCP was thus often one step ahead of other organisations and had the necessary knowledge of the objectives behind the schemes to 'brand' its applications accordingly. Second, their ability to weave their way through the corridors of power meant that they were always able to find a sympathetic civil servant and avoid a more awkward one:

"'X' is the stroppiest, most uncooperative senior civil servant [we've] ever had the privilege to work with; fortunately [we've] managed to bypass him and go to another minister" (Interview 9).

Such abilities made the PCP a highly effective lobbying organisation. Indeed it claimed to have influenced national policy formulation through ensuring the inclusion of flood risk reduction as a secondary objective of Defra's Higher Level Stewardship programme (Land Use Consultants 2005: 20).

The principal pay-off of this lobbying activity was the huge success in accessing additional public funding by the PCP. This applied to the major increase in flood defence funding, new agri-environment schemes and EU funding outlined above. This external funding, though crucial for advancing the PCP, did at the same time create problems which revealed some of the limitations of pursuing a wide range of high-profile funding sources.

Firstly, the availability of major funding focuses attention on meeting the requirements attached, often at the expense of other project activities. For instance, the PCP's involvement in the Joint Approach for managing Flooding (JAF) project, funded by the European Regional Development Fund, brought in the substantial sum of £650,000. However, delivering the JAF project's objectives came to dominate the PCP's work and, in the words of some PCP members, has "diverted attention from strategic thinking" (Land Use Consultants 2005: 21). The PCP attracted criticism for being overly concerned with conducting the demonstration 'pilot' project work outlined by the JAF "with the result that the returns, in terms of flood prevention and regulation, have not justified the high investment in staff time or cost" (Land Use Consultants 2005: 2).

Secondly, national and European funding instruments are often insufficiently flexible to provide the necessary incentives for particular stakeholder groups or local contexts (on the following, see Moss in Chapter 6 of this volume). In the case of the Parrett catchment, the economic incentives available from agri-environment schemes were generally insufficient to persuade farmers not only to change detrimental land-use practices, but to do so on a permanent, or semi-permanent, basis (see the study on the Levels and Moors by Morris *et al.* 2002; Interview 5). Without such an enabling framework, one which allows for the local integration of competences and provides the security of long-term funding, organisations such as the PCP will always struggle to translate their strategy into practice.[9]

[9] Morris *et al.* (2002) propose a Washland Programme to provide strategic direction and management. Tasks would be delegated to member organisations, including the EA in collaboration with Defra, the Internal Drainage Boards and statutory and voluntary conservation organisations.

12.6.3 Promoting policy delivery

The PCP's initial success in uniting actors around the long-term vision of managing the catchment was based on the notions of joint decision-making and common ownership, pooling the resources and support of participants for joint implementation in the common interest.

> "Its strengths have been about building and displaying common ownership across the parties" and "we (now) have a supposition that integration is good and there are efficiencies in integration" (Interview 5).

The imagery of equal partnership and collective action was powerful from the start. But how far did partnership work in practice? In particular, how balanced was the PCP partnership and what effect did power imbalances have on policy delivery?

The issue of power relations within the PCP is best approached through an analysis of the role of the key organisation responsible for implementation, the Environment Agency. The EA was crucial to the emergence of the PCP and was regarded by at least one independent observer as "the single most important player in this situation" (Interview 3). Local government officials, in particular, were keen to ensure the EA was firmly on board. At the same time, the EA itself was interested in improving its image through involvement in the PCP. Previously, the EA had suffered from a perceived inability to deliver improvements on the ground, producing instead a seemingly endless number of reports. This generated a great deal of scepticism amongst other organisations:

> "They keep doing these catchment studies that they keep changing the name of. [You] can get very cynical about how much is written and how little is done" (Interview 7).

This gap in delivery, it is important to note, reflected broader problems confronting the EA in the region and beyond. Despite its wide-ranging statutory authority, the EA is often seen as a "weak institution" (Interview 3). Dependent on other stakeholder groups, the butt of public criticism for flooding events and often limited by internal divisions between departments, the EA is, in the words of one commentator, "a crippled agency in terms of action [...] because it feels weak and exposed to criticism" (Interview 3).

The EA's predicament, its apparent powerlessness to deliver reform, underlines the necessity of building a partnership, pooling resources and providing for joint delivery. To both the EA and others responsible for flood protection the PCP offered something of a new beginning, a chance to break with the inadequacies and unpopularity of the EA's handling of flood management issues in the past. On the one hand, the PCP would allow the EA to better fulfil its statutory obligations to deliver improvements on the ground, creating a forum for the inter-institutional goodwill necessary for improving floodplain management. On the other hand, it provided local government as well as other stakeholders with a chance to reform and complement the EA's approach to catchment management.

How far were these expectations fulfilled? After the emergence of the PCP the EA does appear to have adapted its style of working in the Parrett catchment. According to a member of the Council:

> "one of the big things that we've actually achieved is that we've influenced other organisations or agencies, particularly the EA locally, to actually produce what the EA calls the Parrett Catchment Water Management Strategy Action Plan" (Interview 2).

In the Council's view, the "big difference" with these new plans, and the PCP in general, was that they came with the backing of the local authorities, "so it's not nearly so bound by rule books as any work that comes out from the EA" (Interview 6).

However, this vision of the PCP did not appear to be entirely shared, even by the EA itself. One EA official described the PCP as "an umbrella" organisation which would probably "not deliver very much in terms of the actual activity" but would provide "a focus for that effort, a co-ordination of the different efforts" (Interview 8). The important point here is that the EA still saw its delivery role as being its own, as being something that was determined and performed independently. The PCP was, from the EA's perspective, a means through which their efforts could be combined with others to fill the 'gaps' in the jigsaw of catchment management, "so where nobody is picking up a particular bit of the jigsaw, the Parrett project can do that" (Interview 8).

The implication of this is that although the EA was at the core of the project's management group it did not see 'partnership' as entailing wholesale changes to its own methods of policy delivery. In other words, although there was agreement amongst PCP members in 2005 "that the works conducted by the Environment Agency are a significant improvement on previous practices" (Land Use Consultants 2005: 25) there is also a sense that the PCP was bolstering the work of the EA without necessarily integrating it. For example, it is clear that the PCP's projects under its 'Water' programme were "largely undertaken by the Environment Agency through their programmes of river dredging and flood defence structures" (Land Use Consultants 2005: 24).

At the same time, the EA was undoubtedly benefiting from being part of the PCP partnership. The most obvious illustration was the huge increase in the EA's funding at the catchment level achieved via the PCP. This doubled after the creation of the PCP, from £6.3 to £13.6 million (Interview 1). Local government actors, who determine EA defence budgets at this scale, made a conscious decision to bolster the EA through the PCP, the key proviso being that it was to be a partnership based on the consensus around the long-term vision for the catchment. As an EA official admitted "the reason we have attracted the money is because we have consensus on a way forward" (Interview 8).

Secondly, the EA drew strength from not acting – or not appearing to act – alone. As a PCP participant says, "one outcome of this new set of activities has been that the agency's had fantastic new confidence" (Interview 3). Clearly this has been beneficial for the catchment, as it has eased the development and passage of important strategy plans and funding bids. However, the benefits attained by the EA do suggest that, for all the talk of a new way of working, the PCP did to some extent reinforce existing patterns of policy implementation by strengthening the EA's ability to deliver flood defence improvements. This lack of real change in terms of what is being delivered and who delivers it means that old problems remained intact. Farmers in the catchment interested in changing their land-use practices continued to be frustrated at the EA's inability to assist them through the maze of funding opportunities and regulatory stipulations, or, in their own words, to put "the jigsaw together" and "make the bits gel together" (Interview 5).

Thus, although the PCP created a forum in which all interests in the catchment could have a voice in management issues, inevitably some had stronger voices than others. Power relations, and more specifically power asymmetries, remained intact. As a result the PCP, despite achieving notable improvements, was not able to deliver real progress for all interests within the catchment. It did not re-configure relations on the ground to the extent to which they could be said to have offered genuinely joint decision-making and joint implementation in a common interest.

12.7 CONCLUSIONS AND RECOMMENDATIONS

Three key stories emerging from the PCP's experience have been identified. First, that building a consensus around a long-term vision for the catchment, however much an achievement in the previously volatile Parrett, led to a tendency to avoid implementation of measures which may have jeopardised this unity. As a result, the big issues were avoided and improvements on the ground were slow to transpire. One repercussion of this slow progress was that it undermined the very goodwill and unity amongst participants on which the PCP was founded, as some grew impatient with the lack of action on the ground. This, in turn, jeopardised chances of a truly catchment-based approach.

The second story, leading on from this, is that the PCP's attempts to think 'catchment' and think 'long term' were impeded by the extent to which its work was dependent upon funding from local government, central government and the European Union. This required it to be responsive to the policy priorities and institutional manoeuvring present at these levels and to be highly selective in terms of its proposed actions. As a result, implementation work carried out by the PCP reflected less the strategic, long-term objectives for the catchment and more the requirements of the available sources of funding. This, combined with the need to preserve unity at the catchment level, led to the sense that the PCP was in many ways a branding exercise: the brand being the consensus built around the holistic vision for the catchment.

The final feature of the PCP's experience explored was the very character of the 'partnership' itself and what it meant in terms of actual delivery of catchment management in the area. The PCP's experience seems to point towards the limitations of such an approach in reconfiguring power relations on the ground. Though there was clearly movement towards greater co-operation amongst many actors, with some sharing of resources and competences, a closer examination beneath the veneer of consensus reveals that the old 'power politics', perhaps unsurprisingly, remained present. In the case of the PCP, there was a sense that it was of most benefit to the key institutional actor in floodplain management, the Environment Agency (EA) as the environmental regulator. Indeed some actors within the PCP openly suggested that this was one of the unwritten motives of the key movers behind the establishment of the PCP. Overall, what it does do, however, is dispel the notion that everyone had an equal voice in the partnership; that everyone benefited simply because they had a voice.

12.7.1 Project management

The net outcome of all three analytical perspectives on the PCP is to emphasise the limits of adopting the consensual integrated approach to managing a catchment in the face of the inherently conflicting interests present within the catchment and inadequacies in the institutional framework in the UK. Emphasising this is not to be critical of the PCP itself or of its members who enthusiastically took up the challenge of making the rhetoric a reality. Indeed, if they were guilty of anything, it could be said that some participants expected too much from this new way of thinking and believed too much in the brand that had been created. Nor is it argued here that the PCP should be seen as a failure. In terms of following the guidelines of central government institutions prescribing integrated catchment management, the PCP was highly successful. Instead, we question and challenge some of the thinking on which the PCP was based.

In conclusion, it is suggested that the vigour with which scientific and policy-making communities promote a long-term, holistic approach should be tempered by a realisation that such an approach is not the catch-all cure for floodplain management. Indeed, from the PCP case study it is clear that there are no easy solutions to the complexity of floodplain management and that the pursuit of 'catchment', 'long-term' and 'co-operation' approaches in floodplain management present their own problems and can often be inherently conflictual. To achieve one objective, such as the management of the catchment as a whole, may not always be best served by pursuing the goals of cooperation. It may be necessary to employ more coercive measures or better central government support to promote catchment-level management, given the diversity of interests present within it. Similarly, it may not always be possible to maintain consensus and cooperation around a long-term vision due to the need to provide participants with visible, short-term improvements on the ground as a reward for their support.

Further, on the basis of the PCP experience, it could be argued that too rigid a focus on consensus-building can, paradoxically, undermine a stakeholder partnership in the longer term. If consensus is achieved by systematically avoiding potentially contentious issues then the partnership will fail to develop the skills needed to resolve conflicts in a mutually acceptable fashion. It will be ill-equipped to deal with major issues of conflict when they – almost inevitably – arise, as in the case of the tidal sluice. Research suggests that the suppression of contention also weakens the capacity of a partnership to innovate.

Apart from the importance of confronting contentious issues, the PCP experience suggests that a long-term vision – vitally important as it is – needs to be balanced by activities which can demonstrate policy delivery quickly and effectively. A clear time schedule of activities would have been helpful, ranging from relatively straightforward, but readily visible, short-term improvements, via medium-term actions of a more complex (and often contentious) nature to long-term, substantive changes in land and water use. This would have been important not only to demonstrate effectiveness in implementation but also to engage more actively with some of the more pressing concerns of the partner organisations.

12.7.2 Policy development

What lessons can be drawn from the PCP case for the further development of national and EU policy on floodplain restoration? Creating policies for a more catchment-based, long-term and consensual approach does not necessarily mean that such an approach will necessarily emerge at the regional or local scale. This is not to suggest that the goals themselves are not worthy of aspiration but, rather, that there has to be greater appreciation of the importance of context, the strength of vested interests and the limitations of consensus-based forms of management. Such an approach, more modest in its expectations of success and more realistic in its capacity to meet a variety of interests, may serve to remove some of the burden placed upon actors at the catchment level and to focus the attention of policy-makers and the scientific community on inconsistencies apparent at other institutional levels. What, it should be asked, is the cumulative effect of specific national and European policies on actors in the catchment? How far do the various regulatory, financial and planning incentives from the policy fields of flood defence, agriculture, nature conservation and land-use planning take into consideration the day-to-day practices and perceptions of the intended addressees in their specific contexts of action? By posing questions of this nature a more 'enabling' framework for integrated catchment management may emerge. As the Land Use Consultants Report confirmed, although the PCP had failed to represent its community "in a sufficiently focused way on major issues" this was as much a reflection of the "difficulties that national agencies and local government sometimes have in communicating their policies and proposals" (Land Use Consultants 2005: iii) as it was the deficiencies of the organisation itself. Unless institutional frameworks become more sensitive to the realities of implementation, then the rhetoric of thinking catchment, thinking long term and thinking cooperation will remain just that and the policy delivery gap will persist.

12.8 REFERENCES

Benz, A. and Fürst, D. (2002) Policy Learning in Regional Networks. *European Urban and Regional Studies* **9**(1), 21-35.

Candy, H. and Jenner, K. (2001) Introducing the Parrett Catchment Project. *Parrett People* **5**, 9.

Countryside & Community Research Unit (ed.) (2002) *Socio-economic profile of the southern catchment of the Somerset Levels and Moors*. Report, Cheltenham.

Defra (Department for the Environment, Food and Rural Affairs) (2005) *Making Space for Water: Taking forward a new Government strategy for flood & coastal erosion risk management*, accessible under http://www.defra.gov.uk/environ/fcd/policy/strategy/vision.htm. Accessed on 04.12.2006.

EA (Environment Agency) (ed.) (1999) *Somerset Levels and Moors Water Level Management Action Plan*. Exeter.

EA (Environment Agency) (ed.) (2002) *The Parrett Catchment Water Management Strategy Action Plan*. Exeter.

EA (Environment Agency) (2006) *Effective Flood Management: Position Statement,* accessible under http://www.environment-agency.gov.uk/aboutus/512398/289428/656297/?version=1&lang=_e&lang=_e Accessed on 05.05.2006.

EN (English Nature) (ed.) (1997) *The Somerset Levels & Moors Natural Area*. Taunton.

LAMP (Levels and Moors Partnership) (ed.) (2002) *Introducing REFS*. (s.n):(s.l)

Land Use Consultants (ed.) (2001) *The Parrett Catchment Project: An Action Strategy for integrated Catchment Management*. (s.n):(s.l)

Land Use Consultants (ed.) (2005) *Future Organizational Structure for the Levels and Moors and Parrett Catchment (Draft)*, accessible under http://www.somerset.gov.uk/somerset/ete/pcp/publications/. Accessed on 06.05.2006.

Mayntz, R. and Scharpf, F.W. (1995) Der Ansatz des akteurzentrierten Institutionalismus. In *Gesellschaftliche Selbststeuerung und Politische Steuerung* (eds. R. Mayntz and F.W. Scharpf), Campus, Frankfurt and New York, pp. 39-72.

Morris, J., Vivash, R.M., Alsop, D., Lawson, C.S., Leeds-Harrison, P.B. and Bailey, A.P. (2002) *Economic Basis and Practicalities of Washland Creation on the Somerset Levels and Moors*. Report to Wise Use of Floodplains Project, Cranfield University, Silsoe, Bedford. www.floodplains.org.uk/pdf/somerset/Washlnads%20Report.pdf. Accessed on 04.12.2006.

PCP (Parrett Catchment Project Steering Group) (ed.) (2001a) *Soil management and conservation*. (s.n):(s.l)

PCP (Parrett Catchment Project Steering Group) (ed.) (2001b) *Farming Water - Temporary retention of floodwater in floodplain areas*. (s.n):(s.l)

PCP (Parrett Catchment Project) (ed.) (2001c) *Parrett Catchment Project Business Plan*. (s.n):(s.l)

Scharpf, F.W. (1997) *Games Real Actors Play*. Westview Press, Boulder, CO.

Somerset County Council (2005) *Parrett Catchment Project. Strategy Components: summary description and progress*. Accessible under www.somerset.gov.uk/.../2005%20January%2020%20Item%205%20Parrett%20Catchment%20Project%20Appendix%20B.htm – 46k. Accessed on 13.04.07.

WWF (World Wild Fund for Nature) (ed.) (2002) *Participatory working in Somerset – a review*. (s.n):(s.l)

INTERVIEWS

Interview 1: Project management officer, Parrett Catchment Project, 20 May 2002.

Interview 2: Planning officer, Somerset County Council, Environment and Property Dept., 20 May 2002.

Interview 3: Policy officer, Royal Society for the Protection of Birds, 21 May 2002.

Interview 4: Representative of Council for the Protection of Rural England, 22 May 2002.

Interview 5: Project officer, Somerset Wildlife Trust, 23 May 2002.

Interview 6: Project officer, Somerset County Council, 23 May 2002.

Interview 7: Engineer, Consortium of Parrett Internal Drainage Boards, 24 May 2002.

Interview 8: Project officer, Environment Agency, 24 May 2002.

Interview 9: Project management officer, Parrett Catchment Project, 25 May 2002.

Part E

CONCLUSIONS

13

Coping with complexity: lessons for policy development, project management and research

Timothy Moss and Jochen Monstadt

13.1 INTRODUCTION

At the time of writing this concluding chapter, in July 2007, Britain is undergoing the worst flooding event for 60 years. Together with the floods of the previous month, it is currently estimated that 45,000 properties have been flooded and that costs will exceed £2.5 billion.[1] The immediate responses are targeting the protection of lives and property from damage, the clean-up operation and compensation claims. Floodplains are drawing public attention as locations of vulnerable infrastructures (power stations and sewage treatment

[1] John Vidal in the Guardian, 24 July 2007.

plants) and as areas of critical settlement development in the past. We know, however, on the basis of experience with earlier flood events in Britain, Germany and France, as well as other countries across Europe, that in the medium term disasters of this kind inevitably generate interest in more effective ways of preventing floods and minimising their impact on human settlements and agricultural land use. At this point, the restoration of functional floodplains – as one strategy to this end – receives renewed attention. A 'window of opportunity' presents itself for restoring floodplains, for minimising flood risks and providing a stronger institutional setting for preventive forms of flood management. How far this opportunity is seized, and to what effect, depends heavily on the ability of key stakeholders to address the complex institutional challenges of restoring a floodplain. This book has set out to provide detailed empirical analysis of policies and practices of floodplain restoration in Europe from which such lessons can be drawn for future benefit.

In this final chapter we summarise and reflect on the key messages emerging from the four policy analyses and six case studies of floodplain restoration presented in the book. Looking across the policy analyses of the shifting institutional contexts in the European Union, Germany, France and England and Wales, the principal factors working against, and in favour of, the restoration of floodplains are pinpointed. These are presented in terms of common constraints encountered by initiatives to restore floodplains and of new drivers supportive of such efforts identified across a wide range of policy fields, from flood prevention and water resource management to nature conservation, spatial planning and rural development. As elaborated in Chapter 1, we do not conceive of these institutional constraints or drivers as phenomena extraneous to human action, either circumvented or enrolled as required, but as integral components of the governance of floodplains construed by the stakeholders themselves. Constraints and drivers, in our understanding, not only influence human action but are themselves shaped by political processes and actor perceptions. The basic argument of this first section is that the institutional constraints to floodplain restoration in Europe are considerable and multiple, but that new interest in, and incentives for, using naturally functioning floodplains to help tackle a variety of development problems are creating favourable windows of opportunity to this end.

What our policy analyses also identify, however, is that in all the countries studied there appears to be a substantial gap between favourable policy rhetoric and instruments on the one hand and limited success in restoring floodplains on a large scale in practice on the other. Exploring how far and why floodplains are – or are not – being restored in practice is the subject of the second section to this chapter: a cross-cutting analysis of the six case studies conducted in Germany, France and England. In order to reflect the impact of recent policy changes at EU and national levels, the experiences of two generations of floodplain restoration schemes are contrasted: those conducted in the early to mid-1990s prior to recent shifts in policy thinking and problem awareness, and those being conducted today which are deliberately built around more integrated, long-term and multi-scalar modes of governance. The six case studies reviewed here represent one early generation and one recent generation scheme from each country.

On the basis of these analyses we subsequently reflect on the nature of a persistent policy delivery gap affecting floodplain restoration resting on the over-complexity of today's restoration schemes and problems of vertical interplay between policy making and project management. Finally, lessons are drawn from this for future policy development, project management and research.

13.2 CROSS-CUTTING ANALYSIS OF DRIVERS AND CONSTRAINTS

The task of restoring floodplains poses multi-dimensional challenges to policy-makers and project managers alike. Creating the necessary interaction between a river and its floodplains has far-reaching implications for existing forms of land and water use. Maintaining desirable river flow dynamics demands a catchment perspective to river regulation and cooperation between upstream and downstream users. Striking an optimal balance between the multiple services a floodplain can provide – to humans and the wider natural environment – requires agreement between many actor groups spanning a variety of policy fields. Furthermore, these actors operate according to spatial remits and time scales which only very rarely match those of a floodplain ecosystem. This is a highly complex process. What follows is an elaboration of the principal constraints and drivers of floodplains restoration as identified primarily in the policy analysis chapters on the European Union, Germany, France and England and Wales.

13.2.1 Constraints

What are the principal constraints to floodplain restoration? A simple answer would be: those human activities which have contributed to the loss of functional floodplains in the past. These include primarily settlement development, agricultural land use, technical flood defence and navigation. The institutional framework stimulating these activities continuously poses strong arguments against allowing floodplains to flood naturally and ecosystems typical of floodplains to develop unimpeded.

A closer look, however, reveals many additional factors related to the institutional challenges of changing entrenched practices of land and water use which can frustrate schemes to restore floodplains (on institutional constraints, see WWF 2000). Our policy analyses have identified several of these constraints which, individually or in conjunction, prevent flood restoration schemes taking place or limit the scope and effectiveness of those implemented. Although their intensity and their significance vary between different national and regional contexts, the following constraints are common to all the countries studied.

Unfavourable forms of land use and ownership

Owing to their favourable topography floodplains are generally very attractive sites for development. The quality of the soil makes them amongst the most productive land types for agriculture. The natural scenery of river landscapes and their value for recreational purposes makes them attractive settlement locations. On the other hand, the biodiversity and ecological value of undeveloped floodplains is also very high. Consequently, changes to land use on floodplains usually confronts a variety of vested interests relating primarily to agriculture and forestry, settlement and urban development, nature conservation, navigation and flood defence. Restoration projects, as a result, frequently provoke land-use conflicts. Land ownership on floodplains, partially because of its historically high value, tends to be highly fragmented and complex, requiring the support of a large number of landowners and users. In instances where changes to land use on floodplains have substantial effects downstream, the constellation of affected parties can be huge and highly varied.

Inertia of the traditional flood defence and river regulation regimes

For many decades institutional regimes for flood defence and river regulation have privileged engineering structures over other strategies. These regimes have been highly successful: "By skilful design and construction, hard engineering structures have, by and large, protected urban dwellers from inundation and enabled farmers to cultivate or graze stock right up to the edge of the river" (Werritty 2006: 18) and allowed for navigation. Even if this regime is now being questioned, foremost because of a series of catastrophic floods and the threat posed by climate change, the traditional policy communities of flood defence, navigation and agriculture dispose of enormous financial means, human resources, knowledge and political influence and often try to block attempts that threaten their control. This tendency to institutional inertia is reinforced by the longevity of the material-physical components of the built infrastructure on floodplains. Once dams, dykes and settlements are built, it is complicated, time-consuming and in many cases expensive to remove or relocate them and, thereby, help restore floodplains.

Ineffective policy and planning instruments

Many of the policy and planning instruments designed to protect floodplains from development are, in practice, ineffective. Planning regulations to limit urban development on floodplains have in the past generally failed to halt the loss of existing floodplains, let alone encourage the avoidance or even removal of settlements and the restoration of floodplains. Restrictions that keep land free from building and specify uses of land for nature conservation, water retention etc. have in many cases been circumvented by local authorities in the interest of urban development. Planning permissions have not infrequently been granted on floodplains at odds with state regulations and plans. Nature conservation policies are in many cases oriented towards the protection of existing habitats rather than the ecological development and restoration of human-modified landscapes (cf. Ledoux *et al.* 2005: 130). The planning instrument of protected areas can for this reason act as a constraint on floodplain restoration. Furthermore, the planning and financing of flood prevention measures are still primarily organised within the territorial boundaries of local and regional authorities and not according to the functional scope of floodplain restoration, requiring consideration of multiple spatial dimensions along the river and across the catchment. Due to this territorial fragmentation, conflicts between upstream and downstream communities are generally not well reflected in institutional arrangements. Effective collaboration between the riparian communities across the administrative boundaries of a catchment remains an exception.

Adverse economic incentives

Financial incentives for farmers or landowners to accept changes to land use, for instance via agri-environment schemes, have achieved only meagre results because they generally provide inadequate long-term financial support to cover the economic risks involved. Furthermore, many financial incentives are indeed counter-productive. The emphasis of flood defence funding on building and maintaining defence structures or of the Common Agricultural Policy on increasing agricultural productivity are cases in point. Coverage of flood damage by insurance companies and compensatory payments by national governments and the EU also provide little incentive for home- and landowners in high flood-risk areas to seriously reconsider practices of land use and settlement on the floodplain. Finally, schemes for a river basin-wide allocation of costs targeting those aspects most responsible for adverse impacts on flow regimes (e.g. hydropower, sealed surfaces) are very rare indeed.

Difficulties in coordination

Floodplain restoration requires multi-dimensional coordination between the diverse actor groups affected, which poses a major challenge of strategic and operational complexity. Difficulties apply to the horizontal coordination across diverse policy fields and their respective actor groups. Efforts to restore floodplains are often confounded by the inability to bridge differences in the objectives and modes of action of the various key organisations. Floodplain restoration is also rendered difficult by the need to collaborate across the administrative units of a catchment. Finally, floodplain restoration projects are confronted with inconsistencies between the policy signals of different levels (EU, national, local). Here, the growing importance of EU policies for the management of floodplains and their cumulative incentives for floodplain restoration often conflict with local interests in the use of floodplains for settlement, agriculture etc.

Entrenched practices and mindsets

Perhaps the most severe constraint, and certainly the hardest to change, is the everyday practices and mindsets of key stakeholders. Many people are unwilling to change ways of using the land or the river which have been established for generations, whatever the rational arguments in favour of restoration. Farmers continue to want to keep their land free from flooding at all costs, flood defence managers continue to prefer hard engineering solutions for flood protection, local residents resist floodplain restoration projects in their backyard and nature conservationists continue to seek protection for valuable "natural" ecosystems.

13.2.2 New drivers

Whilst these constraints continue to have considerable effect, there has recently been growing recognition of the fact that the way that rivers are utilised and managed has to change. Current regimes of floodplain management are increasingly regarded not only as expensive and ecologically unsustainable, but also as an important source of many additional problems relating to water and land use today (cf. Maltby and Blackwell 2005: 137). Major flooding events across Europe, growing evidence of climate change, the continued loss and deterioration of valuable habitats, inefficiencies in agricultural production and the cost of maintaining flood defence barriers are prompting growing concern across Europe. Local and regional actors are showing increasing interest in the potential of functional floodplains to address a variety of localised issues, ranging from flood risks and loss of biodiversity to agricultural restructuring. Policy-makers at national and European levels are critically reappraising many existing institutional regulations and incentives with a view to improving their effectiveness in addressing these pressing issues. The emerging new policies at EU and national levels are often not directed at promoting floodplain restoration specifically; indeed they sometimes do not even refer to it as an option. The important point is that, wittingly or unwittingly, a new management and policy framework is developing out of different parallel initiatives, which, together, are creating a powerful case for change in floodplain management (Maltby and Blackwell 2005: 133).

It is difficult to generalise on policy shifts across Europe but certain trends can be observed in the policy areas of nature conservation, flood protection, water protection, agriculture, forestry, land-use planning and rural development on the basis of our policy analyses. These are summarised in Table 13.1 and described in the following section (on the following, Moss *et al.* 2003).

Table 13.1: Recent policy shifts conducive to floodplain restoration.

Policy field	Forces for change	Policy responses
Nature conservation	Growing concerns for the promotion of biodiversity, habitat networks and for riparian ecosystems and wetlands	EU Habitats Directive; funding programmes for functional floodplain ecosystems and wetlands; advanced instruments for the designation and management of protected areas
Flood protection	Flooding events and growing awareness of the limits to technical solutions; climate change; infrastructure costs; progress in, and growing acceptance of, soft engineering techniques; environmental quality	Flood risk management; EU Floods Directive; national legislative reforms; early steps towards river basin institutions in flood prevention
Water protection	Water quality/quantity problems; interest in flow regimes, wetlands, geomorphology	EU Water Framework Directive; catchment-oriented planning approaches and river basin organisations
Land-use planning	Growing awareness of linkage of flooding events to land use; major damage from flood events; growing pressure from insurance industry	Planning instruments for protecting and creating areas for flood retention; legislative restrictions for land use in flood-risk areas
Agriculture	Growing societal concerns over public health and environmental degradation	Agenda 2000; improved agri-environmental schemes; contractual nature conservation
Rural development	Growing spatial disparities and economic disadvantages of rural areas	EU Rural Development Regulation; integrated approaches to rural economic development

Nature conservation

For nature conservationists floodplains represent important habitats, which can contribute to meeting biodiversity targets in accordance with the EU Habitats and Birds Directives. The Habitats Directive obliges Member States to restore, as well as to maintain, favourable status to key habitats and species. In the context of implementation, many Member States have introduced plans to promote biodiversity that include targets relating to the restoration of wetland habitats. As an example, the National Action Plan for Wetlands was introduced in France in 1996 in response to the alarming degradation of wetlands in recent years. In accordance with the plan, the Ministry for the Environment has identified 84 wetlands of national importance. Wetlands with this status are subject to environmental monitoring and measures to enhance their ecological quality. The plan has proved a significant driver behind some notable restoration floodplain projects, e.g. La Bassée on the Seine. Furthermore, new laws make it possible to identify protected areas not only according to their present ecological value, but also regarding their future development and their ecological potential. As a result, restrictions to develop and ecologically restore river ecosystems and floodplains previously transformed by human interventions are being removed.

Flood protection

In the field of flood protection the policy analyses and case studies demonstrate how recent major flooding events in Germany, France and England and concern at climate change have accelerated the willingness of the authorities to entertain catchment-oriented approaches and soft-engineering techniques of flood protection, creating new opportunities for floodplain restoration. Apart from policy reforms at national and European levels the sheer cost of improving and maintaining physical flood defences, in particular in rural areas, has raised interest in alternative strategies. In addition to tightened restrictions for settlement and other land-use forms on floodplains, the insurance industry is treating riverine flooding increasingly as predictable acts of humankind (cf. Ledoux *et al.* 2005: 139). As a consequence, insurance cover for vulnerable property on floodplains may decline or disappear altogether, creating an incentive to downscale or remove settlements on floodplains. The use of functional floodplains for flood storage purposes is gaining credibility in policy-making circles as a consequence. To take an example of a recent policy initiative, Catchment Flood Management Plans (CFMPs), currently being introduced in England and Wales, take a catchment approach to determining areas where flood storage is most suitable. The planned 60-80 CFMPs across England and Wales will provide a general assessment of risks, opportunities and constraints concerning flood management and develop a set of flood-risk management policies for catchments between 1,000 and 5,000 km^2. A flood management strategy developed within the framework of a CFMP for the Severn River, for instance, argues the merits of floodplain woodland for habitat enhancement and flood mitigation as part of a strategic, catchment-oriented approach to flood management (Environment Agency 2002).

Water protection

Water protection agencies, concerned at water shortages and motivated by the EU Water Framework Directive (WFD), are showing growing interest in water flow regimes across whole catchments and in the (disputed) potential of floodplains to improve water quality. The WFD requires surface waters to be restored to a "good status" and advocates river basin management to this end. Implementing the directive will entail paying greater attention than in the past to flow regimes, abstraction levels and the geomorphology of rivers across whole catchments, all factors favourable to floodplain restoration. A guidance document on the role of wetlands in implementing the WFD, a so-called Horizontal Guidance paper, has been produced within the framework of the Common Implementation Strategy process. It includes a programme of measures for better wetland management. The document is, however, not legally binding. Implementation of its recommendations is thus dependent on the discretion of the Member States.

Land-use planning

In response to extensive damage caused by recent flooding events, land-use planning regulations are being modified in many countries to offer more effective protection of existing floodplains. In some instances they are designating land for the future restoration of functional floodplains. In Germany, for instance, the Federal Regional Planning Act (ROG) of 1998 introduced new categories of planning zones to secure natural areas for flood protection. These designated planning zones, termed "Vorrang- und Vorbehaltsgebiete" under Section 7 paragraph 4 ROG, are regarded by some commentators as an important planning initiative towards restoring, rather than simply protecting, floodplains and their habitats. This federal regulation has, for example, created the opportunity for the state of Saxony-Anhalt to identify in its State Development Plan of 1999 a number of flood protection areas behind dykes that could be restored as functional floodplains in the future.

Agriculture

Political pressure is growing for more environmentally-sensitive forms of agriculture and forestry, increasing the funding opportunities for extensive land management practices more suited to functional floodplains. The Agenda

2000 reform of the EU Common Agricultural Policy has given Member States greater powers to attach environmental conditions to the payment of agricultural subsidies that could be used to promote less intensive farming of floodplains. Willingness to consider steps in this direction has been strengthened by growing concern at the consequences of intensive farming practices, in particular for public health, as highlighted by a number of critical incidents in recent years.

Rural development

The EU Rural Development Regulation for 2000-2006 lays the foundations for a more integrated approach to rural development, incorporating agriculture, nature conservation and the diversification of rural economies. Similar thinking at the national level has fostered integrated programmes and projects for rural development that can include floodplain restoration. Land Management Initiatives have been recently introduced in the UK to develop desirable landscapes and agricultural practices with targeted funding programmes. The Severn-Vyrnwy Wetland Land Management Initiative aims to manage water more sustainably in the catchment by taking an integrated approach to the needs of agriculture, conservation and flood defence. In the course of its development it has broadened in scope from a project to create wet grassland as a breeding habitat for wading birds into a collaborative exercise in new ways of managing the rural landscape.

Emergent window of opportunity

Our research of policy contexts in France, Germany, and England and Wales – as well as of the European Union as a whole – reveals many new institutional drivers for floodplain restoration, encouraging the emergence of a larger number of significant schemes. Although the intensity and significance of drivers and constraints vary in time and between different national and regional contexts, these phenomena are common to all the countries studied. Collectively, recent policy shifts in a variety of policy fields are opening a 'window of opportunity' for floodplain restoration projects of a more comprehensive and integrated kind. They represent a general trend towards a multifunctional, holistic, catchment-based approach to the management of rivers and floodplains (cf. also Ledoux *et al.* 2005; Maltby and Blackwell 2005; Werritty 2006). Informed in part by debates on sustainable development and new forms of environmental governance it is widely recognised that schemes of this kind require greater policy integration and stakeholder participation.

More work is needed to understand the nature of this 'window of opportunity' and ways of using it to promote floodplain restoration. Here, a few caveats are in order. We need to bear in mind that, whilst the current 'window of opportunity' is unique in its particular driver constellation, other, less pronounced 'windows of opportunity' have existed in the past. As we shall see, some early generation schemes proved very effective in exploiting the rather different institutional settings of the early 1990s. Secondly, the components of the emergent 'window of opportunity' today do not, obviously, all occur with the same intensity and at the same time in different places. A flooding event has a very different dynamic and spatial scope to, say, reform of the Common Agricultural Policy. Finally, we also need to avoid an overly simplistic, mechanistic understanding of how 'windows of opportunity' work. As Carsten Felgentreff (2003) has illustrated for the Oder flood of 1997, it was not the disaster itself but rather a more long-term and fundamental readjustiment of problem perceptions generated by subsequent floods elsewhere which proved instrumental in changing strategic thinking.

13.3 THE DEVELOPMENT OF FLOODPLAIN RESTORATION PROJECTS

This section turns from the arena of policy development to the task of implementing projects of floodplain restoration. Here, we are particularly keen to explore the distinctive characteristics of past and present schemes to restore floodplains in the three EU Member States studied and to ascertain how far, and to what effect, the recent rethinking in policy described above is shaping today's floodplain restoration schemes. As with the previous section, we look across the empirical findings from the three countries to draw out commonalities and specificities which are – hopefully – of general interest to those engaged in promoting or researching floodplain restoration. While identifying potential schemes for our investigation, we were struck by the significant differences between schemes conducted in the early to mid-1990s (i.e. generally prior to the new drivers referred to above) and those being conducted from the late 1990s onwards. This prompted us to modify our research design so as to be able to explore these differences in detail. We were particularly interested in seeing whether recent shifts in problem perceptions and policies relating to floodplains are generating more successful schemes of floodplain restoration in practice. The following section begins by setting out the principal differences between what we term 'early generation' and 'new generation' schemes (see Table 13.2).

Table 13.2: Main characteristics of early and new generation schemes.

	Early generation schemes	***New generation schemes***
Objectives	* Limited objectives (main focus on nature conservation)	* Multiple objectives / benefits for different stakeholder groups
Spatial scope	* Limited spatial scope / site orientation	* Broad spatial scope / catchment orientation
Temporal scope	* Immediate interventions	* Long-term vision and strategy
Stakeholders	* Limited number of stakeholders	* Strong partnerships and stakeholder participation
Policy fields	* Primarily single-sector orientation	* Inter-policy linkage / high public profile
Instruments	* Limited instruments	* Instrument mix
Management	* Simple	* Complex

13.3.1 Early generation schemes

Up until the late 1990s, our research shows, the severe institutional restrictions described above combined to limit the number, size and scope of schemes to restore floodplains. Schemes from this era in Germany, France, England and Wales – where they existed at all – were small-scale and site-based. They were typically single-issue projects, targeting environmental enhancement as a rule. They involved only a small number of actors and policy instruments, often relying on a single source of funding for the physical interventions. It is generally true to say that these early generation restoration schemes were conducted largely in isolation from national or regional policy initiatives, whether for flood protection, biodiversity enhancement or rural development. The examples studied in depth in this book are: the Rheinvorland-Süd project in Germany, a scheme to improve hydrological and ecological conditions by widening ducts and removing structures in a section of the Rhine floodplain near Rastatt, the Long Eau project in England, in which flood banks were set back for primarily environmental benefits, and the Bourret project in France, a scheme to reconnect an old arm to the river Garonne and restore an alluvial forest – again primarily for environmental benefits.

Still today, many floodplain restoration schemes belong to this early generation of projects of limited scope and scale. It was, and still is, common for a restoration project to begin as an idea to improve habitat quality along a section of a river of low conservation value. The focus of attention is generally on a single issue, such as fisheries or landscape enhancement. In a typical case, a small group of interested people from the region, many representing government agencies and NGOs engaged in nature conservation, would join together to design and implement a scheme on a particular site. Site selection would be strongly influenced by the willingness of the landowner(s) to cooperate. The area of floodplain restored would generally be small. Instrumental behind the effective implementation, and indeed the existence, of the scheme would often be the ready availability of one or more funding sources to cover the capital costs of conducting the physical interventions (e.g. removing flood banks) and also any loss of revenue incurred by the landowner.

Being relatively unambitious and straightforward, schemes of this kind have tended to avoid the most pressing problems associated with high levels of complexity. With the benefit of relatively straightforward administrative procedures, organisational structures and funding mechanisms they have demonstrated how floodplain ecosystems can be restored on a small scale. They have succeeded in restoring floodplain habitats with limited financial means and human resources and, in some cases, within a remarkably short period of time. The Long Eau scheme in Lincolnshire, England, is a case in point. In less than a year between design and completion, two dykes were set back from the river, thereby increasing the standard of flood protection in the surrounding area from a 1-in-10 to a 1-in-30 year flood risk and creating an attractive wet grassland habitat for wintering and migratory birds at a total capital cost of only £70,000 (see Chapter 9).

These schemes do, however, have several critical limitations. They rarely incorporate a catchment perspective on restoration, but concentrate on the site itself. Being predominantly single-issue schemes, they often overlook potential benefits for other policy areas, such as flood protection, recreation or rural development. Little attention is generally paid to cultivating support for the project in the wider policy-making domain, scientific communities or sometimes even in the local community. The performance of many such projects is rarely monitored or evaluated systematically.

13.3.2 New generation schemes

In response to recent institutional drivers, but also to a growing recognition of the inadequacies of current ways of managing rivers, a new generation of floodplain restoration schemes is emerging which are of a quite different scope to those of the early to mid-1990s. These schemes deliberately set out to address some of the complex challenges to large-scale, integrated floodplain restoration (on the following, Moss *et al.* 2003). The three schemes investigated in this book are: a major dyke relocation project at Lenzen on the river Elbe in Brandenburg, Germany, the Parrett Catchment Project in Somerset, England,

and a large-scale flood retention scheme on the Seine upstream of Paris at La Bassée. Distinctive features of new generation schemes of this kind are their multiple objectives, broad spatial scope, long-term orientation, extensive stakeholder involvement, use of instruments from diverse policy fields and linkage to national policy initiatives and policy-making processes:

(1) *Multiple objectives:* Schemes are designed to provide a wide range of benefits affecting several policy areas. Restoring a floodplain – as the case study on the project of dyke relocation at the Elbe demonstrates impressively – can be part of a broader scheme to raise the standard of flood protection, improve groundwater quality, promote biodiversity, create a more attractive landscape, enhance soft tourism and organic farming and strengthen rural development.

(2) *Catchment orientation:* The greater functional scope and the multiple policy targets of floodplain restoration schemes are in many cases related to a larger spatial scale since many of the issues addressed demand catchment-oriented solutions. Restoration sites are selected according to their suitability from a catchment – or at least large-scale – perspective and not primarily because they are available. As an example, the La Bassée project on the Seine River, by virtue of its broad spatial orientation, offers considerable potential benefits for water flows and quality, ecological conditions and land use that go far beyond the boundaries of the immediate project area.

(3) *Long-term vision:* The long time-frame of many of the issues addressed, such as processes of habitat succession, timber production or the impacts of climate change, prompts a long-term vision for restoration projects extending over several decades. The Parrett Catchment Project is a case in point, basing its action plan around a 50-year strategy for the sustainable future of the catchment.

(4) *Strong partnerships:* Great emphasis is placed on building a broad-based partnership of stakeholders from the relevant policy fields that incorporates NGOs and the local community as well as government agencies. To take the Lenzen case, the local manager of a local agricultural holding, together with officers from the regional agency for large protected areas and the regional flood defence authority, formed a strong partnership to promote the project and to apply for funding and institutional support. This core network was then opened to additional stakeholders including officers from relevant policy fields, local authorities, NGOs and the general public. The resulting organisational structures and procedures can, consequently, be quite complex.

(5) *Wide consultation:* Beyond the involvement of those immediately responsible or affected, the support of a wide range of relevant actors at

the regional and national, as well as local, level is actively sought. In the case of the Parrett and Lenzen schemes appropriate consultation was considered to be important for 'selling' the scheme to local communities and relevant regional and national players at an early stage. Moreover, participatory procedures to inform and involve the local communities, such as steering groups and local forums, have been set up to keep the stakeholders involved in the implementation phase (cf. also Ledoux *et al.* 2005: 142).

(6) *Instrument mix:* Schemes draw on a wide range of funding, planning and regulatory instruments from the various policy fields addressed. They often seek effective novel combinations, as for example in the Lenzen project between flood defence funding for capital works, research funding into suitable species, the designation of protected areas for restoration measures, agri-environment payments to cover maintenance costs and private funding for environmental education.

(7) *Policy linkage:* The ambitious objectives and innovative management practices of integrated projects benefit greatly from establishing strong links to national and European policy-making circles. Schemes often act as pilots for new planning or funding initiatives and seek wide publicity for their activities. This is a distinctive feature of the Parrett Catchment Project, which used its influential connections to central government to set itself up as a model scheme of integrated catchment management.

Since these new generation schemes were only launched from the late 1990s onwards it is at present impossible to judge their effectiveness. Most of them – including all three investigated here – have not been fully implemented and many are barely beyond the planning stage. They would at least appear to have the potential to overcome some of the principal institutional constraints to floodplain restoration which have thwarted or curtailed efforts in the past.

13.4 THE PERSISTENT POLICY DELIVERY GAP

Our research findings, however, caution against over-optimistic expectations from the new generation projects. Early signs suggest that the sheer complexity of the tasks they are tackling is posing a major problem for project management (cf. Adams *et al.* 2004: 1938; Ledoux *et al.* 2005: 141). As a direct consequence of the attempt to promote more holistic, integrated, multifunctional and catchment-oriented restoration, project management is becoming much more complex and time consuming than in the case of earlier schemes. The requirement to consult widely with many stakeholders costs time, financial means and human resources. Accessing multiple funding sources obliges restoration projects to adapt their objectives to satisfy different funding agencies and their individual – and often contradictory – expectations of floodplain restoration. Enrolling instruments from different policy fields requires excellent knowledge on how they work and who to involve, and, as experience tells, often reveals incompatibilities and inconsistencies between their objectives and incentive mechanisms. Winning the continuous support of public authorities and private stakeholders from diverse policy fields and levels and from neighbouring territories within a river catchment necessary for planning and implementing such ambitious schemes is an additional factor behind their growing complexity. Our research demonstrates how these innovative features of the new generation schemes, although each justifiable in its own right, are having the combined effect of hindering effective project implementation – and policy delivery. This finding is reflected in a recent study of flood prevention projects in the UK:

> "Experience to date suggests that managed realignment projects are always likely to be more complex and time consuming than "classical" defence schemes. However, it also seems from the review of case studies that they are becoming increasingly costly and take longer to implement. Participants in workshops were broadly unanimous in stating that the current situation is not satisfactory. If managed realignment is to be undertaken on a larger scale and become a central plank in coastal and fluvial flood defence strategy, the current approach has to be streamlined" (Ledoux *et al.* 2005: 141).

Problems of project management are being further aggravated by a lack of sensitivity of national and European policy towards the unintended effects it is having at the operational level. In order to improve the implementation of new policies favourable to floodplain restoration policy makers are introducing procedural specifications which require project managers to collaborate widely, achieve complicated cross-funding of restoration projects by national, regional, local and private investors, seek cross-sectoral synergies and exploit all available opportunities to maximise benefits. These requirements have undoubtedly encouraged greater awareness of policy interconnectivity in general and the emergence of more integrated restoration projects in particular. They are also, in their own right, often highly justifiable as a means to improve the accountability, policy integration and cost-effectiveness of restoration schemes. However, their cumulative impact has been to render more difficult the process of implementing restoration schemes and to limit the scope of project managers to act independently, to use their discretion and to find solutions adapted to local circumstances and situations. Many schemes, faced with growing procedural controls and restrictions, are being seriously delayed, often jeopardising the carefully constructed partnerships, funding opportunities and common objectives essential for their implementation. More disturbingly, we have anecdotal evidence from all three countries studied to indicate that officers responsible at the operational level are, for this reason, increasingly reluctant to get involved in ambitious restoration schemes at all. If a scheme does not have a critical mass of organisational capacity to see it over the bureaucratic hurdles or, alternatively, if it is not small enough to fall below the thresholds for these controls, then the chances are that it will never get started. Given the already small number of floodplain restoration schemes in existence this is a worrying observation.

Related to these issues of over-complexity and excessive bureaucracy are problems of vertical interplay between national or supranational policy makers and local or regional project management. As our expert interviews attest, policy makers are often critical that new policy thinking is not filtering down effectively to the operational level. They suspect a lack of willingness on the part of some project officers and other local stakeholders to entertain changes to established objectives and procedures. Conversely, many project managers feel that policy makers do not fully appreciate how their reform initiatives work on the ground, particularly in combination. This correlates with the observation of Turner *et al.* (2000: 18) that "international and national regulations often fail to address the local subtleties in multiple use of wetland areas". Similarly, Eden *et al.* (2000: 271), referring to the River Cole project, speak of how "the heterogeneity and place-specificity of these river restoration techniques stand in opposition to the uniformity and generality of institutionalised techniques".

13.5 IMPLICATIONS FOR POLICY DEVELOPMENT AND PROJECT MANAGEMENT

Given the persistence of a serious policy delivery gap – albeit for more ambitious objectives than in the past – what recommendations can be drawn for the development of policy and projects for floodplain restoration in the future? How, in particular, can stakeholders at both strategic and operational levels overcome persistent institutional constraints, deal with the 'complexity trap' and problems of vertical interplay described above? In answering these questions we emphasise from the start that we do not dispute the need to strive for policies and projects which are highly integrated, cost-effective and involve a wide range of stakeholders. The answer to over-complexity is certainly not a return to past practices where most floodplain restoration schemes were planned and implemented by a small, select group of enthusiasts with little consideration for connectivity to other policy fields or actor groups. These projects cannot meet the challenges of restoring ecological river environments on a large scale, promoting biotope networks along river corridors, providing more efficient flood risk management or linking floodplain restoration to sustainable rural development. Early generation schemes of floodplain restoration, such as those described in this book, can, however, give us important clues as to ways of coping with the complexity inherent to more ambitious projects.

Building on existing stakeholder structures and local knowledge is one of these lessons (cf. Turner *et al.* 2000: 19). The most successful schemes would appear to be those where local stakeholders are not only involved, but given the scope and opportunity to fulfil their own ideas, individually and collectively. As Adams *et al.* have already pointed out (2004), local "champions" – often respected and influential figures of a community – are important for their knowledge, organisational skills, energy and vision. Examples from this book include, from the Lenzen case study, the local landowner and officers from the state agency for large-scale protected areas (LAGS) and, from the Long Eau study, the three government officers responsible for nature conservation, agri-environment schemes and rural development. In order to work effectively, actors of this calibre need the full support of their own organisations. This can take the form of freedom from excessive bureaucracy but also advice and training in professional project management. Training programmes in communication, negotiation and conflict resolution could strengthen essential management skills. Organising exchanges on project experiences could avoid the duplication of common mistakes and, at the same time, publicise the importance of floodplain restoration. Beyond this, project managers require skills in adaptive management in order to cope with the high degree of flexibility needed for new generation schemes. This should include learning how to identify and exploit windows of opportunity as they present themselves in a particular local context.

Moving from the project level to policy development, the challenges of complexity and vertical interplay call for more critical reflection of the effectiveness of policy instruments and implementation procedures.[2] Considerable progress has been made in all three countries under study here towards generating institutional settings conducive to integrated and catchment-oriented schemes for floodplain restoration which involve a wide range of stakeholders. Without more far-reaching reforms, however, the prospects for the widespread uptake of floodplain restoration schemes are poor. Firstly, policy instruments need to be rendered less contradictory than at present. Classic examples are the existence of funding for both draining and re-wetting agricultural land or of policies in favour of both traditional hydraulic engineering and the conservation of wetlands. Secondly, better ways are needed to link policy instruments. Here, a good

---

[2] Cf. Turner *et al.* (2000: 17-19) who recommend (1) creating awareness for wetland values at all levels, (2) clarifying the division of responsibilities, creating a hierarchy of decisions, and (3) encouraging local arrangements consistent with sustainable wetland use.

example is the case of using capital funding from flood defence budgets to set back dykes where suitable, and combining this with funding from, for instance, agri-environment schemes to pay for the maintenance of the new floodplain ecosystems created. Thirdly, legislation and policy initiatives need to give greater consideration to the improvement – and not just protection – of habitats. This applies in particular to instruments of nature conservation. Fourthly, institutional arrangements for floodplain restoration should be oriented more around river catchments in order to stimulate not only more environmentally sound projects but also a more just distribution of costs. This applies both to policy instruments for planning flood prevention and river restoration and to organisational capacities, such as extending the powers and resources of existing river basin organisations to cover flood risk management and river restoration. Fifthly, economic mechanisms are needed that spread the costs of restoration projects and other flood risk management strategies between the various beneficiaries in a river catchment and penalise activities that cause negative alterations to flow regimes. Examples include taxes on the sealing of surfaces or on hydropower and setting up a catchment-wide fund for restoration projects. Finally, we suggest creating simpler organisational responsibilities at strategic level for the supervision and funding of floodplain restoration projects. This could include streamlining different funding sources to assist co-funding options and naming a first-call agency for coordinating project initiatives across a whole catchment or river basin.

13.6 CONCLUSIONS

We can conclude that recent shifts in problem awareness and problem-solving approaches are opening windows of opportunity for floodplain restoration. In a number of relevant policy fields, notably flood protection, water protection, nature conservation, land-use planning and rural development, policy initiatives are being taken which, wittingly or unwittingly, are creating institutional frameworks favourable to the restoration of functional floodplains. At the same time there is a growing willingness amongst local and regional actors to restore floodplains as one way of addressing a variety of local issues, ranging from flood risks and loss of biodiversity to agricultural restructuring.

This interest at both the strategic and operational levels has combined in several instances to stimulate the emergence of a new generation of floodplain restoration schemes. These schemes would appear to possess the potential to overcome some of the main deficiencies of small-scale, single-issue schemes of the past by taking a longer-term and catchment perspective on floodplain restoration, building strong partnerships, linking in to policy development and making use of a variety of policy instruments. The very complexity of these new generation schemes is, however, posing serious problems for project management. A floodplain restoration scheme of this kind requires a lot of time, money and support from all the stakeholders, which is not always available. The need to meet so many requirements, consider so many factors and satisfy so many interested parties frequently results in ambitious floodplain restoration schemes being seriously delayed, heavily revised or, in extreme cases, shelved altogether. The policy delivery gap continues to exist, therefore, but in a different guise. On the one hand recent policy rethinking within key (national) agencies is often not filtering down effectively to the operational level. On the other hand there is a tendency for policy makers to overlook the implications of policy encouraging integrated, cross-sectoral and multi-agency action for operational practices on the ground.

In the design and implementation of future policy, greater sensitivity towards local contexts of action and the complexities of floodplain management would significantly enhance the prospects for the effective restoration of floodplains. More consideration needs to be given to how individual policy incentives work in conjunction with others; that is, how they alter the existing institutional setting. In addition, policy makers need to be more sensitive to the contexts of action in which their instruments operate. What makes a policy instrument effective is not the assumed preferences of individuals acting according to a rational choice logic but the real scope and willingness of stakeholders to alter their practices. More feedback into policy-making processes is needed about the real-life experiences of project management and stakeholder involvement at the operational level. This will require more monitoring of how and why policy instruments do or do not work in practice. Policy needs to provide not only targets for orientation but also frameworks for developing the necessary economic, social and institutional capital at local and regional level and instruments capable of adapting to the dynamics of a floodplain restoration process. On this basis policy makers and project managers alike should be better equipped to identify and exploit windows of opportunity for the restoration of floodplains in the future.

This process requires the support of the extensive and heterogeneous research community now working on issues pertinent to floodplain restoration. We need to pursue our studies of new generation schemes as they enter the implementation phase in order to understand better how they might be improved and how their example and their experiences might generate more – and more effective – schemes of this kind. We also need better knowledge on the vertical interplay between policy making and project management, in particular on procedures for encouraging cross-scalar learning and striking a beneficial balance between accountability and freedom to innovate. Mechanisms for improving cross-sectoral coordination need further research, especially relating to ways of interconnecting instruments from different policy fields. Beyond this, diverse models of stakeholder participation could usefully be compared to assess their respective value for floodplain restoration schemes. Research into issues of scale could also provide helpful insight into how actors work across scales effectively and which scales lend themselves to which task of floodplain restoration. Seeking ways of linking these research tasks to those of the natural and engineering science communities is, perhaps, the most challenging – but potentially the most rewarding – task of all. Our experience in the FLOBAR2 project encourages us to call on social scientists to pursue this challenge with conviction.

13.7 REFERENCES

Adams, W. M., Perrow, M. R. and Carpenter, A. (2004) Conservatives and champions: river managers and river restoration discourses in the United Kingdom. *Environment and Planning A*, **36**, 1929-1942.

Bernauer, T. (2002) Explaining success and failure in international river management. *Aquatic Sciences* **64(1)**, 1-19.

Eden, S., Tunstall, S.M. and Tapsell, S.M. (2000) Translating nature: river restoration as nature – culture. *Environment and Planning D*, **18**, 257-273.

Environment Agency (2002) *River Severn Strategy. A Flood Management Strategy for the River Severn Corridor*. Scoping Report, January 2002. Environment Agency, Solihull.

Felgentreff, C. (2003) Post-disaster situations as "windows of opportunity"? Post-flood perceptions and changes in the German Odra River Region after the 1997 flood. *Die Erde* **134**(2), 163-180.

Ledoux, L., Cornell, S., O'Riordan, T., Harvey, R and Banyard, L. (2005) Towards sustainable flood and coastal management: identifying drivers of, and obstacles to, managed realignment. *Land Use Policy* **22**(2), 129-144.

Maltby, E. and Blackwell, M.S.A. (2005) Managing riverine environments in the context of new water policy in Europe. *International Journal of River Basin Management* **3**(2), 133-141.

Moss, T., Klaphake, A., Monstadt, J. and Siener, M. (2003) What prospects for restoration? In *The flooded forest. Guidance for policy makers and river managers in Europe on the restoration of floodplain forests* (ed. F. Hughes), University of Cambridge, Cambridge, pp.71-84.

Turner, R.K., van den Bergh, J.C.J.M., Söderqvist, T., Barendregt, A., van der Straaten, J., Maltby, E. and van Ierland, E.C. (2000) Ecological-economic analysis of wetlands: scientific integration for management and policy. *Ecological Economics* **35**, 7-23.

Werritty, A. (2006) Sustainable flood management: oxymoron or new paradigm? *Area* **38**(1), 16-23.

WWF (World Wide Fund For Nature) (2000) *Wise Use of Floodplains. Policy and Economic Analysis of Floodplain Restoration in Europe. Opportunities and Obstacles*. Report, November 2000.

Index

Entries in italics indicate figures.
Abbreviations R., river.

IWA Publishing's authorised EU representative for General Product Safety Regulations is Diane D'Arras, 15 rue Duret, 75116 Paris, France, e-mail: safety@iwap.co.uk.

Printed and bound by CPI Group (UK) Ltd, Croydon, CR0 4YY

12/05/2026

02108864-0008